Ritual bones or common waste

A study of Early Medieval bone deposits in Northern Europe

Groningen Archaeological Studies
Volume 24

Groningen Institute of Archaeology
Poststraat 6
NL-9712 ER Groningen
the Netherlands
gia@rug.nl

Website
www.gas.ub.rug.nl

Publisher's address
Barkhuis
Zuurstukken 37 9761 KP Eelde the Netherlands
Tel. 0031 50 3080936 Fax 0031 87 7844285
info@barkhuis.nl www.barkhuis.nl

Ritual bones or common waste

A study of Early Medieval bone deposits in Northern Europe

J. Thilderkvist

Barkhuis &
University of Groningen Library
Groningen 2013

Book cover design, interior design and typesetting: Johan Thilderkvist

Photographs: ADC ArcheoProjecten, S. Buhmann, J. Jans, O. Magnell, A. Nieuwhof, W. Prummel & J. Thilderkvist

Illustrations: S. Boersma, J-B-S. Chardin & J. Thilderkvist

ISBN: 9789491431319

university of groningen / groningen institute of archaeology

Table of contents

1. Introduction

1.1 The research question

Almost every research endeavour starts with a problem of some sort. This problem can often be summarised into a single question, but such a question frequently gives rise to related issues, some of which have to be answered first in order to address the main research problem, while others could be digressions but still interesting in connection to the theme of the research.

The main problem that I pondered on while writing the research plan for this thesis was a very broad methodological one related to how most archaeozoological research is performed. The problem originated from my interest in taphonomy, the study of the destructive processes that affect, in this case, a bone material from the point in time that an animal is killed and until it is excavated by archaeologists. I had come across several studies that showed that even on sites where the preservation conditions for bone were good and where massive amounts of bones were recovered, only a tiny fraction of the bones that once must have existed on the site, estimated from its size and duration, were ever found by the archaeologists. It has been suggested that the total amount of animal bones found on a typical site where organic material are relatively well preserved make up less than 0.1 percent of the bones that once existed (Noddle 1977: 378-380; Vretemark 1997: 34-35).

I also knew by experience that even if a well preserved bone material is excavated, only a part of it can be identified to species and bone element by the archaeozoologist. Furthermore a vast majority of the archaeozoological studies that I had come across, especially the large bone materials from the Iron Age, Roman Period and Middle Ages that I was most familiar with, had very large scopes and seldom spent much time on taphonomy or small scale (intra-site) depositional patterns. It often seemed to be assumed that all the excavated bones were refuse from consumption of food eaten primarily to provide nourishment. In some cases settlements or parts of settlements of different character could be separated and compared with each other, for example different parts of a town that clearly showed dissimilar characteristics (see for example Noddle 1977: 381-382). But rarely were individual deposits analysed in any detail and often they were not interpreted at all from an archaeozoological point of view. There are of course exceptions to this situation, especially regarding very odd deposits that sometimes received more attention. Ultimately even these deposits however often ended up having the labels "special" or "ritual" slapped onto them without much further discussion.

What was bothering me is mainly that if it is clear that we only find a tiny fraction of the animal bone material that once existed on a site and if we at the same time care little or not at all about the reasons behind the individual deposits, is it not possible or even likely that some of our large scoped studies are severely biased? I furthermore think that the interpretation of individual sites, often the primary aim of commercial excavations, is even more susceptible to this danger. If we find a few tens of kilos of animal bones from an excavated settlement (not an uncommon amount) this is often used to give a general view of the diet, economy and hunting/husbandry of the community. But if half of this bone material for example was found in a feature that was the result of an activity not related to butchery, cooking or food refuse, interpretations regarding diet and economy in the settlement might turn out wrong.

In theory the answer to the problem described above is simple, the archaeologists and archaeozoologists only have to identify the activities that formed the individual deposits in order to know which parts of the bone materials they should use for answering which research questions (O'Connor 1996: 6-12). A bonus of such an approach would be that human activity involving bones within a site would be more visible than is usually the case. A simple theoretical answer is unfortunately not equal to a simple practical solution. Every archaeologist is familiar with the often huge difficulties involved in interpreting any feature, and the fragility of bones does not make this job easier. I therefore came to the conclusion that the main task must be to improve the possibilities for such interpretations of individual deposits. This should be done by both discussing the theoretical framework of archaeological interpretation of features and by reviewing different methods already used for such tasks, or if it is necessary develop new methods.

When reading about this and thinking of different scenarios where a broad archaeozoological analysis might be biased, ritual deposits repeatedly kept coming up. This was partly because it is a category of activity that, when represented by a sufficiently odd bone assemblage or in combination with certain artefacts, is actually sometimes identified, but also because such deposits and sites, even if they are identified, still are regularly included in purely economic or dietary analyses. Some archaeozoologists are very aware of the problematic nature of such broad views, as can be seen for example in O'Connor's *A Critical Overview of Archaeological Animal Bone Studies*:

> *"...animals are an essential part of the cognitive mindscape of a human population, not merely some useful resource to be exploited on an objective, rational basis, and bones are often encountered in settings which indicate a non-rational or 'ritual' significance... An interpretation of the bones from an archaeological site of any period cannot ignore this point."*
> (O'Connor 1996: 12)

Many archaeozoologists however seem not to be concerned by this problem and with this in mind I concluded that ritual deposits would be suitable as an example subject for my methodological assessment. Some of the specific methods evaluated or developed might be useful for identifying deposits of varying types both within and outside the time frame and geographical area chosen for the examples in the thesis. Nevertheless the main goal is not to find specific methods, but to formulate a systematic methodological approach to the problem that in turn could be used for similar contexts. This means that the thesis should not be seen as purely archaeozoological research, or that the circumstances discussed here only apply to animal bones. Much archaeological research faces similar problems and even if the perishable and biological nature of bones creates some unique challenges, a lot of the discussion will be relevant for other fields of study within the discipline, especially when dealing with other mass materials such as ceramics and flint. Both human bones and various types of artefacts are also repeatedly discussed in this thesis since they are both practically and theoretically intermingled with the animal bone assemblages. With this in mind the research aim is formulated as follows:

> **The main aim of the thesis is to test and evaluate different methods in order to formulate a broad methodological framework for approaching the identification of ritual deposits of bones in settlements.**

This aim will be achieved by describing and discussing various archaeozoological and archaeological methods used for identifying ritual deposits. A structured system for approaching materials using these methods will then be formulated. Finally this system will be applied on a number of case studies made up of bone materials from various sites within a specified time period and geographical area.

1.2 Research background

The early interest in ritual

It is easy to suppose that the interest among archaeologists for religion and ancient beliefs in the supernatural is a recent phenomenon connected with the increased awareness of non-functional activities during the post-processual movement in the 1980s and onwards. This is however far from true. Already during the infancy of the discipline a curiosity and awareness about these things existed, as can be seen for example among some of the most influential scholars of early archaeology such as Thomsen, Worsaae and Montelius:

> *"2. Store ringe, som man anser for hellige. De synes ikke at kunne have været brugte om handledene, hvortil tvende mod hinanden vendende udbugninger, hvori de ende sig, gjøre dem mindre bekvemme; de ere derfor for smaae for halsen eller for hovedet.*
> ...
> *Da de derfor ofte ere af puur guld og meget massive, saa at de i oldtiden maae have været store kostbarheder, har man troet i dem at gjenfinde de hellige ringe, som omtales brugte ved eds aflæggelse i den hedenske tid.*
> ...
> *4. Flade, store fade eller kar af bronse, som sædvanlig have en breitet fod; de antages for at være de så kaldte Offerboller, hvori offerblodet heldtes."*

> *"2. Large rings, which are assumed to be sacred. They do not seem to have been used around the wrists since two extensions opposite each other at the ends of them make them less comfortable, and they are too small for the neck or head.*
> ...
> *Since they often are of pure gold and very heavy, so that they must have been great treasures in ancient times, they are believed to be the holy rings that are said to have been used for the swearing of oaths in heathen times.*
> ...
> *4. Flat, large plates or trays of bronze, that usually have a wide foot; they are supposed to be the so called offering bowls, into which the offering blood was poured."*
> (Thomsen *et al.* 1836: 44-45, translation from Danish by the author)

> *"In complete agreement with this the gold bracteates and larger ornaments of gold with pictures of the gods and sacred signs which had been worn as amulets, like the Thor's hammers in the latest Iron-Age are seldom found in graves, especially in Denmark, but almost always in the hoards deposited in bogs and field, most probably by way of offering."*
> (Worsaae 1886: 162)

> *"The great Danish hoards so often mentioned before seem also to have been originally offerings made by a victorious army to the god who gave them victory."*
> (Montelius 1888: 122)

The subject might not have been as widely discussed as today, but it did exist and archaeologists were not afraid to use this type of interpretations to explain deposits observed during excavations.

Occasional use of ritual in interpretations continued to appear during the whole era of culture-historical archaeology. One prominent example is Ole Klindt-Jensen's splendid book covering the Migration Period of Bornholm, *Bornholm i Folkevandringstiden*, in which large sections deal with interpretations of suspected ritual deposits of varying types (Klindt-Jensen 1957).

Another albeit more superficial example is Albert van Giffen's article *Het bouwoffer uit de oudste hoeve te Ezinge* that, despite the fact that the text focuses on the building rather than the house offering, is still a ritual interpretation of the finds of large parts of a cattle, a horse and a dog (or possibly a sheep) found in the walls of an Iron Age house (Van Giffen 1963).

The New Archaeology-era

The spread of the new theoretical paradigms often called New Archaeology or Processual Archaeology in the early 1960s did however change the situation dramatically. These new ideas turned against what was perceived as the unscientific practices of the earlier cultural historians. Archaeology as a discipline became more systematic, more deductive, functionalistic and positivistic, and generally moved towards the natural sciences. In such a research environment the difficulty to prove and categorise (apparently) non-functional deposits made ritual activity less interesting. Subsequently this type of activity was not used as often anymore in interpretations of archaeological contexts.

The very existence of ritual activity within ancient societies was however seldom put in doubt and if a deposit was convincing enough it could still be interpreted as such, but little attention was generally given to these features. Some examples from this period where ritual deposits are still the main focus are Anne Ross' article *Shafts, pits, wells – sanctuaries of the Belgic Britons?* and C. J. Becker's *"Mosepotter" fra Danmarks jernealder - Problemer omkring mosefundne lerkar og deres tolkning* (Ross 1968; Becker 1972).

In some countries, such as former West Germany that was not very influenced by the theories of processual archaeology, ritual was more often used in interpretations. This can be seen for example in Müller-Wille's article *Pferdegrab und Pferdeopfer im frühen Mittelalter* and in Haarnagel's book about the excavation at Feddersen Wierde *Die Grabung Feddersen Wierde. Methode, Hausbau, Siedlungs- und Wirtschaftsformen sowie Sozialstruktur* (Müller-Wille 1971; Haarnagel 1979).

This was possibly even more the case for the former East Germany and, to some degree, the whole of Eastern Europe, as can be seen for example in Müller's archaeozoological report on the excavations at the temple fortress Arkona on Rügen *Die Tierreste aus der slawischen Burganlage von Arkona auf der Insel Rügen* where ritual interpretations play a prominent role (Müller 1974).

Post-processual archaeology and a renewed interest in ritual

In the 1980s a counter movement against processual archaeology appeared. This movement actually contained a number of diverse theoretical views that often, very broadly, is called post-processual archaeology. The rigid scientific framework and pursuit for objective knowledge was challenged and as a part of this, non-functional (or at least not only functional) models of interpretation regained attention.

One of the best examples of this change of interest is the interpretation of the finds from Danebury. This Iron Age hillfort in Southern England was the location of extensive excavations from 1969 until 1988 led by Barry Cunliffe. During the project a large animal bone assemblage was excavated and it soon became clear that a number of deposits contained bones with unusual characteristics compared to the rest of the assemblage. Pits, trenches and ditches with articulated parts of animals, strange combinations of species and objects and a large number of skulls were noted and generated specialized research into the subject, mainly by Anne Grant. Danebury eventually became a mile stone for the research into ritual activity on settlements, especially regarding animal bones. Grant's work includes extensive chapters in the excavation reports (see for example Grant 1984a) and several articles such as *Survival or sacrifice? A critical appraisal of animal burials in Britain in the Iron Age* in *Animals and Archaeology: 4 Husbandry in Europe* (Grant 1984b), *Animals and ritual in early Britain: The visible and the invisible* (Grant 1989) and *Economic or Symbolic? Animals and Ritual Behaviour in Sacred and Profane. Proceedings of a Conference on Archaeology, Ritual and Religion. Oxford, 1989*, that dealt both with this material and other related Iron Age sites (Grant 1991).

Other researchers also showed interest into this subject and time period, for example Wait with his large survey of ritual deposits from the British Iron Age in *Ritual and Religion in Iron Age Britain* (Wait 1985), while Merrifield wrote a more general book, *The Archaeology of Ritual and Magic*, dealing will all kinds of unusual deposits in Great Britain from various time periods (Merrifield 1987).

With his thesis *Ritual and Rubbish in the Iron Age of Wessex – A study on the formation of a specific archaeological record* dealing with the interpretations of Iron Age deposits in Britain, Hill made an important contribution to the field (Hill 1995). In his thorough study Hill questioned the strict division between ritual deposits and rubbish, and found many structured deposits containing what at first sight could seem to be normal refuse.

During this upswing of research into the subject of ritual activity, especially in Great Britain, some critical voices could be heard. The archaeozoologist Bob Wilson was one of these and in his article *Considerations for the Identification of Ritual Deposits of Animal Bones in Iron*

Age Pits he criticised what he considered vague criteria for identifying ritual deposits and the lack of attention for taphonomic factors (Wilson 1992). Wilson never expressed doubt about the existence of ritual deposits, but considered the methods used to identify them as unclear or based on faulty assumptions. A few years later he formulated his own thoughts about identifying activity within settlements, not only regarding ritual action but many types of activities such as slaughtering, cooking and cleaning (Wilson 1996).

While all this was going on in Great Britain the research in Germany basically continued as before. Torsten Capelle for example published a large overview of building offerings *Eisenzeitliche Bauopfer in Frühmittelalterliche Studien. Jahrbuch des Instituts für Frühmittelalterforschung der Universität Münster*, while Müller-Wille continued his work with the book *Opferkulte der Germanen und Slawen* and in many articles (Capelle 1987; Müller-Wille 1999).

Important recent works in Scandinavia

As could be seen above research about ritual activity has been of some interest in Scandinavia since the beginning of archaeology as a formal discipline. This interest continued throughout the 20th century (see for example Kjær 1928: 19; Hatt 1938, 1960 as well as Klindt-Jensen 1957), even if a decline in connection with the popularity of New Archaeology can be seen here as well. For most of this long time period the research into the field was however almost exclusively focused on graves, sacrificial bogs and hoards, with the volumes dealing with the sacrificial bog Skedemosse on the island of Öland as a prominent example (see for example Hagberg 1967). The reason for this focus was at least partly because actual finds of buildings and settlement activity areas were rare in Scandinavia before the large scale excavations in the 1960s (Carlie 2004: 9-10). Occasional books also discussed rituals, magic and religion from a more general perspective, mainly with reference to the Early Medieval written sources. Actual archaeological research about ritual deposits on settlements was however uncommon before the late 1980s.

One important event within this research in Scandinavia was the conference *Arkeologi och Religion* in Lund in 1989 that resulted in several articles about the study of ritual and religion within archaeology. Many of these focused on the traditional areas of hoards, graves and cult sites but some dealt, at least partly, with settlement finds (Larsson & Wyszomirska 1989).

In *Brogård – ett brons- och järnålderskomplex i södra Halland. Dess kronologi och struktur* a couple of years later Lennart Carlie showed, with examples from the Bronze and Iron Age settlement of Brogård in the province of Halland (Southern Sweden) that deposits of relatively small amounts of ceramic fragments in features could constitute the remains of ritual activity (Carlie 1992).

In 1993 Tove Paulsson produced a student thesis, *Huset och lyckan: en studie i byggnadsoffer från nordisk järnålder och medeltid*, that a few years later was reworked and published as an article in *Fornvännen*, *Iron Age building offerings. A contribution to the analysis of a die-hard phenomenon in Swedish preindustrial agrarian society* (Paulsson 1993; Paulsson-Holmberg 1997). In this text Paulsson focused on building offerings and its long consistency in Scandinavia. Even though the article was rather short it resulted in a renewed interest and attention for this find category within Scandinavian archaeology.

In the later part of the 1990s and up until today a great number of publications dealing with ritual activity on settlements from various time periods have been published in Scandinavia and the awareness about this aspect of human behaviour have also seeped over into commercial archaeology during this time period.

In 1994 Per Karsten published his thesis *Att kasta yxan i sjön - En studie över rituell tradition och förändring utifrån skånska neolitiska offerfynd* that dealt with ritual deposits in the province of Skåne in Southern Sweden during the Neolithic Period, but that also included thorough theoretical and methodological parts (Karsten 1994).

In 1997 *Kungens Gods i Borg – om utgrävningarna vid Borgs säteri*, the synthesis of an excavation in the province of Östergötland, Central Sweden, was published. Of special interest was a small house with a surrounding animal bone layer and a large catch of amulet rings that was interpreted as a ceremonial building with the remains from various ritual activities (Lindeblad & Nielsen 1997).

Around this time several more articles dealing with ritual activity in connection with buildings were published, for example *Guden under gulvet - ofringer under fynske huse fra ældre jernalder*, that was a short review of Iron Age building offerings on the island of Funen, Denmark (Henriksen 1998). Several articles also described ritual activity on large farms and on central places during the Roman Period and Early Middle Ages (see for example Jeppesen & Madsen 1997; Jørgensen 2002; Larsson & Lenntorp 2004).

In 2004 Anne Carlie published her book *Forntida byggnadskult. Tradition och regionalitet i södra Skandinavien* that is a comprehensive review and discussion of all finds in or near buildings that have been interpreted as ritual in Southern Sweden and Denmark, throughout all time periods (Carlie 2004).

Finally one recent example of archaeological research concentrating on ritual deposits within settlements is Gavin Lucas' and Thomas McGovern's work on the Viking Age settlement Hofstaðir on Iceland. Here the finds of a great number of cattle heads were interpreted as the remains of ritual decapitations (Lucas & McGovern 2008).

Important recent works in the Netherlands

While the interest for ritual deposits in settlements has not been as intensive in the Netherlands during recent years as in Scandinavia, a number of interesting articles and books have nevertheless been published. In 1999 the animal and plant remains, including a number of animal bone deposits interpreted as ritual, from the Late Roman Period settlement of Heeten in the province of Overijssel was published in the article *Between ritual and economics: animals and plants in a fourth-century native settlement at Heeten, the Netherlands* (Lauwerier *et al.* 1999). Despite the unfavourable preservation conditions on the site, the article showed that a thorough analysis can reveal a lot about animal bones beyond the usual lists of bone elements and species.

In 2004 Linda Therkorn published her thesis *Landscaping the Powers of Darkness & Light. 600 BC – 350 AD settlement concerns of Noord-Holland in wider perspective.* The book deals with unusual deposits at settlements in the province of Noord-Holland during the Iron Age and Roman Period and the interpretations connecting features in the ground with star constellations have been much debated. Despite the criticism of these interpretations, the very existence of this thesis has undoubtedly led to discussion and interest for this category of finds.

Maaike Groot published a less controversial thesis *Animals in Ritual and Economy in a Frontier Community* in 2008 that dealt with the bone material from Roman Period settlements in the Dutch river area (Groot 2008). Among the settlement finds presented in her book were a number of unusual deposits of which some were interpreted as the remains of rituals.

In addition to Groot's thesis the number of publications from excavations in the Netherlands where ritual deposits have been identified and discussed have increased in the last few years, as can be seen for example in the publications from the Englum *terp*-excavation in the province of Groningen and the Midlaren-project in the province of Drenthe (Nieuwhof 2008; Prummel 2008; Prummel *et al.* 2008).

1.3 Essential definitions and terminology

Clear definitions of the important terminology used in a text are crucial in order to prevent misunderstandings, especially when dealing with the rather ambiguous vocabulary of symbolism, ritual, magic and religion. It is therefore quite surprising that this vital part is lacking in many archaeology books and articles dealing with these matters. While clear definitions are fundamental and have to be either commonly accepted or logically formulated, their exact content will always be open for discussion.

I will keep the definitions presented here short and concise; partly because this thesis is not primarily an exercise in the theory of ritual, and partly since long and detailed definitions, however accurate from a more theoretical point of view, run the risk of increasing misunderstandings instead of avoiding them. This might lead to somewhat simplified definitions, but will on the other hand be closer to the meanings of these words used in everyday archaeological research.

Special

The term special used in this thesis is a purely constructed external definition that basically has nothing to do with the actual meaning of the human activity involved. It is applied to deposits that in one or several ways differ from the characteristics of a normal assemblage. The normality in this case is usually defined by the complete material on the site in question, but could also include comparisons with materials from similar sites.

Ritual

A ritual is a set of symbolic actions used to convey some form of meaning. The purpose and content of a ritual might vary and include, but are not limited to, religion, magic, social interaction, maintaining of traditions, education, upholding of hierarchies and projection of power.

A wide definition of ritual could theoretically include any set of symbolic human actions, regardless of their meaning or even the performers' awareness of them. Such casual or unconscious actions, however interesting for understanding human behaviour, are extremely difficult to identify in an archaeological context. I have therefore in this thesis chosen to use the term ritual to describe only the actions that could best be called *explicit rituals.* With an *explicit ritual* I mean conscious ritual acts that are not considered mundane events by the performer.

It is important to note that the etymologically and in meaning closely related word rite is sometimes used either as a synonym for ritual, to describe parts of a ritual, or to represent a specific religious act, often associated with formalized religions (see for example Steinsland 2005: 261-273). Because of this ambiguity I have chosen not to use the word rite unless it is a part of an already established definition, as for example in *rite de passage.*

Symbolic

Another word often used in texts dealing with special deposits and ritual activity is symbolic. Its use is often very broad and its definition vague. Paul Mellars has in his search for symbolic activities among Neanderthals presented what he describes as a widely accepted definition (see also for example Hodder 1982a) of the word symbol as being:

> *"...anything, be it object, sign, gesture or vocal behaviour which in some way refers to or represents something beyond itself..."*
> (Mellers 1996: 369)

The consequence of such a broad definition is that most things in life are at least partly symbolic. In this thesis the word will however be reserved for deposits and human actions that mainly have a symbolic purpose. They may very well have had a functional meaning as well, but the symbolic significance was the most important one.

Offering and sacrifice

With an offering, and the sub-category sacrifice, I mean the giving of something, usually an object, food, drink, plant or animal, to a supernatural entity in order to communicate with it. The goal of the communication can be to appease the entity and make it favourably inclined, to obtain something from it, to make something holy, or as a thanksgiving for favours already received, often with the ultimate goal to support the cosmic order in general (Van Baaren 1964). In contrast to magic an offering is not an act by which the entity is supposed to be tricked or forced into performing the desirable action, but a show of worship and possibly a humble request for favours.

In the terminology of religious studies an offering is any act of presenting something to a supernatural being while a sacrifice is an offering accompanied by the ritual killing of the offered object (Van Baal 1976: 161). Since this separation is often difficult to observe in the archaeological record the terms are often used interchangeably. In this thesis I have chosen to use offering in all cases where an act of killing the offered object can not be clearly identified.

Religious and magical rituals

One common purpose for rituals is to deal with the supernatural. This can be done in a number of different ways, with religion and magic being the most common. Both types of actions try to create some form of positive effect for the performer by affecting the supernatural powers. The most common definitions of these two are that the performer of magical rituals actively tries to persuade, trick or even force the supernatural entity to do his or her biding, while the performer of religious rituals more humbly asks for favours. It is however important to note that there is a gradual distinction between the two, and how a ritual is viewed often depends on the subjective performer or observer, as Steinsland points out:

> *"Often the attitude reveals that «we» perform religion, while «the others» deal with magic."*
> (Steinsland 2005: 267, translation from Norwegian by the author)

1.4 The choice of materials

In accordance with the discussion in section 1.1, four bone materials were selected as case studies on which various methods should be applied with the aim of identifying special deposits and, to some extent, interpret these further in order to find possible ritual deposits.

The materials used as case studies were chosen according to a number of criteria. To limit the scope of the research and make comparisons between the various materials possible, the period for sites to be included was set to between ca 400 AD and until the definitive introduction of Christianity in the area, around 800 AD in the Northern Netherlands and around 1000 AD in Southern Sweden (Heidinga 1997: 22-23; Steinsland 2005: 424-430). The time period was chosen to avoid the direct influence from the Roman cultural sphere and to ensure that transition of ritual customs in connection with the introduction of Christianity would be minimal. I certainly do not suggest that the tradition of ritual deposits vanished after the introduction of Christianity in Northern Europe; many sites indicate that this is not the case (see for example Ersgård 2002). It is however clear that the frequency and form of ritual deposits changed quite dramatically during the religious transition. Despite these considerations it is important to note that a different time period could have been chosen without dramatically altering the nature of the research, which primary aim is to test methods and not to research ritual practices of a specific time or society.

Other criteria for the materials were that they should contain a well dated bone material, excavated and recovered with relatively modern archaeological methods. It was furthermore essential that both the bones and the field documentation from the excavations were accessible for first hand examination.

Initially a selection of materials evenly spread over Northern Europe was envisioned, but ultimately the accessibility to suitable materials as well as practical and economical considerations restricted the choice. In the end three bone materials from the Northern Netherlands and one material from Southern Sweden were selected. Since the aim of the thesis is not to give a survey of ritual deposits or to study regional variation this uneven spread of sites should not pose any significant problem. Since three of the materials had been excavated in collaboration with the Groningen Institute of Archaeology at the University of Groningen and the final one came from an ongoing excavation supervised by the Department of Archaeology and Ancient History at Lund University, where I was an undergraduate student, I have had virtually unlimited access to the sites, materials, documentation and persons involved in the excavations.

The four assemblages represent three very different types of settlements (see further section 1.5). Two of the materials come from artificial dwelling mounds (*terps*) in the then undiked salt marches of the North of the Netherlands, one comes from a settlement in the inland areas of the Northern Netherlands and the final one is a settlement of central place-type from Southern Sweden. These varied settlement types were chosen in order to give a range of different deposits, but in the case of the *terp*-sites and the central place also because these have good preservation conditions for bone.

As can be seen in table 1.1 the size of the bone mate-

Table 1.1. The bone materials from the four sites used as case studies.

Name	Location	Weight	Dating	Archaeozoological researchers
Dongjum	The Netherlands (Friesland)	17.6 kg	VMEA- VMEC	Roblićková, M. & Soetens, L.
Leeuwarden	The Netherlands (Friesland)	33.9 kg	VMEA- VMEC	-
Midlaren	The Netherlands (Drenthe)	13.2 kg	ROML-VMEC	Prummel, W., Holl, J., Jans J.E.A. & Huisman, N.
Uppåkra	Sweden (Skåne)	57.7 kg	ROMLB-VMEB	Magnell, O.

rials chosen for analysis is quite small with archaeozoological standards, having a total fragment weight of only slightly more than 120 kg. The reason for this was time constraints in connection with the unusually detailed process of examination. During a normal archaeozoological analysis the bone fragments are counted, weighed and identified to species, element, part and side. The bone fragments are during this process usually examined in search of various marks, age and sex indicators, and in some cases they are also measured. In addition to these standard procedures I applied a number of often very detailed analysing methods on each bone fragment. These methods, that are further described in chapter 2, and the recording of them turned out to be very time consuming which is the reason for the relatively small bone sample sizes used in the thesis.

1.4.1 The four bone materials

The four bone materials that were chosen as case studies were the *terp*-sites Dongjum and Leeuwarden (Oldehoofsterkerkhof) both in the province of Friesland (the Netherlands), Midlaren located on the sandy soils in the province of Drenthe (the Netherlands) and finally a part of the central place Uppåkra located on the Southern plains of the province of Skåne (Sweden).

In figure 1.1 the geographic locations of these sites are shown. In table 1.1 the location, weight and dating of the selected assemblages, as well as the names of the archaeozoological researchers that, in addition to myself have analysed parts of the materials are presented for the four sites.

Table 1.2. The Dutch time period abbreviations used in the thesis.

Abbreviation	Dutch name	English translation	Time period
IJZ	IJzertijd	Iron Age	800 - 12 BC
IJZV	IJzertijd vroeg	Early Iron Age	800 - 500 BC
IJZM	IJzertijd midden	Middle Iron Age	500 - 250 BC
IJZL	IJzertijd laat	Late Iron Age	250 - 12 BC
ROM	Romeinse tijd	Roman Period	12 BC - 450 AD
ROMY	Romeinse tijd vroeg	Early Roman Period	12 BC - 70 AD
ROMVA	Romeinse tijd vroeg A	Early Roman Period A	12 BC - 25 AD
ROMVB	Romeinse tijd vroeg B	Early Roman Period B	25 - 70 AD
ROMM	Romeinse tijd midden	Middle Roman Period	70 - 270 AD
ROMMA	Romeinse tijd midden A	Middle Roman Period A	70 - 150 AD
ROMMB	Romeinse tijd midden B	Middle Roman Period B	150 - 270 AD
ROML	Romeinse tijd laat	Late Roman Period	270 - 450 AD
ROMLA	Romeinse tijd laat A	Late Roman Period A	270 - 350 AD
ROMLB	Romeinse tijd laat B	Late Roman Period B	350 - 450 AD
ME	Middeleeuwen	Middle Ages	450 - 1500 AD
VME	Middeleeuwen vroeg	Early Middle Ages	450 - 1050 AD
VMEA	Middeleeuwen vroeg A	Early Middle Ages A	450 - 525 AD
VMEB	Middeleeuwen vroeg B	Early Middle Ages B	525 - 725 AD
VMEC	Middeleeuwen vroeg C	Early Middle Ages C	725 - 900 AD
VMED	Middeleeuwen vroeg D	Early Middle Ages D	900 - 1050 AD
LME	Middeleeuwen laat	Late Middle Ages	1050 - 1500 AD
LMEA	Middeleeuwen laat A	Late Middle Ages A	1050 - 1250 AD
LMEB	Middeleeuwen laat B	Late Middle Ages B	1250 - 1500 AD

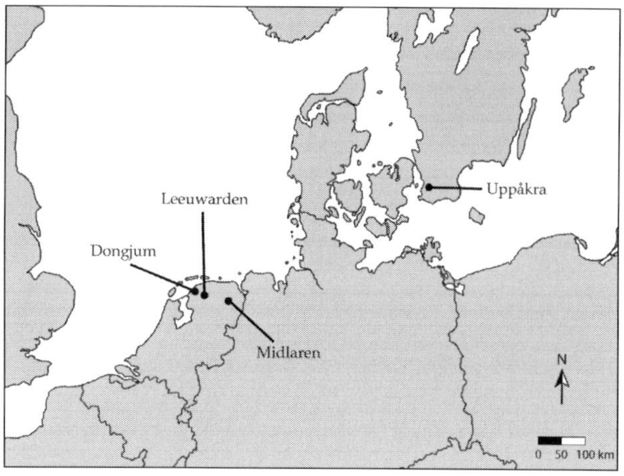

Figure 1.1. The geographical location of the four sites used as case studies.

1.4.2 The Dutch time periods and their abbreviations

Since the majority of the bone materials used in the thesis originate from the Netherlands I have selected to use the standardised Dutch time periods for dates. I have used English translations for these in the text and the Dutch abbreviations in tables and for find references. The abbreviations, period names and what time periods they encompass are presented in table 1.2.

1.5 Special aspects of the case study-sites

1.5.1 *Terps*

The sites Dongjum and Leeuwarden located in the northern parts of the Netherlands are so called *terps*, a unique form of settlement. A *terp* (also regionally called *wierde, woerd, warf, Warft, werf, Wurt* or *værft*) is a type of artificial dwelling mound that was constructed in the salt marsh landscape of North Western Europe from the early Iron Age until the Middle Ages (figure 1.2).

The *terp*-region stretches along the coast from the north of the Netherlands up till Southern Denmark (figure 1.3).

Before the building of the protective dikes along the coast the majority of this area consisted of low laying flatland that was regularly flooded. The lowest laying areas located close to the sea were flooded every high tide, while slightly higher areas were only flooded at spring tide or during storm tide. The recurring floods created large salt marshes where for long periods of time no human habitation was present. During the 7th century BC this changed and small scale colonisation can be observed. These early settlements were built on natural heights in the landscape, mainly salt march ridges. Such features were however always at risk of being flooded during extreme circumstances. Therefore during the 6th century BC the construction of settlements on artificial heights was initiated (Taayke 2005: 155, 161).

Figure 1.2. The *terp* on the island Hallig Hooge in Germany (photo by S. Buhmann 2005).

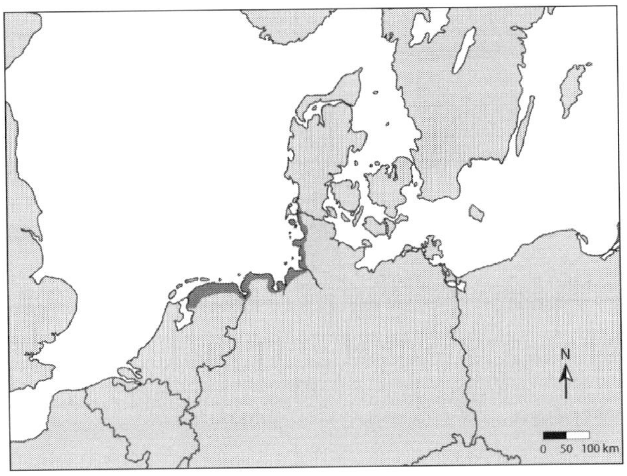

Figure 1.3. The *terp*-area, with present day coastlines (after Heidinga 1997).

At first small podiums with enough space for only a single building were constructed using sods or manure. Later these house platforms were expanded, or several closely situated podia were merged, into a larger *terp*-settlement. The raising and expanding of the *terps* became an integrated part of the life in the settlements and this process often continued throughout the sites' existence. This occasionally resulted in very impressive mounds measuring up to several hundred meters in diameter and sometimes close to ten meters in height, although the majority of the *terps* seldom reached more than half this height and a fraction of that size (Taayke 2005: 161).

An explosive population growth can be observed all along the North Sea coast in the first century AD and it has been estimated that as many as 1500 *terp*-settlements existed at this time in what is present day Netherlands (Taayke 2005: 155). This poses the question of why so many people decided to settle down on dwelling mounds in the seemingly inhospitable salt marshes. The answer is probably multifaceted, but one important aspect is that the landscape that at first glance seems to be only barren wetland, did in fact provide excellent and virtually unlimited areas for cattle and sheep grazing (Dijkstra *et al.* 2008: 315). This meant that even if the possibilities for crop cultivation were perhaps limited (see for example Zeist 1974: 364-366), a stable subsistence system could be based on live-stock farming and trade. The archaeological record shows that the economy of the area flourished during extended time periods due to extensive trade with neighbouring areas, as well as through long range networks (Heidinga 1997: 27-32).

These periods of wealth were however interrupted by periods of decline. In what are the present day Dutch provinces of Friesland and Groningen the habitation density of the *terps* slowly dwindled during the 2nd century AD, and dramatically decreased during the 3rd century. During most of the 4th century Westergo (the western part of the present day province of Friesland) was totally abandoned while the only place in Oostergo (the eastern part of the province) where settlements still existed was a small area near Ferwerd and Hallum. In the area that is the present day province of Groningen some scattered habitation also remained along the Hunze River.

The reason behind this break in habitation is not clear but three important factors have been presented, all of them most likely contributing to the situation (Bazelmans 2000: 49-52; Taayke 2008: 195):

- The deterioration of the drainage from the gullies and rivers into the North Sea due to sedimentation, and the subsequent flooding of areas close to watercourses where *terps* were often situated.
- An increasing political pressure from the east.
- The collapse of the Roman *limes* and the consequences of this for trade and exchange networks.

The situation eventually became more favourable and at the end of the 4th and beginning of the 5th century the abandoned parts of the *terp*-area were resettled, possibly by people from the northeast. The *terp*-area did not only recover, but became the core region in the Frisian kingdom that emerged in the 7th century. This Frisian kingdom prospered for some time, but was soon put under pressure from the expanding Frankish empire in the south and was conquered during the 8th century. The inhabitants of the region were converted to Christianity and the economic centre was moved south to the river area of the present day Netherlands (Heidinga 1997: 16-22). The *terp*-settlements however remained densely inhabited until the large protective dikes were built along the coasts, from the 12th century onwards. After the building of the dikes the farms were spread out in the landscape, but the *terps* often continued to be habited, although no longer out of necessity (Dijkstra *et al.* 2008: 314).

From an archaeological point of view the *terps* are remarkable treasure troves. Their thick and highly organic soil layers create superb preservation conditions for usually perishable materials such as plant remains and animal bones, and the long lasting habitation within the same restricted area generate high concentrations of finds. Unfortunately the very fact that the *terp*-soil has such a high organic content made it commercially interesting as fertilizer in the second half of the 19th century. Countless *terps* were partially or completely dug away and the soil transported inland to the sandy areas for use in land improvement (Arjaans 1991). This practice continued up until the beginning of the 20th century, when the archaeological potential of the *terps* were brought to public knowledge, much thanks to the pioneering work of the archaeologist Albert Egges van Giffen. During the first decades of the 20th century he directed several of the earliest large scale archaeological investigations of *terps* in the north of the Netherlands, such as Tuinster Wierden and the now classic excavation in Ezinge (Van Giffen 1936).

Since then many important *terp*-excavations have been carried out, including the complete investigation of Feddersen Wierde in Lower Saxony, Germany, between 1954 and 1963 that has been of great importance for our knowledge about the spatial organisation of *terp*-settlements. In more recent years the excavation between 1991 and 1993 of parts of one of the *terps* near Wijnaldum in Friesland produced many interesting finds. This excavation also became one of the catalysts for the Frisia Project, a joint multidisciplinary project of the Groningen Institute of Archaeology (GIA) of the University of Groningen, the Institute for Pre- and Protohistoric Archaeology of the University of Amsterdam and the State Service for Archaeological Investigations (formerly ROB, now the RCE). The project had the explicit aim to arrive at a coherent view of the Frisian area during the first millennium AD (Heidinga 1997: 9-11; Taayke 2008: 189-190). This ambitious project unfortunately never evolved in the way that was envisioned and did eventually fade out after a couple of excavations and the publication of the first volume covering the Wijnaldum excavation, a sad thing considering the great archaeological potential of the region (Besteman *et al.* 1999; Prummel *et al.* 2011). Fortunately the archaeological research of the *terp*-area has not stopped. In the last years several *terp*-excavations have been carried out, most notably in Wierum 2004 (Nieuwhof 2006), Anjum 2006 (Nicolay 2010), Oldehoofstekerkhof in Leeuwarden 2004-2006 (Dijkstra & Nicolay 2008), Achlum 2009, Jelsum 2010, Firdgum 2011 and Oosterbeintum 2011.

1.5.2 Central places

One of the bone materials used in the thesis comes from the settlement of Uppåkra in the province of Skåne. This region of present day Sweden is dominated by fertile plains separated by smaller areas of low forested hills. The region has been inhabited since the end of the last ice age (ca 12,000 BC) and especially the fertile southern and western parts have for long periods been densely populated. The area had strong cultural, economical and political links with what is present day Denmark during prehistory and was a part of the Danish kingdom from its consolidation in the 10th century until the mid 17th century.

Uppåkra has been categorized as a central place, a term used since the 1980s to describe certain Roman Period and Early Medieval settlements with special functions, especially in Southern Scandinavia, but in recent years in other parts of Northern Europe as well. Since the definition of a central place, especially in relation to the contemporary so-called workshop sites and the later towns, is unclear and has been debated for nearly two decades, a short introduction of this type of settlement and its function in South Scandinavia during the first millennium is necessary.

It has been known for a long time that sites whose function differed from that of the mainly agriculturally focused villages and farms did exist in Scandinavia at least since the Roman Period. At places like Helgö in Central Sweden and Dankirke on Jutland small settlements with signs of large scale specialised craft production as well as long range exchange existed during the 3rd and 4th centuries. These trade and workshop sites were undoubtedly controlled by the elite of the society in their respective areas and are often seen as direct precursors of the first towns. In the case of Dankirke and Helgö the towns of Ribe and Birka appeared nearby and seem to have taken over their role as centres of power, production and exchange (Jensen 1991: 73-76; Holmqvist 1961).

Beginning in the mid 1980s however another large type of settlement, contemporary with and even sometimes predating the trade and workshop sites, became the focus of attention of researchers interested in how society was organised before the first towns. Some sites had been excavated before the 1980s but received renewed interest due to numerous new finds of prestigious objects, mainly found with metal detectors. These sites are often called central places and include settlements such as Gudme on Fyn, Sorte Muld on Bornholm, Tissø on Sjælland and Uppåkra in Skåne (Thrane 1991; Watt 1991, Helgesson 1998; Jørgensen 1998). The earliest of these sites appeared in the Late Iron Age (according to Dutch chronology) and some of them continued to be in function until the 10th century (figure 1.4).

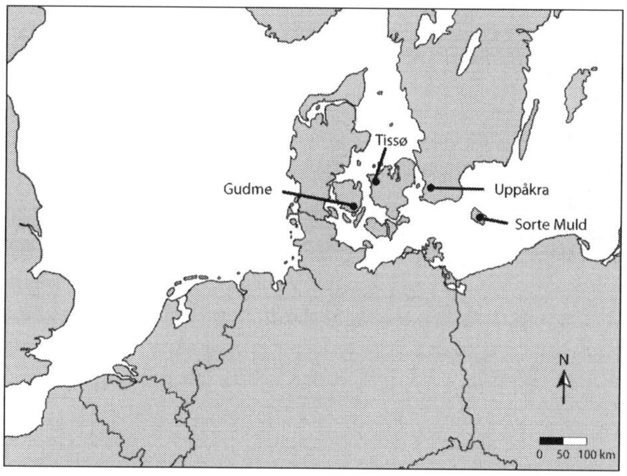

Figure 1.4. Central places in Southern Scandinavia mentioned in the text.

The exact definition and function of a central place has been widely debated. It is clear that they had the same functions as the workshop-sites, but it also seems as if they in addition had many of the same roles as the later Early Medieval towns in Scandinavia, albeit on a smaller scale. A central place is therefore, at the very least, a settlement that strongly influenced a surrounding area of varying size, politically as well as religiously, that shows signs of extensive exchange and that had local high quality craft production (Jensen & Watt 1993: 195; Grønnegaard 1997: 6). The last two characteristics are in fact intimately connected with the first, if we accept the widespread idea that power in Scandinavian societies during the first millennia was based on a system of gifts and exchange between different layers of the society in an area, and between the leaders of different territories in return for loyalty and alliances (Jensen 1995: 232-234).

Sites such as Sorte Muld, Gudme, Tissø and Uppåkra have produced thousands of archaeological finds, mainly from the top soil, including numerous high-status finds such as fibulas, figurines, bracteates, gold-figure foils, glass and weapons, but are these finds a good indication of the function of these sites in relation to their surroundings? The answer may be yes, as long as our field methods are sound (especially with regard to the investigations of the so called normal agrarian settlements where few metal finds have been found even when extensive metal detecting have been carried out), and as long as we also include qualitative aspects on the sites. The massive numbers of finds must have a cause and these sites indeed differ from the majority of settlements from the first millennium in Scandinavia. Even relatively large and long lasting villages such as Trabjerg and Vorbasse on Jutland do not show such a numerous and varied material culture (Jensen 1995: 214-218; Skibsted Klæsøe 1995: 111-119). The existence of the artefacts and what they indicate in the form of wealth, religious activity, excellent quality local craft and highly organised long range exchange networks tell us that these sites had central functions in their society.

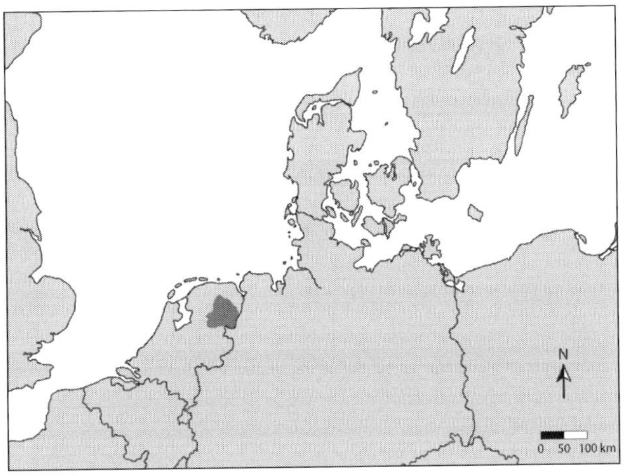

Figure 1.5. The province of Drenthe.

1.5.3 Settlements in Drenthe

The third type of site that is represented in this thesis is a settlement in the province of Drenthe in Northern Netherlands (figure 1.5).

The province is an inland area dominated by the Drenthe plateau. This region is situated well above sea level, in contrast to the low-lying coastal areas and mostly has a sandy top soil that does not produce favourable preservation conditions for metal or organic materials. Such objects are only found during special circumstances, usually in features such as wells that are deep enough to reach below the groundwater level.

Finds of human presence in Drenthe are known from the Middle Palaeolithic onwards. During the Neolithic Period permanent population is indicated by grave fields and a number of dolmens (*'hunnebedden'*), but finds of actual settlements with house remains have however not yet been found from before the Bronze Age (Waterbolk 2009: 40-41).

During the Roman Period and Early Middle Ages Drenthe seems to have had a stable, if not very large population concentrated on the sandy plateau (Harsema 1980: 67; Krol & Bos 2005: 103; Groenewoudt *et al.* 2006: 6-7). In contrast to the *terp*-sites and the Central places described earlier, the settlements of inland Drenthe are unexceptional and hardly differ from other inland settlements focused on agriculture in Northern Europe during the same periods. The habitation was made up of hamlets with a few farms that each normally consisted of a farm house divided into living quarters and a stable area with boxes for cattle and horses, and a few additional buildings and sunken huts used for storage, craft activities and for keeping sheep, goats and pigs (Harsema 1980: 7; Waterbolk 2009: 122). Notable excavations of settlements from these periods in the area are Gasselte, Midlaren, Noordbarge, Odoorn, Peelo and Wijster.

Politically and economically Drenthe has often been described in early written sources, and sometimes by modern scholars as well, as a peripheral area during the Roman Period and Early Middle Ages. The settlements were in most cases small and lacked the high status material and imported goods found on sites that had a central function or a well developed trade network (Nicolay & Den Hengst 2008: 587). There are however two important exceptions to this situation. One of these is the Roman Period settlement of Wijster that seems to have had both a central function and long distance connections, especially with the Roman Empire, something that is shown in the import goods, and in a couple of rich graves that contained Roman artefacts (Waterbolk 2009: 108). The settlement was larger than most villages in Drenthe from the same period and had an especially large, centrally located farm that included an unusually sturdy building that has been interpreted as having a hall function (Waterbolk 2009: 159).

The other example is Midlaren, the site used as a case study in this thesis. In comparison with Wijster, the settlement is small but from a regional perspective it is a site with special functions during the Roman Period. Midlaren was strategically situated on the outskirts of the Drenthe plateau, close to the Hunze river that functioned as a connecting route to the coast. The village was a small trading centre with contacts both with the nearby coastal *terp*-region, but also further away to the north and south, something that is shown by import goods and luxury products found at the settlement (Nicolay 2005: 60; Nicolay & den Hengst 2008: 577). It is however also important to keep in mind that Midlaren has some of the very best preservation conditions for metal and organic material in Drenthe, something that might contribute to the image of a special settlement compared to other sites in the area (Nicolay & Den Hengst 2008: 590).

Around 800 AD Drenthe was integrated into the expanding Frankish kingdom with subsequent Christianisation, a process that could be considered as finished some time during the 10th century. These events lead to many social and economic changes, including the end of the old religious beliefs and of many ritual activities deemed as pagan by the church (Nicolay & Den Hengst 2008: 577; Waterbolk 2009: 176).

2. Archaeozoological methods

2.1 The material analyses

The four bone materials used as case studies in this thesis have been analysed using acknowledged archaeozoological methods, but with special emphasis on the registration of taphonomic agents. In this chapter the primary archaeozoological methods that have been used in the bone analyses will be presented, while the methods and theoretical framework for identifying ritual deposits will be presented in chapter 3.

2.1.1 Identification

The identifications of the bone fragments was performed with the help of a bone reference collection, for the Dutch materials at the Groningen Institute of Archaeology (Rijksuniversiteit Groningen) and for the Uppåkra material at the Department of Archaeology and Ancient History (Lund University). The access to an extensive and well organized reference collection is vital to any archaeozoological study. It is through comparisons with the reference skeletons that the often fragmentary bones can be identified to element, side and species.

In addition to the reference collections several manuals and books were used, primarily complementary during the identification and for the correct naming of the different bone elements. The pictorial manuals include *Atlas of Animal bones/Knochenatlas For Prehistorians, Archaeologists and Quaternal Geologists* (Schmid 1972) and *Bildkompendium Historisk Osteologi* (Eriksson & Iregren 2002). The Latin terminology that was used in the thesis is based on the book *Lehrbuch der Anatomie der Haustiere* (Nickel *et al.* 1968).

2.1.2 Measurements

Measurements on bones from the materials have been taken with dial callipers according to the guidelines given by Angela von den Driesch in her book *A guide to the measurement of animal bones from archaeological sites* (von den Driesch 1976).

2.1.3 Registration

The registration of the collected observations were made in writing on forms that included both the standard information categories used for archaeozoological analyses by the Groningen Institute of Archaeology and the Department of Archaeology and Ancient History respectively, with the addition of the extra taphonomic information described in section 2.2.

Every material, phase and deposit was furthermore described in text during the analyses and especially interesting bone fragments were drawn or photographed. Afterwards all the information was registered in MS Access databases that are stored at the Groningen Institute of Archaeology.

2.2 Taphonomic processes

2.2.1 Introduction

The study of the different taphonomic processes, both natural and human, that affects and changes a bone material from the time an animal dies until the bones are rediscovered is considered a vital part of most osteological analyses. Nevertheless many archaeozoologists deal with the subject in a very general way, presenting the problems involved in studying such a fragile material as bone, but mostly without going into detail and without evaluating the actual effects of the different taphonomic processes on the assemblage in question (Magnell 2006: 1).

Because of this, taphonomy is seldom used to extract more information from an assemblage, but only as a way to explain biases in a material. I think that the writing of a taphonomic history for every site and deposit is a vital step towards understanding them, and therefore also a useful method for identifying human activities behind deposits.

In dealing with taphonomy some terminology first has to be addressed. The terms *taphonomic agent* and *taphonomic process* are often used in archaeozoological literature and have been defined by several authors. In the article *Current Developments in Bone Technology* Eileen Johnson for example defines the term *agency* as:

> "...*the general group to which the phenomenon causing change in the original state of the bone or a bone assemblage can be classed (e.g., biological, geological, hominid).*"
> (Johnson 1985: 158)

She differentiates between this and the term taphonomic process, which she regards as:

> *"...the 'how' it became modified (e.g., scored by a tooth scratching the cortical surface) and the 'what' that caused the modification (e.g., large carnivore)."*
> (Johnson 1985: 158)

In his book *Vertebrate taphonomy* Lyman uses the terms taphonomic agent and taphonomic process in similar ways:

> *"A taphonomic process is the dynamic action of an agent on animal carcasses and skeletal tissues, such as the downslope movement, gnawing, or fracturing..."*
> (Lyman 1994: 3)

Of these two terms the taphonomic process will be most frequently mentioned in this thesis. In the analysis of my materials I decided to systematically register six different types of taphonomic processes (in addition to the standard data collected for this type of material at the mentioned institutes): *weathering, trampling, heating, fracture freshness index, animal processing* and *animal manipulation*. Many more processes were registered in the analysis of my earliest materials, that were later discarded. The reason for this was either that the categories were impossible to identify, because their value for identifying special and ritual deposits were found to be non-existent, or because the frequency of the specific process was so low in the materials that a comment in the remarks column was sufficient.

For some of the taphonomic processes that I choose to record there is a multitude of registration systems in use, while others lack any consistent system at all. The different systems have advantages and disadvantages, and much of their usefulness depends on the aim of the analysis and the time available. As was mentioned above I decided to put great emphasis on the registration and quantification of taphonomic processes in this study and therefore decided to use relatively detailed registration systems for these.

2.2.2 Weathering

Anna Behrensmeyer defines the process of weathering on bone as:

> *"...the process by which the original microscopic organic and inorganic components of a bone are separated from each other and destroyed by physical and chemical agents operating on the bone in situ, either on the surface or within the soil zone."*
> (Behrensmeyer 1978: 153)

Most research on the subject deal with bone weathering on the soil surface, but weathering caused by chemical and biological agents also affect the bones in the ground (White & Hannus 1983). These processes result in bone surface deterioration that is often impossible to distinguish from weathering of bones lying on the ground surface. It is therefore practical to use one word to describe the collective result of both processes.

I decided to use the weathering stage scale that Behrensmeyer presented in her article *Taphonomic and ecologic information from bone weathering* (Behrensmeyer 1978). These stages illustrate not only a continuous process of destruction by weathering, but actually describe real stages that have comparably short intermediary transition phases. The system with its six stages is simple to use and has become a standard within taphonomic studies (table 2.1).

According to Behrensmeyer (1978) the most severe signs of weathering must cover more than one square centimetre of the bone's surface to be registered. Edges or areas with physical damage should be ignored. Since all soft tissue and marrow is usually long gone in an archaeological bone material (and not necessarily as the result of weathering), these parts of the descriptions were not taken into consideration in my registration.

2.2.3 Trampling

The process of stepping on and kicking bones lying on the ground or in the top soil performed both by humans

Table 2.1. Weathering stages (Behrensmeyer 1978). Phrases in parenthesis are additions from Johnson (1985).

Stage	Description
0	Greasy, no cracking or flaking, perhaps with skin or ligament/soft tissue attached (marrow edible, bone still moist)
1	Cracking parallel to fibre structure (longitudinal); articular surfaces perhaps with mosaic cracking of covering tissue and bone (split lines begin to form, low moisture, marrow sours and is inedible)
2	Flaking of outer surface (exfoliation), cracks are present, crack edge is angular (marrow decays, split lines well developed)
3	Rough homogeneously altered compact bone resulting in fibrous texture; weathering penetrates1-1.5 mm maximum; crack edges are rounded
4	Coarsely fibrous and rough surface; splinters of bone loose on surface, with weathering penetrating inner cavities; open cracks
5	Bone falling apart in situ, large splinters present, bone material very fragile

Table 2.2. Trampling stages.

Stage	Description
0	No sign of scratch marks
1	Very few, but clear traces of scratch marks have been identified
2	Several and clear examples of scratch marks
3	Most of the bone fragment's surface is covered with scratch marks

and animals is called trampling. The process of trampling can affect bones in three principle ways. Through stratigraphical observations and experiments it has been shown that trampling can move bone fragments both horizontally and vertically. The distances the bones are transported depend on several factors such as the intensity of the trampling activity, the density and moisture of the soil, and the size, shape and orientation of the bone fragments. Horizontal movement can be several meters for small fragments while vertical transport seldom exceeds a few centimetres, unless the soil is very sandy (Gifford-Gonzales *et al.* 1985).

Trampling can also result in the breaking of bones into smaller fragments. This effect is especially common on slightly weathered bones, since fresh fragments have a much higher flexibility. Fresh bone can still break as the result of trampling, but this is probably rare (see also section 2.2.5 Bone fragmentation) (Lyman 1994: 379-380).

The third effect trampling can have on bones is the production of scratch marks. These can occasionally be difficult to distinguish from stone tool marks and animal tooth scoring, however the trampling marks are usually shallow and tend to be very randomly distributed and concentrated on the shafts of long bones (Lyman 1994: 180-181). As a contrast animal scoring is usually concentrated on the epiphyseal ends of long bones, while cut marks often are repeated side by side.

Since to my knowledge no established method for recording trampling exists I decided to register this taphonomic process through a simple new system. Every bone fragment was given a trampling stage where 0 indicate that no trampling traces are observed and 3 means that most of the bone is covered with scratch marks. The different stages are presented in table 2.2.

2.2.4 Heating

If a bone is affected by extreme heat its basic physical and chemical properties will change dramatically. This is mainly manifested through a decrease of weight and size, and a change of colour. During the process of heating water is first evaporated after which the destruction of the bone's organic component follows, resulting in a lighter and eventually smaller fragment. The colour also changes with raising temperature from yellowish to red-brown, blue-black and grey-white. The colour changes are equivalent to the maximum temperature the bone had been exposed to according to table 2.3 (Lyman 1994: 384-392).

2.2.5 Bone fragmentation

The physical characteristics of bone breaking are very complicated since the specific bone elements have complex shapes and structures. Bone as a material is also very changeable, from a piece in fresh condition that can absorb quite some force before failing, to a brittle and dry state that easily breaks if force is applied to it. The details of these characteristics are however not of immediate concern for my analyses, and have been thoroughly described elsewhere (see for example Johnson 1985).

Of more interest is the actual identification of different types of fractures on bones, the relative size of the resulting fragments and what these things can tell us about the different events a bone fragment has gone through. The way a fracture is formed is the result of taphonomic processes induced on the bones, naturally or by human actions. By identifying these processes it is possible to describe a part of the taphonomic history for the specific bone and its assemblage.

There exists a myriad of possible fracture sequences for bones, and often many different taphonomic processes have affected the same fragment, making the identification of the cause of a fracture difficult. The solution to this problem is a simplification of the methods used in the analysis. In general simplification means that some information is lost, in this case this loss is nevertheless justifiable since the high complexity of fragmentation probably cannot be solved by "looking closer" at the material, and since the time spent on one specific part of an analysis always has to be weighted against the information that will be produced.

In my analyses I use two systems for registering bone

Table 2.3. Bone colour as the result of heating (partly from Lyman 1994: 386).

Stage	Temperature	Description
0	<250°C	Yellowish (only possible to distinguish on fresh bones)
1	250-525°C	Red-brown
2	525-675°C	Blue-black
3	>675°C	Grey-white

Table 2.4. Criteria for calculating fracture freshness index (from Outram 2001: 405-409).

Criteria	0	1	2
Fracture angle	No more than 10% of the fracture surface is perpendicular to the cortical surface	Between 10 and 50% of the fracture surface is perpendicular	More than 50% of the fracture surface is perpendicular
Fracture surface texture	Surface is entirely smooth (apart from stress relief features)	Some roughness but texture mainly smooth	Largely rough edges
Fracture outline	Only helical (spiral) breaks	Partly helical form	Total absence of helical form

fragmentation. The first is a system to distinguish fresh from dry bone fractures with the help of something called a fracture freshness index (see table 2.4), while the second is a simple estimation of a fragments size compared to a complete bone of the same type.

Fracture freshness index

There are many systems that categorize the fracture surface of a bone according to a number of different breakage templates (see for example Shipman *et al.* 1981; Marshall 1989). The problem with these systems is for one part that it is difficult to simplify something as complex as a fracture surface into a few type cases, but also that the categorizing itself not necessarily help in identifying the actual process that created the fracture. The lengthy discussion surrounding the causes behind early bone assemblages with so called spiral fractures is a good example of this (see among others Dart 1956; Hill 1976; Binford 1981; Haynes 1983b).

Instead of breakage templates I decided to use a system devised by Alan Outram that attaches a fracture freshness index (FFI) to all bone fractures on dense diaphysis bone in order to discern whether the fracture was made on fresh or dry bone. The exact cause behind the fracture is not revealed with this system, but it will indicate whether the fracture happened while the bone was still fresh or later in a dry state (Outram 2001: 405-409). If a dry fracture is identified it could be assumed that this in most cases is the result of non human activity (or at least not of conscious human activity).

The system uses three criteria: *fracture angle*, *fracture surface texture* and *fracture outline*. Each of these criteria is given a score between 0 and 2 where 0 indicates a very fresh bone and 2 a very dry one (table 2.4). By adding the scores for the three categories a fracture freshness index is produced that can vary between 0 and 6. A fracture freshness index of 0-2 indicate a bone that was fractured in a fresh state, while 4-6 indicate a bone that was fractured in a dry state.

Only fractures on bone shafts can be determined with this system, and after some testing it was obvious that some further specifications had to be made. I decided to record a fracture freshness index for every easily separable fracture surface on the bone, but did not apply the system at all on fragments that weighed less than 5 g since the fracture surface texture was the only criterion that could be reliably determined for such small fragments.

Fragmentation degree

One of the main problems involving quantification within archaeozoology is the varying fragmentation degrees between different bone assemblages and even between different features and layers within the same material. The processes that result in a more or less fragmented material are of great interest from a taphonomic point of view. To give some indication of the fragmentation degree in my materials I chose to use a slight modification of a system devised at the Groningen Institute of Archaeology.

All bone elements except the cranium were divided into three fragmentation zones (numbered 1 to 3 beginning from the cranial direction, if possible). The cranium was because of its complexity and its tendency to fragment along its sutures, instead divided into 6 different zones. The zones for a number of common bone elements are depicted in figure 2.1.

The numbers of all the zones that were observed in a fragment was registered. Only complete or virtually complete zones were recorded. A humerus with all but the proximal end remaining would thus be recorded as having the zones 2 and 3. Fragmentation zones were never recorded for unidentifiable fragments or teeth.

2.2.6 Animal processing

An animal that is slaughtered is usually processed in order to extract its valuable resources. Under normal circumstances an animals' hide, meat, fat, marrow, bone and horn are extracted and used for food, fuel or as raw material. When searching for human actions behind deposits of bones, understanding animal processing is important for two reasons. First of all it is vital to know the taphonomic history of a bone deposit if any conclusions about the reasons behind it should be drawn, and the way an animal is processed and how the bone refuse is handled during and after this process is in its own right a very important part of the taphonomic history of a material. Secondly it is also essential to know how the processing of animals in a settlement was usually performed, in order to make it possible to identify irregularities in this regard. It is likely that the processing of an animal that was for example killed for a ritual purpose differs from other forms of butchery and only by observing what is normal practice in a society is it possible to find the special (see for example Wilkens 2002: 75).

There are several aspects of a material such as ele-

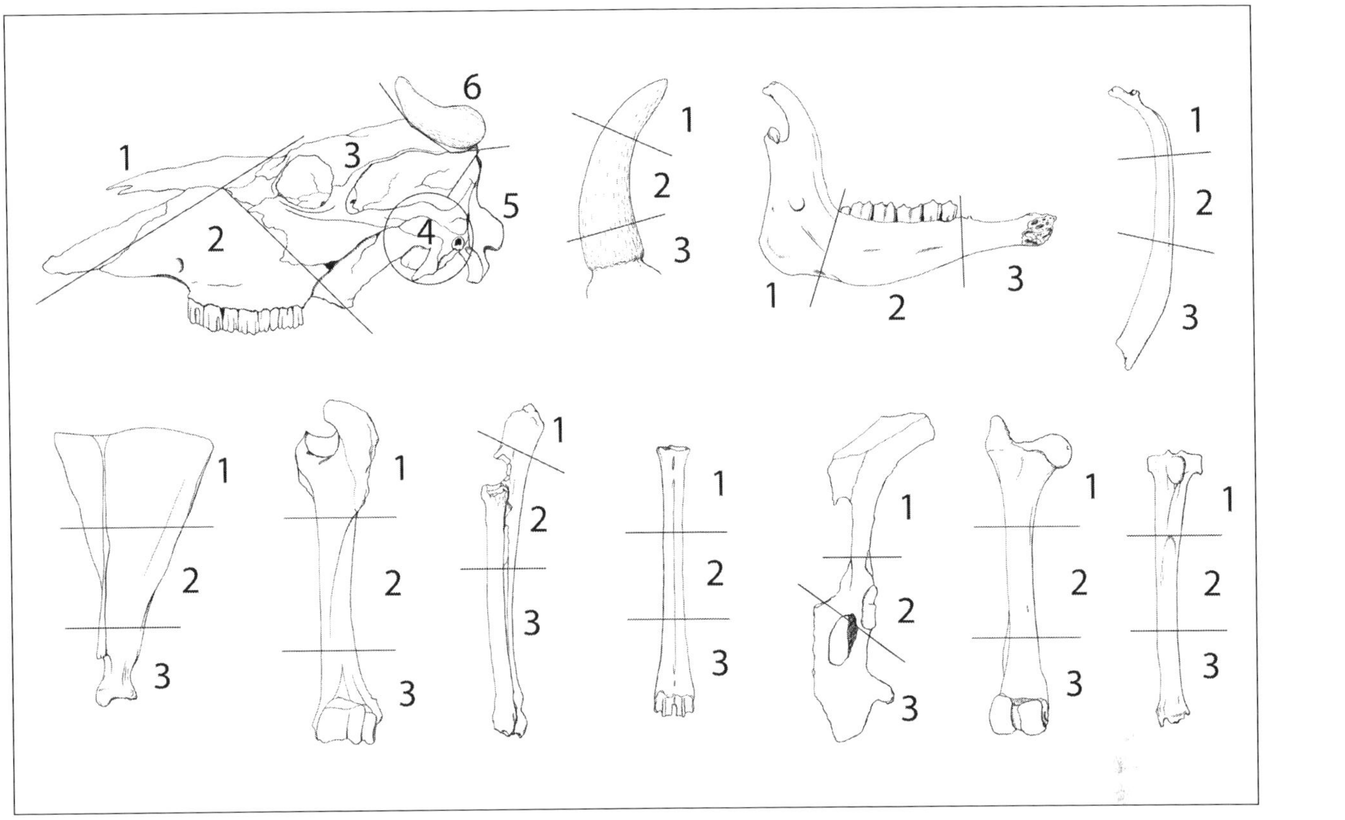

Figure 2.1. Fragmentation zones for some common bone elements (the bones depicted from the top left are a cattle cranium and horn core in addition to a sheep mandible, rib, scapula, humerus, radius/ulna, metatarsus, pelvis, humerus and tibia).

ment distribution and bone fragmentation that can give information regarding the way animals were processed. The most frequently used tool for understanding animal processing is however the recording of the different cut, chop, scraping and saw marks that derive from such activities as disarticulating and filleting. In the course of the recording of the taphonomic processes it is both important to correctly identify which marks are the results of animal processing, as well as to register their frequency and placement.

For my materials the marks' character and placement were registered according to the system presented by Roel Lauwerier in the book *Animals in Roman times in the Dutch eastern river area* (Lauwerier 1988: 40-42, 181-212). This registration system is built around a catalogue of drawings and definitions for different types of butchery mark. Every bone element has a series of different possible marks that each has an individual number beginning with 1. The number indicates where the butchery mark is located on the bone, and what type it is. If a new type or location of mark is found during the analysis, this is given the first free number for the specific element (see table 2.5 and Appendix I for the additional butchery marks found in the analysed materials, but not identified by Lauwerier in his Roman materials).

A mark was registered as one instance of force application if it was more than one centimetre from the nearest other mark (compare with Lyman 1994: 304).

The processing of an animal is usually described in different stages that cover the specific steps that humans take to retrieve mainly skin, meat, marrow, horn and bone from an animal. The different stages are mainly a terminology that an outside ethnographical observer or archaeozoologist use to structure the process. These divisions are influenced both by arbitrary decisions from the observer's side and on actual differences in the animal processing. The different stages might also sometimes flow seamlessly into each other, or vary somewhat depending on cultural habits and specific circumstances at the time and place that the animal was killed and processed. It has for example been observed that the butchery process among different contemporary hunter-gatherer populations is not the same, and that there is even variability within the same group. Nunamiut Eskimos for instance butcher an animal with a slightly different technique when the body is frozen, as compared to when it is fresh (Binford 1981: 109-110, 126).

Despite these variations the main processing sequence seems to be roughly the same. Among the above mentioned Nunamiut Eskimos Binford has observed the sequence of *skinning, dismembering, filleting* and *marrow consumption* (Binford 1981: 106). Noe-Nygaard has the same stages in her description of Danish Mesolithic materials but uses the expressions *skinning, butchering, filleting* and *marrow fracturing* (Noe-Nygaard 1995: 182-186). Lyman uses a slightly different terminology with his *kill-butchery stage, secondary butchery stage* and *final butchery-consumption stage*, a system maybe even more

Table 2.5. Additional butchery marks found in the four materials but not already included in Lauwerier's system.

Element	Number	Type	Description
Cranium	13	chop	on zygomaticum
Cranium	14	cut	on zygomaticum
Cranium	15	chop	on the rear part of os frontale
Cranium	16	chop	inside the dental row
Cranium	17	chop	chopped through
Cranium	18	chop	on frontale
Cranium	19	chop	incisivum chopped through
Cranium	20	cut	on incisivum
Cranium	21	cut	As 18, but cut
Horn-core	9	chop	as 7/8, but not cut through
Mandibula	34	chop	on lateral side of the diastema
Mandibula	35	cut	on medial side of the diastema
Mandibula	36	chop	on medial side of processus coronoides
Mandibula	37	cut	as 36, but cut
Mandibula	38	chop	on lateral side of processus coronoides
Mandibula	39	cut	on lateral side of the ramus
Mandibula	40	chop	as 39, but chop
Mandibula	41	chop	mandible chopped through near the incisors
Atlas	19	cut	on cranial part
Atlas	20	chop	chopped through
Atlas	21	chop	chopped through
Atlas	22	cut	on cranial surface
Vertebrae	20	chop	on processus transversus
Vertebrae	21	cut	as 20, but cut
Vertebrae	22	cut	on ventral corpus vertebrae
Vertebrae	23	chop	as 22, but chop
Vertebrae	24	chop	as 8, but not chopped through
Vertebrae	25	cut	on lateral side of the corpus
Vertebrae	26	cut	on processus articularis cranialis
Vertebrae	27	chop	as 26, but chop
Scapula	41	chop	as 29, but chopped through
Humerus	37	chop	as 18, but chopped through
Ulna	16	cut	on volar-lateral side of the proximal end
Ulna	17	chop	on lateral side of the distal part
Ulna	18	chop	as 16, but chop
Ulna	19	cut	on medial side of the distal part
Metapodia	31		shaving mark
Phalanges	13	chop	on distal surface
Pelvis	30	cut	on medial side of the ischium
Astragalus	14	cut	on middle part of the planar side
Astragalus	15	chop	on proximal part of the dorsal side
Astragalus	16		as 15, but carved out
Sacrum	1	cut	corpus chopped through
Sacrum	2	cut	processus articularis chopped through
Centrotarsale	1	cut	on corpus

aimed towards hunter-gatherer populations than the others (Lyman 1994: 300).

For my material studies I chose to use Binford's stage division with two exceptions. Since I am working with societies that mainly exploited domesticated animals, the addition of a *killing* stage seems reasonable, especially since ritual practice often involves special ways of killing the animal in question (see for example Cope 2004: 26). I also added a sixth processing category *craft* that includes activity connected with bone and antler craft. These additions result in a system with the following processing stages:

Killing

This stage includes the actual killing of an animal, whether it is in the form of the killing of a domesticated animal or the hunting of wild game.

Skinning

This stage deals with the removal of the skin from an animal.

Disarticulation

This stage involves the disarticulation of an animal and the partitioning of the body into large pieces.

Filleting

This stage includes the removal of meat from the bones, and the partitioning of the already disarticulated sections into pieces ready for cooking or storage (e.g. smoking, drying, salting etc.).

Marrow extraction

This stage deals with the preparation for and actual extraction of marrow.

Craft

This stage includes the bones that have been used (or prepared for use) as raw material for bone and antler craft.

There are of course other man-made marks on the bones that do not fit into any of the above mentioned categories. Semi-finished artefacts do qualify for the Craft stage even if no marks were identified according to Lauwerier's system (which is quite natural since the system primarily was created to register butchery marks). Even more specialized marks such as rune carvings or decorations may appear and would then be noted in the remark column of the registration sheet.

It is clear that the processing stage system is a rather schematic and crude tool that does not take into consideration all the different possibilities and variations of real animal processing. Despite this the system serves its purpose and gives an indication of what general stages a bone deposit, or even an individual bone, has gone through. Something to keep in mind is that since none of the above mentioned stages is guaranteed to leave marks on the bones and since complete or almost complete skeletons are seldom found, the system has to be quite general if any results should be expected.

In order to identify which processing stages a bone material has been through, the registered butchery marks have to be associated with one or more of these stages. Doing this is however a very delicate task. Many researchers have tried to identify what type of butchery activity produced a specific mark on the bones, but the consensus is often lacking (compare Ekman 1973: 66-67; Von den Driesch & Boessneck 1975: 18-20; Binford 1981: 98, 107, 110-116, 119-126, 132-133; Luff 1982: 30, 103-104; Lauwerier 1988: 154-160, 182; Maltby 1989: 78, 83, 86-88; Rixson 1989: 50-57; Noe-Nygaard & Richter 1990: 179-181; Hüster-Plogmann 1993: 225-230; Noe-Nygaard 1995: 184, 190-191, 194-195, 198, 202-205, 212; Wilson 1996: 34; Nilsson 2001: 93; Magnell 2003: 674-676).

The placement of a mark described by a researcher does also often pose a problem. In many cases it is virtually impossible to figure out exactly where the described mark is placed, or how exact this placement must be to still represent the same activity.

Another problem is that many of the texts referenced above either deal with Palaeolithic or Mesolithic materials, or are ethnographic studies of present day hunter-gatherer populations. This brings forward the question of how the medieval butchery practices differed from these early examples. One important question here is the use of chopping instruments. In the Medieval Period it was common to use both chopping instruments and knifes in the butchery process, something that is shown in the bone materials as well as in medieval paintings (Hodgson & Jones 1982: 238; Van Wijngaarden-Bakker 1990: 172; and figure 2.2). The use of both cleavers and knifes is evident in the materials I have examined as a part of this thesis.

According to several ethnographic sources cleavers or choppers are sometimes also used in the butchering performed by contemporary hunter-gatherer as well, but just as often, as is the case of for example the Nunamiut Eskimos, they are not used at all. Despite the belief of earlier scholars the use of chopping instruments seems to have been quite rare in prehistoric times, at least until the Bronze Age (Von den Driesch & Boessneck 1975; Binford 1981: 142-147). To complicate things further Maltby observes clear differences in butchery practice between urban and rural contexts in English materials from the Roman Period. Cleavers and choppers were used in both type of sites but were a lot more frequent in the towns (Maltby 1989: 89).

Because of the above mentioned problems I have chosen to base the interpretation of the different processing marks on two main sources. The interpretations are

Figure 2.2. Medieval butchery with a meat cleaver (after the Luttrell Psalter 1325 AD).

primarily based on the personal comments from Ola Magnell, an archaeozoologist with good experience from butchery experiments, and on the information provided by Lauwerier himself in the book where he presents his registration system. In this way confusion regarding the placement of the marks is avoided, and the interpretations will also be made with cleavers taken into consideration since it deals with Roman Period materials. In addition to this, information from the researchers referenced above has been used in uncertain or ambiguous cases.

In table 2.6 the interpretations are presented for the processing marks according to Lauwerier's number system divided into the six different processing stages. Marks appear in more than one column if they could be the result of more than one type of activity.

There are some problems one should be aware of when using this table. The division of a number of different animal processing marks into stages is obviously a simplification of a fairly complicated process. I have also chosen to deal with one single system of interpretation for animal processing, regardless of what species that is being studied, which is of course yet another simplification. Differences in butchery patterns, especially between large and small animals should be expected. It is likely that for example smaller animals were more seldom disarticulated to the same degree as bigger ones, and that these marks therefore will be less frequent. Obviously the system is aimed towards mammals, and most of the references has dealt with the domesticated species (mainly cattle, sheep/goat, pig and horse) and some of the more commonly hunted animals (mainly wild boar, red deer and caribou), and should be used with caution for other species than these.

2.2.7 Manipulation by animals

A multitude of practical experiments, ethnographical studies and observations of animals in the wild have shown how different species can manipulate bones, mainly by moving, gnawing, crushing, licking and swallowing them, if given the opportunity. It has also been shown that these processes can seriously affect a bone assemblage. The vast majority of the animal manipulations we observe in settlements in Northern Europe were done by dogs but under special circumstances it is possible that wild carnivores or animals such as rodents, birds, ungulates and pigs were responsible (see for example Brothwell 1976; Greenfield 1988; Berryman 2002).

The identification and quantification of animal manipulation is an important part of the study of the taphonomic history of a bone assemblage. The existence of this taphonomic process in a deposit can among other things explain fracture patterns and help determining how long bones have been lying exposed.

Several researchers have discussed the identification and registration of animal manipulation activities, but a single system has not yet been accepted as standard (Haynes 1980, 1983a; Binford 1981: 44-49; Lyman 1994: 206-212; Thilderqvist 2001).

Since a detailed registration of animal processing beyond appearance and frequency is of less interest in this research I have only registered the existence of animal manipulation, with the addition of a frequency indication in the form of bone destruction. This column in the form was filled in if parts of the bone were missing and this destruction could be attributed to animal manipulation. Such a connection is usually indicated by the existence of other animal related marks such as pitting, furrowing and punctures on the remaining parts of the bone fragment.

2.3 Sex and age determination

2.3.1 Introduction

Both the sex and the age at death of animals found in a deposit can be of importance in the search for human activity, such as rituals. It has been shown that animals of a specific sex were preferred for specific ritual activities in some societies, for example in the Roman *suovitaurilia* offerings (Lauwerier 2002: 68; Wilkens 2002: 74). In the same way animals of a certain age, especially young ones, were preferred in some ritual activities in various cultures (see for example Merrifield 1987: 34; Lentacker *et al.* 2004: 92; Lauwerier 2002: 65, 2004: 68). In addition to this a meticulous age determination is necessary if seasonal analyses are to be performed on a material (see section 2.5).

Table 2.6. Bone marks according to processing stage (mainly after Magnell, personal comments and Lauwerier 1988: 154-160, 182). The marks within parenthesis can belong to multiple stages while question marks indicates uncertainty.

	Processing stage					
Element	**Killing**	**Skinning**	**Disarticulation**	**Filleting**	**Marrow extraction**	**Craft***
Cranium	1, (2), (18)	3, 8, (14), 20-21	(2), 4-7, (9-11), 12-13, 15-16, (17?), (18-19)	(9-11), (14?)	(17), (19)	-
Horn core	-	6	1, 3, 5	-	-	2, 4, 7-9
Antler	-	-	-	-	-	1-10
Mandibula	-	6, 8, 31, 33	(4), (7), 11, (12), 14-24, (27), 29-30, 32, 34-41	(12), 13, 26, 28	1-3, (4), 5, (7), 9, 25, (27)	10?
Hyoideum	-	-	-	1-2	-	-
Atlas	-	-	1-4, (5-9?), (10-11), 12-17, (18), 19, (20), 21-22	(5-11), (18)	(20)	-
Axis	-	-	1, (2), 3, 4?, 5-10, (11), 12-14	(2), (11?)	-	-
Vertebra	-	-	2-8, (9?), 10, 15-19, 23-24	1, (9), 11-14, 20-22, 25	-	-
Costa	-	-	1-4, (5), 7	(5), 6	-	-
Scapula	-	-	1, 4-25, 27-32	2-3, 33-39	-	26?
Humerus	-	-	1-6, (7), (9), (11), 16, (17), 18-20, 23-36	8, 10, 12-15, (17?), 21	(7), (9), (11), 22	-
Ulna	-	-	1-10, 12-13, (14), 15, 17-18	11, (14), 16, 19	-	-
Radius	-	-	1-8, (9), 10-12, (13), 14, (15), 16, (17), (19), (21), (23), (26-27), 29-33, (34?)	(9), (13), (15), 18, 20, 22, 24, 28	(17), (19) , (21), (23), (26-27), (34)	-
Metapodia	-	(13**), 20	1-10, 12, (13***), (21), 22-23, 25-30	24	14-15, (19)	11?, 16-18, (19), (21), 31
Phalanx I	-	10	1-9, 11-13	-	-	-
Pelvis	-	-	1-14, (15), 16-17, 19-20, 22, 24-29	(15), 18, 21, 23?, 30	-	-
Femur	-	-	1-13, (14), 15-16, (17-18), (21-22), (24), 29-35, (36)	19-20, 23, 25	(14), (17-18), (21-22), (24), (36)	-
Tibia	-		1-6, (7-8), (12-13), (15), (17), (19), 25-31, (32-33), 34-38	9-11, 14, 16, 18, 20-24	(7-8), (12-13), (15), (17), (19), (32-33)	-
Calcaneus	-	(6), 8	1-5, (6), 7, 9-16	-	-	-
Astragalus	-	-	1-15	-	-	16
Sacrum	-	-	1	2	-	-
Centrotarsale	-	-	1	-	-	-

* Only the marks in Lauwerier's system is presented here, but many other traces on the bone fragments can of course be interpreted as the results of craft.

** If it is placed lateral/medial on the bone.

*** If it is placed dorsal/planar on the bone.

2.3.2 Age determination

The age at death of an animal can be determined from criteria on different parts of the skeleton. The epiphysis and diaphysis of different bone elements fuse with each other at varying ages for different species. Teeth erupt at various ages and are gradually worn down with time. These different facts can be used to give an indication of how old an animal was when it died. The age can be presented as an exact length of time, or as a time range, depending on how precise the characteristics for determining the age at death are. It is important to remember that the suggested age at death of an animal always contains a degree of uncertainty. The different systems for determining age at death do, both because of this uncertainty, and due to methodological problems, not always produce the same results (Vretemark 1997: 35-36).

In the material descriptions I have chosen to translate the different age criteria into seven broad categories taken from the bone registration manuals used at the Groningen Institute of Archaeology, something that makes it possible to compare the ages of different animals regardless of the specific method used. Of course this is a simplification, but in the situations where detailed questions regarding the age at death for animals in the materials were to be answered, for example in connection with seasonal analysis, more detailed criteria were used. In table 2.7 the seven age categories are presented together with their abbreviations and corresponding real ages for sheep, cattle and pig.

The category subadult/adult/senil is included as a complement to the others in an effort to increase the number of bone fragments that can be used in age assessment. It is often clear that a bone fragment is subadult or older when the epiphyses are completely fused, but since it is in most cases still impossible to differentiate between subadult, adult and senil, they are usually not included.

Fusion of epiphyses

The use of fusion of epiphysis for determining the animals' age at death has been described by many authors (e.g. Silver 1969; Habermehl 1975). The possibility to use the method on many different bone elements and not only teeth is useful, but unfortunately the method is not very precise due to both the problem of using modern observations as the basis for old materials, and the existence of castrates since the castration affects the time of fusion. For modern comparative animals the time of fusion has mainly been registered through the use of x-ray photographs. These are in themselves difficult to analyse and it is furthermore likely that the time of fusion in some cases is not the same for modern as for historic and prehistoric animals (Silver 1969).

Epiphyseal fusion has been recorded according to the registration manual in use at Groningen Institute of Archaeology for the Dutch materials and according to the standard at the Department of Archaeology and Ancient History in Lund for the Uppåkra material (see table 2.8).

In the case where two epiphyses can be observed two symbols have been used with the first one representing the proximal epiphysis and the second symbol the distal one.

In those cases when an age in years or months from the epiphyseal fusion is presented in the text, this has been derived from the publication *Die Altersbestimmung bei Haus- und Labortieren* (Habermehl 1975).

Tooth eruption

The fact that the teeth of animals erupt from the jaws in different order and at different ages has been known for a long time and tooth eruption data for animals have been available at least since the beginning of the 19th century (see for example Girard 1810; Schwab 1833). There has also been an extensive discussion about the reliability of this method. Some researchers have suggested that the intensive breeding in modern times have changed the eruption times of the teeth and that the oldest possible data should therefore be used. Recent research does however criticize this opinion, partly on the grounds that the oldest available data contain both problems with definitions and obvious errors, and partly because the change in eruption times due to intensive breeding cannot be shown with certainty (Vretemark 1997: 37-38; Legge *et al.* 2000: 153).

The age according to tooth eruption for animals in the materials used in this thesis has therefore been determined according to the modern data presented by Habermehl (1975).

Table 2.7. Age categories and suggestions for corresponding real ages for sheep, cattle and pig.

Age category	Abbreviation	Sheep	Cattle	Pig
Foetal/neonatus	fo/nn	Unborn/newborn	Unborn/newborn	Unborn/newborn
Infantil	in	0-6 months	0-8 months	0-6 months
Juvenil	ju	6-12 months	8-18 months	6-14 months
Subadult	sa	12-24 months	18-30 months	14-21 months
Adult	ad	>24 months	>30 months	>21 months
Senil	se	Old	Old	Old
Subadult/adult/senil	sa/ad/se	>6 months	>18 months	>14 months

Table 2.8. Registration systems for epiphyseal fusion.

GIA symbol	Lund designation	Fusion stage
-	fui	Unfused epiphysis
±	fuc-	In the process of fusion
+	fuc	Fused
0		No information available since the part is missing

Tooth wear

After a tooth has fully erupted the enamel will start to wear down. The intensity of the wear depends on different circumstances such as what type of food the animal was eating, but mainly on how long the teeth were in use. If several molars and premolars from the same jaw are present, it is possible to determine a reliable relative age at death for an animal with this method. The diagrams from Grant's *The use of tooth wear as a guide to the age of domestic ungulates* have been used for determining a tooth wear score for mandibles from cattle, sheep/goat and pig (Grant 1982).

2.3.3 Sex determination

There are several methods available for determining the sex of animals from different species. These methods can be divided into two groups: specific and metric. The specific methods deal with traits that can hardly be mistaken, such as the shape of pig's canines or the existence of red deer antlers. The metric methods involve detailed measurements of traits that differ slightly in size between males and females. The problem with these traits is that the size of the sexes often overlaps, especially for domesticated animals. To complicate things the existence of castrates among domestic animals has to be accounted for (Davis 1987: 44-45; Vretemark 1997: 42; Wigh 2001: 51).

Since sex distribution is only one of many tools used for explaining animal bone deposits I have chosen to use only established methods that are relatively quick and easy to apply. Only the methods that were actually used in my analysis are presented here and the presentations are deliberately short. For more detailed information and criticism of the different methods see the referenced literature.

Pelvis from cattle, sheep and goat

The cattle, sheep and goat pelves show clear sex dimorphism already from one year of age and are thus a very good and reliable attribute for sex determination. There are three traits on the cattle pelvis that can be used, the shape of the pubic bone, the thickness of the edge on *fossa muscularis* on the *musculus rectus femoris*, and the thickness of the medial edge of *acetabulum* (Vretemark 1997: 43-44).

Horn cores from cattle, sheep and goat

The size and shape differences for the horn cores from male and female animals, especially the length and base circumference or diameter, can be identified. This can sometimes be done visually, and in more ambiguous cases through measurements and correlation diagrams (Vretemark 1997: 45-46).

Metacarpus from cattle

Measurements of cattle metacarpus-bones are a very good sex-indicator since the difference in size especially affects the front part of the animals. It has been shown that especially the breadth-measurements are of interest here. Another advantage with the use of this element is that the bones are often found complete in archaeological assemblages. The only real problem with the method involves castrated males, at least if the castration took place at a relatively early age (Vretemark 1997: 46-48).

Canines from pig

The canines from pig differ distinctively between the sexes. The boar's tusks have open roots and thus grow all their lives. The size, curving and cross-section of the canines both in the lower and upper jaw make it easy to distinguish between the sexes (Vretemark 1997: 48).

Canines from horse

Horse stallions have canine teeth which the mares usually lack. This means that the absence of such a tooth shows that the animal is a mare, while their existence only suggests that the animal is a stallion, without constituting any clear evidence (Davis 1987: 44; Wigh 2001: 51).

Deer antlers

All deer found naturally in Europe have antler, and with the sole exception of reindeer (*Rangifer tarandus*) only the male carry them. This makes sex-determination simple for antlers and cranial fragments with antler attached. The number of deer finds in my materials was however very small and did mostly consist of loose antler fragments that most likely are raw material for craft.

2.4 Quantification methods

2.4.1 Number of identified specimens

The most frequently used quantification method in archaeozoology is NISP or number of identified specimens. This method is a straightforward count of all bone fragments for every species. The NISP is easy to use and is the basis for many more advanced quantification methods. Used on its own it is however not without problems. Especially a varying degree of fragmentation in different deposits can create skewed results if only the NISP is used. NISP is therefore often studied together with fragment weight or other quantification methods to spot such anomalies (O'Connor 2000: 54-58; Eriksson & Magnell 2001: 164).

2.4.2 Fragment weight

Another common quantification method within archaeozoology is the weight of each fragment. The method is quick and easy to use, but the weight of a fragment does unfortunately also depend on several external factors that can bias the numbers. The amount of moisture in the bones, how well the material has been cleaned and in what soil type the bones were deposited are all such aspects that obscure the information actually sought. All these things make weight comparisons between different sites, and sometimes even different deposits very difficult (O'Connor 2000: 57-58).

The weight is however an easy measurement to take and is often used to give quick estimates of the meat contribution of different species, and can also be used for intra-site comparison and to find the correct fragment after the initial analysis, something that is often called for when old materials are re-examined. As was mentioned above it can also be used to support NISP-quantifications regarding the problems with varying fragmentation degrees.

2.4.3 Minimum number of individuals

The most widespread of the more advanced quantification methods is MNI (minimum number of individuals). The method gives an absolute minimum number of animals from the examined species that existed at the site, phase or feature. The number is derived through the calculation of the amount of a unique element from one species (from one side of the body). Sometimes this basic calculation is refined through differentiation according to sex, age and size, as well as spatial and temporal distribution of the bone elements in question.

There are unfortunately several serious problems involved when using the MNI-method. First of all it is very important to remember that the MNI has little to do with the actual number of individual animals that once existed on a site. The fact that a single bone of a species results in an MNI of 1 for example results in a situation where rare species tend to be overrepresented compared to common species.

The frequency of different species can also be biased by the method depending on preservation conditions for different types of elements and different meat and refuse handling practices. Four left upper hind legs from an animal brought to a site does as an example give an MNI of 4, while the bones of a complete animal that was butchered on the same site only counts as MNI 1 (compare O'Connor 2000: 59-61).

These problems are further accentuated when the method is applied to relatively small materials, such as most of the individual deposits analysed in this thesis. The MNI-scores for small materials tend to get low and are thus sensitive to vary small changes. If only a couple more of the "correct" fragments shows up, the whole ratio between the different species will be affected dramatically. With these problems in mind I have only used MNI for deposits with a large number of bone fragments, and for the complete case-study materials as comparison.

2.4.4 Element distribution

It has been shown on many occasions that irregularities in the element distribution among animal bones are useful indicators for identifying varying human behaviour (Reitz & Wing 1999: 203). This is for example the case with ritual deposits such as the so called "head and feet"-deposits found in several locations across the north of Europe, as well as deposits with an unusually high percentage of cranial fragments (Schmid 1965; Grant 1989: 79; Dickmann 1997: 60-62; Carlie 2004: 124-126).

In an analysis such as this where irregularities in a material are actively sought after, one important quantification tool is the use of element group distribution diagrams. In such a diagram all the bones of an animal are divided into a number of logical element groups. This division is a form of interpretation and the reasoning behind the categories can always be discussed. I have chosen to use the 16 different categories presented in table 2.9.

After the element distribution of an assemblage, deposit or species has been established, the percentage of NISP for the element groups, as compared to the total number of identified fragments, is calculated. Usually the resulting NISP-percentages are presented in a diagram containing all the element groups.

In such a diagram it is also possible to present an expected element distribution. This curve is based on the frequency of the complete bones that actually exists in the body of the specific animal species. This could be considered the "ideal" element distribution if no taphonomic processes affected the material. Such a curve for cattle can be seen in figure 2.3.

This situation does of course never exist in reality and such a curve is therefore of limited use. The fragmentation of cranial and long bones into several identifiable parts often makes these groups somewhat overrepresented

Table 2.9. Bone element groups

Element groups	Description
Cranium	All cranial fragments with the exception of loose teeth
Mandibula	All mandible fragments with the exception of loose teeth
Dentes	Loose teeth
Vertebrae	All the vertebras including the sacrum
Costae	All the ribs and the sternum
Scapula	
Humerus	
Radius/ulna	All fragments from both radius and ulna
Carpalia	All the carpal bones
Metacarpus	
Coxae	The pelvis
Femur	
Tibia/fibula	All fragments from both tibia and fibula
Tarsalia	All the tarsal bones including the astragalus and calcaneus
Metatarsus	
Phalanx	Phalanx-bones regardless whether they come from the fore or hind legs

in archaeological materials, while for example the fragmentation of vertebras and ribs into smaller pieces that are often difficult to identify, makes these groups underrepresented. Furthermore, in archaeological situations many of the smaller bones such as *carpalia* and *tarsalia* are underrepresented. If these fragmentation and find circumstances are taken into account the result is a more usable "expected element distribution". In figure 2.4 the modified curve for cattle is presented.

As with all the quantification methods it is important to recognise what the expected element distribution really is, a tool that in itself is already an interpretation.

2.5 Seasonal analysis

The time during the year when a site was in use or even when a deposit was created can sometimes be of interest when interpreting deposits. First of all, butchery practices, including possible seasonal slaughter periods, are always important to investigate in order to understand the norm from which unusual human activities can be separated. It has been shown that ritual activities in many cultures are connected with the seasons of the year. There are examples both in literature and in the archaeological record of holy feasts or special offerings that were only

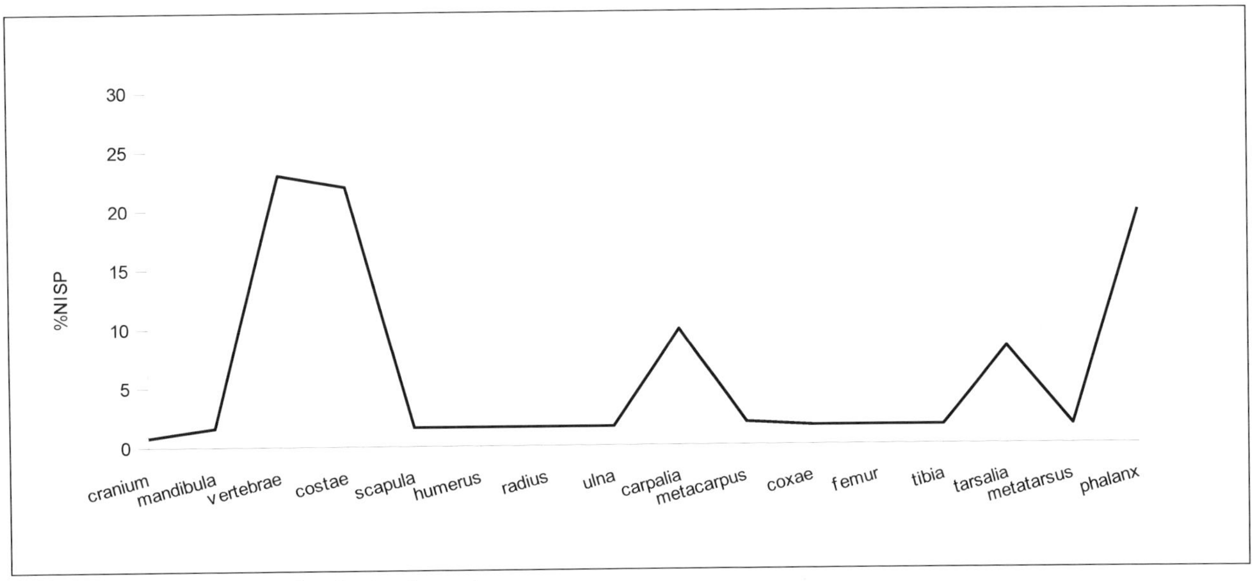

Figure 2.3. The ideal element distribution for cattle.

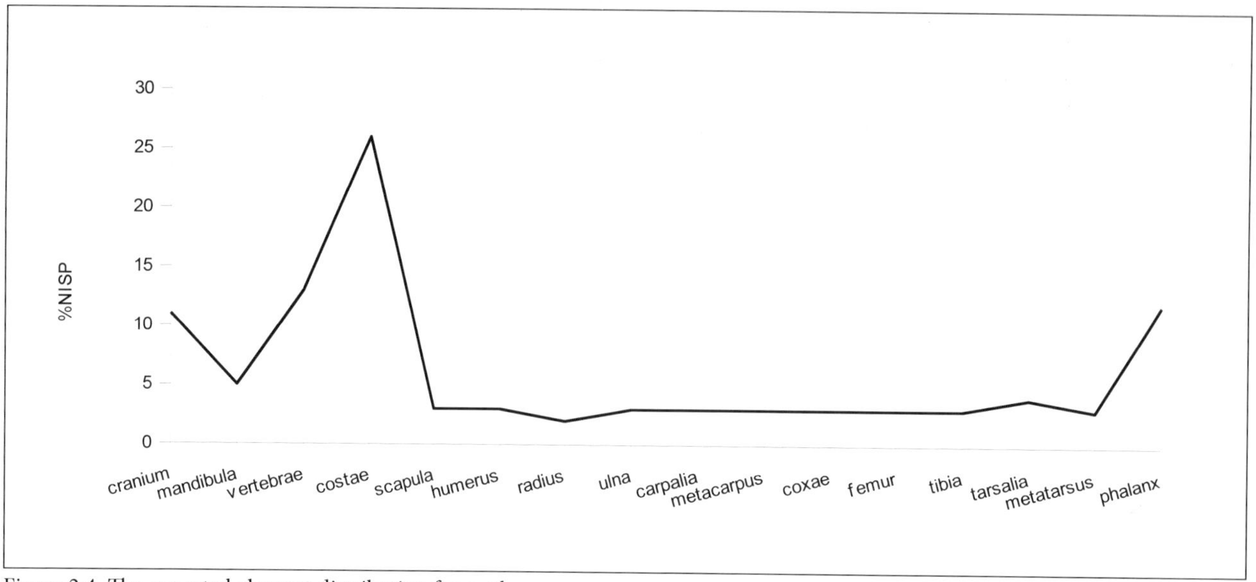

Figure 2.4. The expected element distribution for cattle.

performed during specific time periods, and that these can be identified partly by seasonal studies (see for example Legge *et al.* 2000: 153-154; Raudevere 2002: 52; Therkorn 2004: 31-32).

There are several methods that can be used to identify seasonality in an animal bone material. Direct methods use the presence or absence of specific animals in an assemblage that either occur in the area only at certain times of the year, or that for some reason are more desirable to catch or harvest during a specific season, as an indication for during what time of year a site was used or a deposit was made (Monks 1981: 180-181; Serjantson 1998).

Certain physiological events in the life of animals can also be used to determine seasonality. These methods involve for example the shedding of antlers during a specific part of the year and the incremental structures that can be observed in for example teeth, molluscs and fish scales (Monks 1981: 185, 193-194; Martin 1998).

Finally the use of animals' tendencies to give birth only during a small period of the year is one of the most versatile methods for determining seasonality, since it can be applied on many of the most common mammal species. Mostly the animals give birth in the spring to give the offspring sufficient time to grow before the winter. If the age at death of an animal can be determined with precision the season in which it died can be calculated.

The method has been frequently utilized for identifying seasonality of settlements and hunting stations during the Palaeolithic and Mesolithic, but has fairly seldom been used for post-Neolithic periods. One of the reasons for this is that seasonal use of sites seldom is of interest when the population has become sedentary. But as a major objection to using the method on domestic animals it has been stated that they can be controlled to give birth at "unnatural" times of the year, thus undermining the primary requisite of the method. In the last years it has however been convincingly argued that such a control of the time of birth, although possible, does not fill any practical purpose on domestic animals. The fact is that such a control system actually requires special efforts in the form of extra food resources during the winter to be successful. Since this is the case seasonal analyses should basically be possible to perform on every type of material (Legge *et al.* 2000:1 52; Ervynck 2005: 153-154).

2.5.1 Problems with seasonal analyses

There are unfortunately some problems concerning seasonal analyses. Even if the relevant indicators are available several possibilities have to be addressed. Transport from areas with different seasonal circumstances, or the preservation of food may produce erroneous results and must always be taken into consideration. The amount of the evidence in question is vital here and several different examples of the same indicator are therefore preferable when seasonality is determined (Monks 1981: 184-185).

It is also important to critically view the other methods that are used as the basis for the seasonal analysis. The reliability and level of detail for the methods used to determine the age at death of animals is for example very important when seasonality is based on the birth season of the animals. Because of this the different problems mentioned in connection with the ageing of animals are essential to address when seasonality is regarded (O'Connor 1998).

When single deposits are investigated, as is the focus in this thesis, the problems are even more important. The sample sizes are then in most cases small and fewer indicators are thus likely to be identified. There is also always the possibility that the deposit was produced over a long period of time, or during several distinct phases, but that seasonal indicators only show up from part of the whole time of use. As has been remarked: *"Absence of proof is not proof of absence."* (Davis 1987: 75). Because of this a critical mind is vital when this type of method is used.

3. Identifying ritual deposits

3.1 Theoretical background

3.1.1 Introduction

The categories *sacrifice*, *offering*, *special deposit* and *ritual deposit* are often used interchangeably within archaeology to describe certain unusual dumps of artefacts or biological material. In recent years such non functional interpretations of deposits have been more widely accepted, but similar explanations have actually been used for a long time in Northern Europe, albeit not very often (see for example Worsaae 1886: 162; Montelius 1888: 122).

I will in this text mainly use the two words special and ritual, as they were defined in chapter 1, with special indicating anything out of the ordinary and ritual representing explicit ritual action.

3.1.2 The essence of the ritual deposit

The non-functional interpretations have thus been around for quite a while in archaeology, at least well over a century. The problem has for a long time been, and still is in many analyses, the lack of a structured framework for the methods used to decide if such a non-functional interpretation is valid in a specific case. The question that is too seldom asked is what really should be used as indicators for special or ritual deposits and why?

The question is not easy to answer since the methods commonly used often seem to be *ad hoc*-solutions not clearly presented or even worse, plain "common sense"-based criteria. Generalizing a bit I think that special or ritual interpretations in archaeology commonly are based on the, perceived or real, existence of one or more of the following indicators.

Symbolism

Something concerning the situation is recognized as having a symbolic meaning. The notion of what is symbolic most often comes from contemporary analogies, or more rarely from historical or ethnographical sources. One example of this could be the burial of dogs near the entrances of houses that have been interpreted as symbolic guardians of the doorways (Carlie 2004: 112).

Irregularity

Something in the composition of the deposit is perceived as odd. In this case the objects themselves might be ordinary, but the manner in which they are deposited, where they are positioned, or the combination of objects seems out of the ordinary for some reason. Finds of upside down turned ceramic vessels that appear occasionally over the whole of prehistoric Northern Europe are an example of this, since this is a clearly irregular way of burying ceramics (Carlie 2004: 65; Becker 1972: 16-17).

Impracticality

If, for one reason or another, the deposit is difficult to explain in any functional way. That is to say that the deposit does not fit into any strictly economic or biological model of interpretation. Deposits of for instance complete articulated, and hence not butchered, limbs of animals are often interpreted as a waste of resources, given that the animals in question were usually eaten in that period and society (Grant 1989: 81).

Repetition

There is a tendency to distrust unique deposits within archaeology (see for example Merrifield 1987: 6). If a type of deposit is repeated it can however not be dismissed as a freak incident anymore and is thus more credible (Groot 2007: 100). This is the only one of the indicators that in itself is never used as a sign of symbolic or ritual activity since a great variety of archaeological deposits are of course repeated.

3.1.3 General problems

In connection with all analyses of unusual deposits a number of problems seem to be recurring, especially in the *ad-hoc* or "common sense"-type of studies. I will here briefly present six general problems I consider important to address or at least keep in mind when analyses of special deposits are made or evaluated.

Generalization

Even though the opposite is virtually always assured there is a tendency to apply categories that have been suggested for a certain place over very wide areas and time periods (see for example Piggot 1962; and to some extent Klindt-Jensen 1957: 84-87). It is of course fully possible that a particular type of deposit could have the same meaning at two completely different places or periods without any contact. Such situations could emerge either as the result of surviving traditions, biological constraints or pure coincidence, but every such case has to be assessed on its own merits and from its specific contexts and should not be accepted without criticism (Henriksen 1998: 209-210; Paulsson-Holmberg 1997: 172-173).

Ignorance of the normal

The identification of the remains of explicit rituals or symbolic activities, is per definition a search for the abnormal. As was mentioned in chapter 1 some everyday activities could definitely be defined as both ritual and symbolic, but they are in reality seldom looked for and are not the subject of this thesis. When trying to find the activities that we hope could shed light on explicit rituals we are however almost always seeking the irregular. In such an enterprise the definition and study of the normal material from a site, society or geographical area during the relevant time period is vital. All too often the normality is not studied sufficiently, or at least not taken enough into consideration. Complete animal crania deposited in the soil could be seen as the remains of explicit rituals, but we must first try to understand the normal butchery practice and refuse handling of the society we are studying. It could for example be that the skulls as a rule were not processed and deposited as refuse in this specific culture.

Lack of source criticism

The debate on the reliability of the information we extract from the archaeological material is a difficult topic and much a matter of opinion, both in a practical sense and on a more theoretical level regarding what actually can be deduced from the material culture. A thorough discussion about the limitations and possible gaps in the specific material that is studied should be a mandatory part of any archaeological analysis. Unfortunately this discussion is not always presented, something that is especially risky when dealing with individual deposits instead of larger assemblages, as is the case in the study of special and ritual activity (Wilson 1992).

The over-emphasis of contemporary symbolism

Again despite claims of the opposite the symbolism that we deduce from the circumstances of a deposit plays a large part in whether we interpret it as something out of the ordinary. There is nothing especially wrong with this, but we must be careful to not <u>only</u> see the symbolism we ourselves recognise from our society and time. Some symbols might very well be of a general nature, but we must still reflect upon why we for example so readily interpret deposits containing even tiny amounts of human bones as special.

The over-emphasis of context

Archaeological finds cannot be studied in isolation, neither from the society that produced them, nor from their physical context. Many of the strongest indications of ritual activity come from the place where a deposit is found. It is however incorrect to take only the context into consideration, and forget the material itself. If a certain type of feature for example is shown to contain offerings, it is too easy to interpret everything found in similar features as offerings, regardless of other possible circumstances. Examples of this is the tendency to interpret structures situated close to "established" types of ritual features (such as for example stone formations, graves or, in Scandinavia, early wooden churches) as special regardless of their actual form or content, or to interpret all finds in post holes, however small, as building offerings (see for example Carlie 1992: 58-59; Artelius & Nordin 2007: 40-41).

The paradox of repetition

As was mentioned in section 3.1.2 repetition is often used as a supportive indication for special and ritual deposits, especially to distinguish them from assemblages that look odd out of pure coincidence. It is however noticeable that there is a minor paradox connected with this idea.

There is nothing to indicate that a unique deposit could not be of a special character (uniqueness it is even a part of the definition of speciality), but everyday functional human behaviour is on the other hand in most cases considered to be very repetitive. The bottom line is then that if a type of deposit is not repeated it will usually not be interpreted as ritual, but if it is repeated too often it might not be considered a ritual either.

3.1.4 The sources at our disposal

If the paragraphs above could be regarded as a view of how most earlier analyses of special or ritual deposits have been performed, the question is of course what can and should be done differently to get better results. At this point it is important to take a step back and see what we actually have at our disposal to work with.

Primarily we have the archaeological data collected from excavations and surveys that is sometimes further analysed and interpreted by archaeologists and various specialists. This is by definition the basic dataset available that we are trying to interpret. To do that however we are forced to use various types of analogies. The use of these are most obvious when more elaborate explanations of the archaeological finds and sites are presented, but in reality they are utilized in all phases of the archaeological process. The different types of analogies that are usually employed within the discipline are contemporary, historical and ethnographical. It is of course possible to use written or pictorial sources as primary information if they come from the relevant place and time period, and deal directly with the subject. This is however a rare occurrence for early medieval Northern Europe.

These sources and their advantages and problems will be discussed later in this chapter when the different methods and analogies relevant for identifying special and ritual deposits will be presented.

3.1.5 The five step process

Now that the different sources available have been presented the next stage is to deconstruct the actual practical process of identifying and studying the ritual deposits. There are a number of different paths that can be taken when a material is studied with this aspect in mind depending on the approach and aim of the research project. In this thesis the materials have been analysed with the aim of finding special and ritual deposits. Most archaeological excavations and archaeozoological analyses will however have more general aims and requirements and then the unusual deposits will presumably be treated as one of many interesting aspects of the material. In these cases it is however even more important to analyse the material in a structured way.

The manner of structuring the procedure that evolved during my research was to divide the process into five consecutive steps. These steps should be seen as a framework that could be used in the process of identifying special or ritual activity, or in any similar search for human activities responsible for deposits. The five steps I chose to use are *definition*, *description*, *identification*, *interpretation* and *explanation* (see figure 3.1).

Definition

The definition of what is sought for in the analysis is doubtlessly the most important step of the whole process. It is really the most important step in any study that aims to identify human behaviour in some form. In this case ritual deposit is quite a wide term and is given various meanings by different researchers. The definition I have used has already been presented in chapter 1.

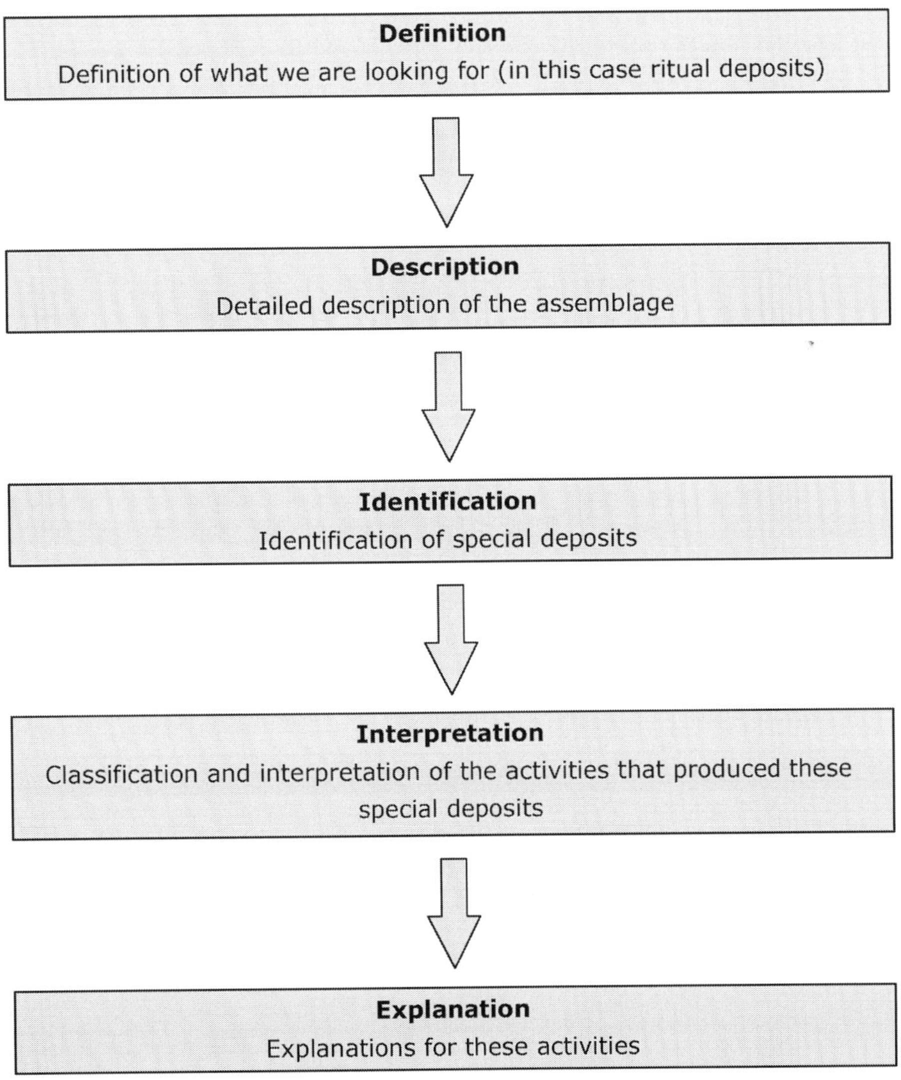

Figure 3.1. The five step process.

Description

That a detailed and methodical description should be produced for the studied material both in general and for every deposit might sound like an obvious step in any analysis, but is far from always done in a satisfactory way. It is however of vital importance, and what is overlooked in this step can never be repaired later.

In the description all those aspects of the material or sample that possibly could help to understand the activities behind the deposits should be recorded in a systematic way. These aspects include a detailed archaeological and archaeozoological investigation, but should also contain as much contextual information as possible.

The realities of archaeological projects are unfortunately often such that a complete detailed analysis cannot be performed. In such cases the researcher has to choose what is most important. This type of prioritizing is encountered in all fields of practical research and does always pose a challenge to the researcher.

Identification

The third step in the process is to identify which deposits differ from the normality of the material, site and region that is investigated. The aspects of how the deposit is unusual can vary a lot. For animal bones these may include species frequency and combinations, element distribution, taphonomic history and find contexts. It is in this step not yet necessary to interpret what type of activity the special deposits are the result of, but this step should instead be seen as a selection process in which everything out of the ordinary is put aside for further study.

Unfortunately many researchers never go beyond this step in their analysis, either because it is too time consuming, too difficult, or because it (in their opinion) would only produce vague or speculative results. In some cases this is perfectly acceptable, for example if the material is small or the time available is short. It is however important to realise that the identification of special deposits is only the half-way point in the process, and not a very interesting result in itself. It is obviously precarious to try to extract more information out of a material than there is a potential for, but in the end the archaeologist must also try to explain and not only observe!

Interpretation

The fourth step of the process is to determine which of the special deposits actually can be interpreted as, in this case, ritual according to the definition from step one. The material is still the main focus but the researcher also has to use analogies in order to produce a convincing interpretation. The activities that can be interpreted in this step are rather general and can consist of categories such as building offerings, remains of ceremonial meals, food offerings, protective magical deposits etc.

Explanation

The final, and without doubt most difficult step in the process is the explanation of the deposits that in the fourth step were interpreted as ritual. In this step some understanding of the reasons and logic behind the ritual activities that produced the deposit is gained. This is a very difficult task that requires a thorough understanding of the culture of the geographical area and period in question. Availability of written sources as well as iconography and the use of ethnographic analogies are of huge importance at this stage.

3.2 Archaeozoological methods

3.2.1 Introduction

For reasons of clarity I have chosen to distinguish between archaeozoological and archaeological material and methods in this chapter. Essentially I see little difference in how archaeozoological material should be treated in contrast to other archaeological material. Both are the product of human activities and have the potential to convey information of past societies and human behaviour. Despite this the specific methods are, at least when viewed in detail, different to some degree and are based on originally dissimilar research traditions.

3.2.2 Species selection

It has often been assumed that certain species should be more prominent than others in ritual activity. This theory is most likely based on the few written sources we have that deal with pre-Christian rituals in Northern Europe. As an example the description by Adam of Bremen concerning the great religious feasts that were supposed to happen every ninth year at the pagan temple in Old Uppsala in the province of Uppland (Sweden), has frequently been used as a strong indication that especially dogs and horses where important sacrificial animals.

> *"They hang the bodies in a grove close by the temple. This grove is so sacred for the pagans that they believe that every single tree in it has become divine through the death and decay of the victims. Dogs and horses hang here and together with them humans. A Christian has told me that he has seen 72 such bodies there, animal and human all mixed up."*
>
> (Adam of Bremen's Deeds of Bishops of the Hamburg Church, book IV, own translation from various sources)

Tacitus tells us in his book Germania of offerings to the gods Mercury, Hercules and Mars (usually interpreted as

Odin/Wotan, Tor and Tyr) among the Germanic tribes. He mentions that Mercury was given human sacrifices while the other two gods received certain typical animal species.

> *"Of all the Gods, Mercury is he whom they worship most. To him on certain stated days it is lawful to offer even human victims. Hercules and Mars they appease with beasts usually allowed for sacrifice."*
> (Tacitus' Germania, chapter IX, translation by Gordon 1737)

The 10th century Arabic geographer Ibn Rusta describes the offerings of either humans or horses among the Rus of Novgorod (for a discussion about the relevance of the Arabic sources of the Rus for Scandinavia, see section 3.4 Analogical reasoning and the use of written sources below).

> *"Sie haben gewisse Wahrsager, von denen einige sogar über ihre Könige Urteile fällen, als ob sie deren Herren wären. Sie befehlen ihnen, dem Schöpfer Frau, Mann oder Pferd als Opfer darzubringen, wie sie [die Wahrsager] es wollen. Haben die Wahrsager das Urteil gefällt, so bleibt ihnen [den Rus'] nichts übrig, als deren Befehl auszufüren: der Wahrsager nimmt ihnen den Menschen oder das Tier, wirft einen Strick um dessen Hals und hängt ihn an einem Holz auf, bis der seine Seele aufgibt, und er [der Wahrsager] sagt: Dies ist ein Opfer für Gott."*
> (Kitāb al-alāq an-nafīsa by Abū Alī Ahmad ibn Umar ibn Rusta, chapter 6 ar-Rūsīya, translation by Kmoskó)

Another 10th century Arabic scribe, Ibn Fadlān wrote a famous text about the Rus he met in the Volga river area. He among other things describes a Rus trader that makes sacrifices of certain animals to his gods after a quick and fruitful trade bargain has been made.

> *"He produces a number of sheep or cows and slaughters them, donating a portion of the meat to charity and taking the rest and casting it before the large piece of wood and the small ones around it."*
> (Ibn Fadlāns Kitāb, Rūsiyyah, translation by Montgomery 2000)

As a final example the 10th century Icelandic poet-warrior Egil uses a horse head put on a stick in his curse on king Erik and queen Gunhild in Egil Skallagrimson's Saga.

> *"He took in his hand a hazel-pole, and went to a rocky eminence that looked inward to the mainland. Then he took a horse's head and fixed it on the pole. After that, in solemn form of curse, he thus spoke: 'Here set I up a curse-pole, and this curse I turn on king Eric and queen Gunnhilda.' Here he turned the horse's head landwards. 'This curse I turn also on the guardian-spirits who dwell in this land, that they may all wander astray, nor reach or find their home till they have driven out of the land king Eric and Gunnhilda.'*
>
> *This spoken, he planted the pole down in a rift of the rock, and let it stand there. The horse's head he turned inwards to the mainland; but on the pole he cut runes, expressing the whole form of curse."*
> (Egil Skallagrimson's Saga, chapter 60, translation by Green 1893)

There is however also an obvious danger involved in this way of reasoning. The finds of specific species, for Northern Europe primarily dogs and horses, in ritual contexts from excavations are used as a proof that the texts by for example Adam of Bremen and Tacitus are correct. The question is of course how strongly the interpreting archaeologists were influenced by these well known written sources when they chose to interpret the deposits in the first place? There exists a risk of circular reasoning here, something that is added to the many problems already involved in using the written sources.

Had the search for special and ritual deposits only been about the existence of certain species in irregular numbers, or even as complete skeletons, it would indeed be a weak indication. Fortunately there are more methods available that in association with species selection could indicate speciality.

The wild animals are worth a short note here. Non-domesticated mammals are usually rare on settlements after the Neolithic and are also not a frequent part in special deposits. When they do appear, mainly in the form of complete animals, they can however be considered a strong indicator of speciality, since it is unlikely that anyone would bring a dead wild animal to the settlement to bury it there for any practical reason (see for example Lauwerier *et al.* 1999: 185).

3.2.3 Age structure and seasonal analysis

Age structure and seasonality are linked together through the biological reality of fixed birth periods among animals. I will here disregard other possible indicators for seasonality (such as the existence of migratory wild birds) beside age at death since a majority of the bone material from my sites contained bones from domestic mammals.

As was told in chapter 2 most of the mammals we encounter in Northern Europe are born in spring, mainly to take advantage of the summer months to grow before facing the harsh winter conditions. It was also shown that even if it is possible to change these breeding patterns for domestic animals, there is usually little sense in doing so, at least in a pre-industrialized society, because a higher fodder input has to be made with modest or no gain in return. Because of this mammals killed at a certain age are also most likely killed at a certain time of the year. Both seasonality and age can however each in their own right be indications of speciality.

It has been suggested that young animals were especially sought after for certain rituals. One example of such a practice is the Roman Period tanning pit at Lullingston (Great Britain), where a suckling pig and the skull of a lamb were found together with 34 leather sandals. In several Roman villas in England, finds of lambs and piglets deposited in pits have been made, indicating at least a partial preference for young animals in rituals (Scott 1991: 116-117).

Another good example of the use of young animals in rituals comes from the Roman temples of Harlow and Great Chestford in Essex (Great Britain) where large scale slaughtering of lambs were observed. Here however the selection of lambs as a main part for the rituals or feasts also indicated seasonal events. The slaughtering seems to have taken place in large numbers every autumn at Harlow, and both in spring and autumn at Great Chestford (Legge et al. 2000: 153-154, 157).

As with most of the archaeozoological methods described in this chapter, the use of age and seasonality does not constitute very strong indications of speciality on their own. In combination with other indications they can however be of great interest. A lot of what we know about contemporary religions and about pre-Christian practice in Northern Europe points towards recurring seasonal ritual events, something that might be observed in the bone material if young animals are present.

Seasonality of offerings is sometimes described in the written sources. In Adam of Bremen's book Deeds of Bishops of the Hamburg Church there is for example a comment (possibly by the author himself) saying that the great feast in Uppsala took place at or around the vernal (spring) equinox.

> *"Feasts and sacrifices of this kind are solemnized for nine days. On each day they offer a man along with other living beings in such a number that in the course of the nine days will have made offerings of seventy-two creatures. This sacrifice takes place about the time of the vernal equinox."*
> (Adam of Bremen's Deeds of Bishops of the Hamburg Church, book IV, translation by Tschan 2002)

In Ynglinga saga Snorri Sturluson also mentions on several occasions what time during the year the offerings should take place.

> *"On winter day there should be blood-sacrifice for a good year, and in the middle of winter for a good crop; and the third sacrifice should be on summer day, for victory in battle."*
> (Snorri Sturluson, Ynglinga saga chapter 8, translation by Laing 1844)

> *"Onund's district-kings were at that time spread widely over Sweden, and Svipdag the Blind ruled over Tiundaland, in which Upsal is situated, and where all the Swedish Things are held. There also were held the mid-winter sacrifices, at which many kings attended."*
> (Snorri Sturluson, Ynglinga saga chapter 33, translation by Laing 1844)

The tendency to have seasonal recurring feasts and religious holy days is an almost universal trait in cultures around the world. Since it is supported by several written sources there is little reason to suggest otherwise regarding the cultures of Northern Europe. To exactly pinpoint the dates or even the existence of individual events is however another matter.

3.2.4 Sex selection

As with selection of animals according to age it is a reasonable thought that the sex of the animal could be of importance when animals where chosen for rituals. This is confirmed in some of the written sources. In the text already partly quoted above describing the great feast in Uppsala Adam of Bremen states that the humans and animals chosen to be sacrificed should be male.

> *"The sacrifice is of this nature: of every living creature that is male they offer nine..."*
> (Adam of Bremen's Deeds of Bishops of the Hamburg Church, book IV, translation by Tschan 2002)

In the Saga about the Ynglings, Snorri Sturluson writes about a bull that was obviously selected to be sacrificed some time before the actual event, with eventual devastating consequences since it escaped and killed the king that tried to hunt it down.

> *"It happened in Sweden that an old bull, which was destined for sacrifice, was fed so high that he became dangerous to people; and when they were going to lay hold of him he escaped into the woods, became furious, and was long in the forest committing great damage to the country."*
> (Snorri Sturluson, Ynglinga saga chapter 26, translation by Laing 1844)

As with most of the methods described in this chapter the sex of the species in a deposit is only a weak indication of speciality in itself, but can be used as one of several criteria.

3.2.5 Element distribution

An unusual element distribution in a deposit is the most frequently used indicator for speciality among the archaeozoological methods described in this chapter. Finds of complete animals, heads, articulated bone groups or odd combinations of different animal bones have often been taken as prime indicators for ritual deposits (see among others Grant 1984a, 1984b, 1989, 1991; Wait 1985; Hill 1995: 27-29; Müller-Wille 2002: 155-156; Carlie

2004: 104). This idea is to some extent supported in the written source material. There are of course many cases without descriptions of the actual treatment of the sacrificed animals, but in other places there are hints about the element distribution.

The animals (and humans) in the already mentioned text by Adam of Bremen, seem to have been hanged in the trees as complete bodies. In the case of Egil Skallagrimson's curse of king Erik and queen Gunhild a horse head was used. The special treatment of heads can also be seen in Ibn Fadlāns description of the sacrifice made by a merchant where the heads of the animals were not thrown on the ground with the rest of the animals but instead attached to a wooden pole.

> *"He ties the heads of the cows or the sheep to that piece of wood set up in the grounds."*
> (Ibn Fadlāns Kitāb, Rūsiyyah, translation by Montgomery 2000)

These and similar sources could, and often have been, taken as signs that complete animals and heads were of special interest. This can of course be true, but there are also written sources that suggest more complex divisions of animals in ritual situations.

In the same text as above Ibn Fadlān describes the intricate customs involved in the ship burial of a Rus chieftain. Beside the very noticeable sacrifice of a slave girl, the ritual involves several examples of animals and food used as grave gifts and offerings.

> *"Then they brought alcohol, fruit and herbs and placed them beside him. Next they brought bread, meat and onions, which they cast in front of him, a dog, which they cut in two and which they threw onto the ship, and all of his weaponry, which they placed beside him. They then brought two mounts, made them gallop until they began to sweat, cut them up into pieces and threw the flesh onto the ship. They next fetched two cows, which they also cut up into pieces and threw on board, and a cock and a hen, which they slaughtered and cast onto it."*
> (Ibn Fadlāns Kitāb, Rūsiyyah, translation by Montgomery 2000)

This is a very interesting section of text that both shows what was given to the deceased, but also something about how these gifts or offerings were treated. The fact that meat as food together with fruit, alcohol, bread, herbs and onions are mentioned, but that whole animals in the form of a dog, two horses, two cows, a cock and a hen were also sacrificed indicate that these two types of gifts/offerings had slightly different meanings. It is particularly interesting that all the animals (with the possible exception of the fowl) were divided into parts before they were thrown onto the boat. This indicates that not only complete animals but also articulated parts of animals should be examined when ritual deposits are sought after or suspected. The fact that articulated limbs or body parts can be of significance is also something that several scholars have concluded from archaeological finds (see below).

If we leave the written sources and instead examine a bit more in detail which criteria regarding element distribution researchers chose to use when they identified speciality, we find a few element combinations that turn up repeatedly in analyses of this type.

Complete animals

Finds of complete and unbutchered (or at least mainly unbutchered) animals have by many researchers been identified as something out of the ordinary. Annie Grant has as an example presented this type of find as one of her main groups of special deposits from the Iron Age, especially if they are found in pits or ditches (Grant 1984a: 553; 1984b: 221).

> *"...a group of bones, certainly or very probably found in articulation, representing a whole animal or at least a substantial part of a single animal, and usually showing no evidence of having been butchered."*
> (Grant 1984a:534)

Grant originally identified these deposits in the hillfort of Danebury, but later the same type of deposit was found in other hillforts and settlements in England (Wait 1985: 125). This type of find is however not limited to Iron Age England, but appears in virtually all time periods all over the world, something that is also the main problem with the criteria. It is very likely that complete animals were deposited untouched in many cultures if they died of illness and therefore could not, or were not allowed to, be eaten. The question is which deposits really are the result of special activity and which ones only represent ordinary carcass removal?

According to Grant the answer to this problem lies in simple statistics. If the deposits of complete animals in a society would only represent animals that died of illness, they should show roughly the same species and sex ratio as the normal bone material. Furthermore should two or more animals, especially if they are not of the same species, be found together it is very unlikely that they died of natural causes (Grant 1989: 82).

Another way to tackle the problem of carcass removal is presented by Haarnagel regarding the many complete animals deposited in the Roman Period *terp*-settlement of Feddersen Wierde in North-Western Germany. His view is that carcasses would never be buried inside the settlement, and those that were found in Feddersen Wierde could therefore safely be interpreted as being special or even ritual in nature (Haarnagel 1979: 228). I would consider this view slightly simplistic, since we know that cultural attitudes towards what is seen as clean and dirty can vary a lot between different societies, and over time. It is however still a logical starting point for a discussion.

Skulls and mandibles

The appearance of loose but complete or at least virtually complete skulls have also been defined as a category of special deposits for Iron Age England, again primarily if found in pits and ditches. Related to these deposits, loose horse mandibles were also interpreted in the same way for the same features and time period (Grant 1989: 81; 1984a: 553; 1984b: 221; Wait 1985: 125).

> *"Special deposits of skulls were thus defined as including skulls that had clearly been deposited whole and skulls where a substantial amount had survived, any damage having probable or possibly occurred post-deposition."*
> (Grant 1984a: 538)

> *"Another special category was defined that related only to horses – isolated complete mandibles, where the left and right halves of the mandible were still joined and the mandibles were substantially undamaged."*
> (Grant 1984a: 538)

The importance of loose animal skulls is of course not confined to the Iron Age or to England. Such loose skulls appear in many places and time periods over the world. The category also appears repeatedly in the written sources (see above) and in ethnographic descriptions of offerings (see section 8.2.2 for several references to archaeological and ethnographic sources). The complete skull of an animal seems to have had a strong symbolic function that made it a recurring ritual object in a variety of societies.

"Head and feet" or "Heads and hooves" deposits

A type of deposit that is somewhat related to the loose skulls and mandibles is the combination of animal skulls (or parts of these) and legs (usually the lower legs, often with the phalanges still remaining). These are usually called "Head and feet" or "Heads and hooves" deposits and are known from large parts of the world in different time periods. The type of deposit appears in the Neolithic and continues until the Bronze Age in many parts of Europe and Central Asia. Many of them are found on burial sites, either in the form of deposits in pits near cemeteries, or as parts of the actual graves (Grant 1989: 79; Rech 2006: 7-8).

This type of deposit has sometimes been interpreted as the remains of a skin with the head and hooves still attached, and even if this is an interesting explanation, it is not the only possibility. The bones are often deposited close together and not spread out or neatly placed as if part of a folded skin. In some cases the pit in which the bones were deposited is furthermore too small to ever have contained the skin of the animal. Therefore more complex explanations might be needed, even if the skin may very well have played a part in the activities we are observing the remains of.

Even if these deposits seem to disappear in most areas after the Bronze Age it is of specific interest for this study that examples from the Netherlands, Germany and Scandinavia indicate a much later practice in Northern Europe. Here they appear both at settlements, offering sites and burial grounds at least as late as the Early Middle Ages (Klindt-Jensen 1957: 83-88; Piggot 1962; Schmid 1965; Dickmann 1995, 1997: 60f; Müller-Wille 1999: 32-33, 2002: 156; Carlie 2004: 124-126).

Articulated body parts or limbs

The final category of elements that frequently has been interpreted as special in nature is the articulated limbs, articulated legs, articulated bone groups or articulated body parts as they are called by different authors. The main idea is that a part of an animal is deposited, usually in pits or ditches, in an anatomical correct position without signs of butchery, consumption or other forms of processing, except of course possible marks produced during the separation of the part from the rest of the body (Wait 1985: 125).

This type of deposit was first repeatedly observed in pits and ditches at Danebury and later on in other Iron Age hillforts and settlements in England. The English finds mainly come from horses, but also cattle, sheep and occasionally pigs are represented. As with the other categories the articulated body parts appear in many parts of Europe in different time periods (Grant 1984a: 553, 1984b: 222, 1989: 81, 1991: 110).

> *"The category of special deposits defined as articulated legs consists of several bones from a single animal leg, found in articulation within a pit."*
> (Grant 1984a: 539)

> *"In this class a considerable amount of the animal was clearly dismembered and butchered, but these limbs did not have their meat removed..."*
> (Wait 1985: 125)

There are two main problems involved in using this category. It is first of all by definition quite a vague category. It is rarely defined exactly how many bones the bone group should contain, what parts of the animal (if not all) can qualify, and if they really have to be placed in complete articulation or whether it is enough if it can be assumed that they come from the same body part and have not been further processed.

The second problem is just as with the deposits containing individual complete animals, that the body part for some reason could have gone bad or was forbidden to be eaten. As with the complete animals above this is a possible explanation that has to be addressed. If however several articulated skeletal elements were found together, if there were other special circumstances surrounding the deposits, or if the species frequency is not the same as in ordinary refuse on the site, these explanations are less convincing.

Despite these problems, this category seems to be a common type of special deposit in many cultures, something that is also supported by several written sources, for example the description of the Chieftain burial by Ibn Fadlān quoted above for which at least the dog, the two cows and the two horses were partitioned without further butchery.

3.2.6 Taphonomic history

The final archaeozoological method to be described is also the one most seldom used, at least in the positive sense discussed here. As was mentioned in chapter 2 the field of taphonomic research primarily deals with the biasing processes that affect a bone material from the time an animal is killed, until the bones are excavated. This is a very important part of any archaeozoological analysis, and especially so for those that deal with individual bone deposits where poor preservation and representation has a strong impact on the results. It is also a field where analyses of ritual behaviour are often lacking. Many studies do not include any taphonomic discussions, while others include only superficial ones. This has also been strongly criticised by for example the British archaeozoologist Bob Wilson. Wilson has, maybe because of this, a very sceptical attitude towards symbolic and ritual interpretations of for example the already mentioned special deposits in Danebury and similar sites (Wilson 1992).

> *"Even more importantly, the full taphonomy of Iron Age and Romano-British deposition in pits and other features needs further investigation. The part of natural and mundane processes needs elaboration so that the part of ritual or other complex phenomena may be elucidated as much as possible without confusing ritual processes with other taphonomic events."*
> (Wilson 1992: 347)

Beside this important aspect it is however also possible to use taphonomy as a method that goes beyond the procedure of analysing the degree of destruction that the bone material has gone through. It is just as much the aim of taphonomy to identify and understand the specific processes that affect the bones, the taphonomic history of the material, and by doing so it becomes to a large part a study of human handling of bones (Magnell 2006: 1).

The main idea behind writing the taphonomic history of a deposit or assemblage is to collect as much information as possible about those processes, both of human origin and others, that have affected the bones from the time of death of the animal until the recovery (Lyman 1994: 5-6).

In practice this is achieved through meticulous visual observation of the bones and the registration of these observations according to different categories. When the analysis of a deposit or an assemblage is used later in the interpretive phase, a taphonomic history is produced. The level of detail for this will depend on the size of the material, its general condition and what processes could be identified. In many cases little can be said about the taphonomic history of a specific deposit, but even this is important information. The lack of traces might warrant a discussion in itself.

In table 3.2 a purely hypothetical, but still quite possible example of the activities involved in a ritual deposit and the taphonomic traces these activities might produce on the bones is presented. The deposit is supposed to come from the post hole of a roof bearing post in a house and consists of a cow head together with a few pieces of pottery.

The taphonomic traces on the bones could, if correctly identified and recorded, be used to write a taphonomic history somewhere along the following lines:

> **The cow was killed by a blunt blow to the forehead. The cranium was then severed from the rest of the body with an axe or a cleaver. At least one blow missed its target and struck the neck. There is some very light gnawing, probably from a dog, on the left mandible. This indicates that the bones were not immediately covered after butchery (but the total lack of weathering and trampling also indicate that the time it was exposed in the open was short). At some point a part of the nose was broken off. The head was finally deposited in a post hole adjacent to a roof bearing post together with a few pieces of pottery.**

As can be seen from this hypothetical example the exact order of events is sometimes quite difficult to uncover, but most important of all is that virtually nothing can be said about the ritual activity that the bones were subject to. The very notion that this made-up deposit was the result of ritual activity can almost exclusively be attributed to the element distribution (the unprocessed complete cranium) and the archaeological circumstances (the placement together with pottery in the post hole of a roof bearing post).

Taphonomic histories do in fact virtually never give any indications of the reasons behind activities that affected the bones, but are still important in so far as to give a lot of information about the practical side of the actions performed, and also give important clues about what activities were not performed. In this case the fact that the head was not processed further in order to obtain the meat, fat and brain of the cranium is an indicator that something is out of the ordinary (unless of course all cattle heads on the site, or in the area, were found complete in this specific time period, in which case this might only be a normal butchery and processing procedure).

As can be seen in the example above, taphonomic history is even more than the other described methods to be regarded as a supplement, but an essential one as such. It is undoubtedly time consuming to analyse the

Table 3.2. The hypothetical taphonomic history of a cow cranium found in a post hole of a pre-historic house somewhere in Northern Europe.

Event	Taphonomic traces on the head	Bone composition after event
A cow is slaughtered somewhere on the settlement.	The forehead of the cow is smashed as a result of the killing with the neck of an axe.	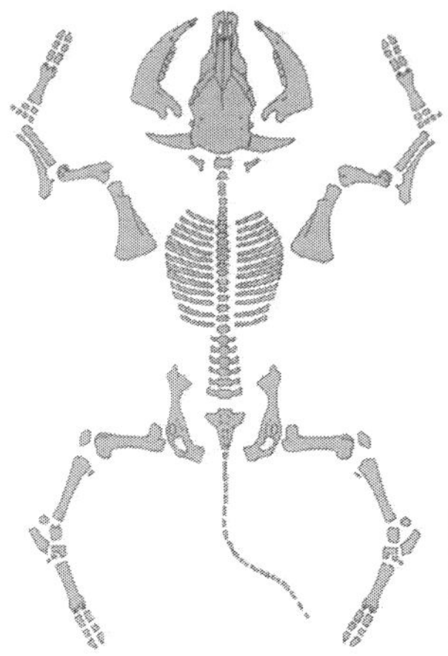 The whole skeleton
The animal is disarticulated. The limbs and parts of the trunk are divided among several families on the settlement while the head, ribcage and vertebra column are set aside.	None	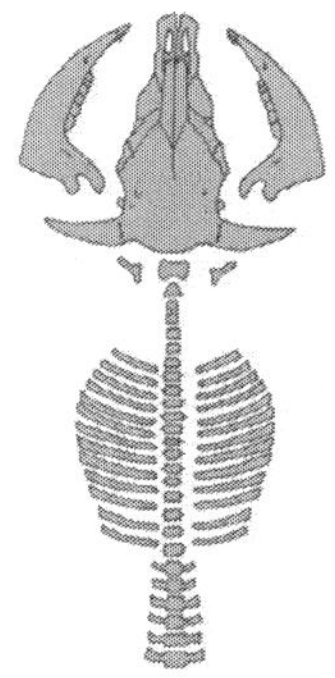 The skull, mandibles, ribs and vertebra column.
The head of the cow is separated from the trunk.	Chop mark on *condylus occipitalis*. This is a miss since the butcher wanted to sever the head slightly lower at the axis vertebra. This is then performed with the next few chops.	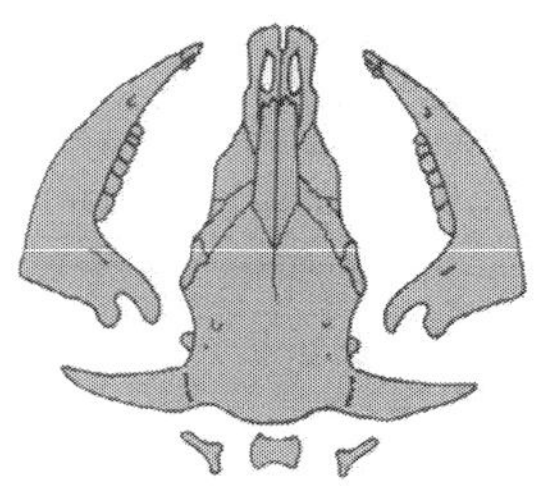 The skull, mandibles and atlas vertebra.

Event	Taphonomic traces on the head	Bone composition after event
Most of the remaining meat is filleted and cooked for the feast that is to be held as a celebration of the nearly finished house. The head is put aside for now.	None	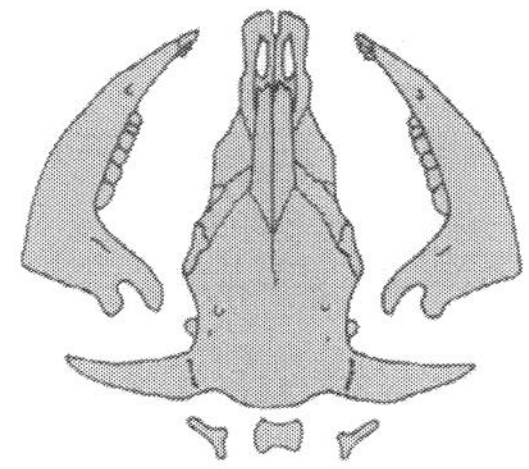 The skull, mandibles and atlas vertebra.
After the feast the skull is taken to the new house and placed in the post hole of one of the roof bearing posts, together with part of a ceramic pot, as a foundation offering.	None	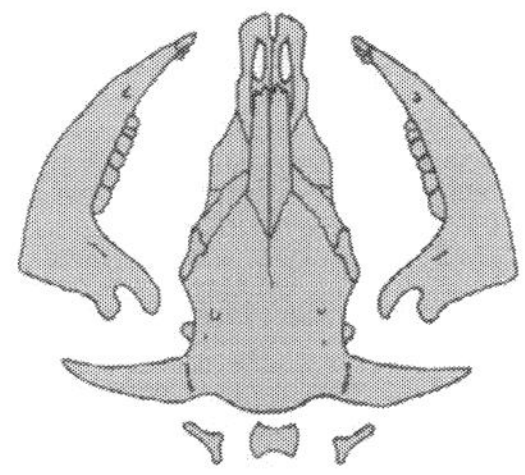 The skull, mandibles and atlas vertebra.
Before the ritual is finished a dog manages to sneak up and starts to gnaw on one of the mandibles and gets hold of the atlas vertebra before being chased away.	Superficial gnaw marks on the left mandible.	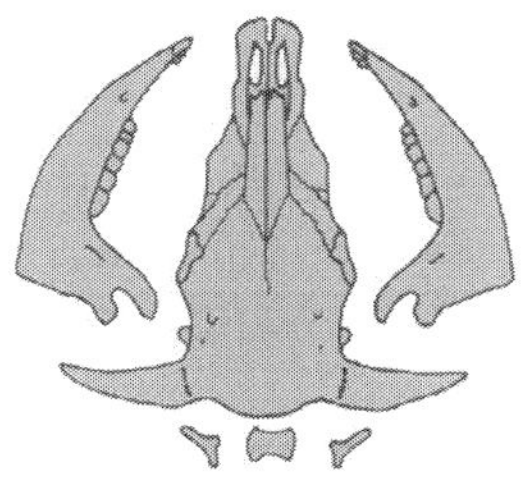 The skull, mandibles and atlas vertebra.

Event	Taphonomic traces on the head	Bone composition after event
The offering is covered by soil.	A part of the skull is accidentally broken off in the process, producing a fresh fracture.	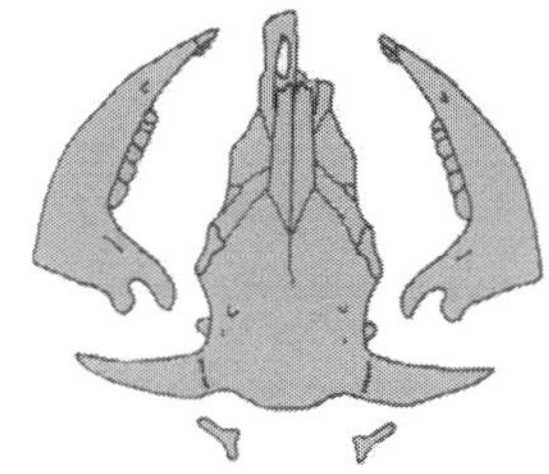 Part of the skull and mandibles.
The deposit lies buried for centuries.	Many of the bones break as a result of the pressure from the soil and during excavation.	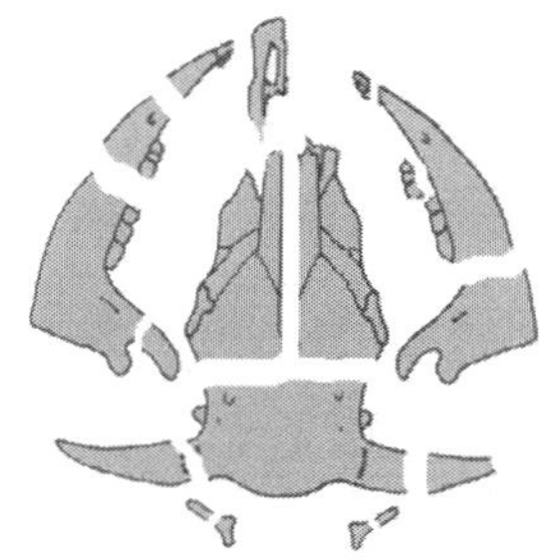 Fragmented part of the skull and mandibles.

taphonomic processes at such a detailed level that is required to write a useful taphonomic history, but since the general loss-oriented taphonomy in any case is of such significance two birds are in this case actually caught with one stone.

Figure 3.2. Gold-foil figure found in Uppåkra, Sweden (after B. Almgren LUHM).

3.3 Archaeological methods

3.3.1 Artefact context

One of the most commonly used archaeological methods for identifying special deposits is the artefact context, in other words what type of objects the bones were deposited together with, if any. For various reasons different artefacts were regularly used for ritual purposes in the past. Some of these, for example the Scandinavian early medieval gold-foil figures or "*guldgubbar*" (figure 3.2), have long been interpreted by archaeologists as having apparent ritual and religious meanings (see among others Watt 1999; 2004).

In other cases ordinary but valuable objects or things with (at least according to some) inherent symbolic value, such as weapons, are interpreted as being connected to rituals. This type of interpretation is especially popular if the objects have been deliberately destroyed (see among many others Worsaae 1886: 126; Müller-Wille 1999: 38-40; Jørgensen 2003; Ilkjær 2003: 59-60).

In other cases the main focus has been the sheer amount of things deposited. During the excavation of the early medieval farm of Borg in the province of Östergötland (Sweden) roughly one hundred rings used for keeping amulets attached together were found in a pit in the middle of an animal bone layer with large amounts of cranial fragments. In this case the deposit was interpreted as the remains of offerings connected to local iron working, partly because of the sheer amount of rings (Lindeblad & Nielsen 1997: 111).

Most objects are not as straightforward to interpret as special, but do nonetheless seem to have been repeatedly used in rituals. This is for example the case for ceramics that were used in ritual acts on many places and during various time periods (see for example Becker 1972: 5-6, 51-52; Carlie 1992: 58-59; Carlie 2004: 41-48).

Other objects often connected with special deposits are grinding stones and iron tools, but even quite obscure things such as flint nodules appear in odd deposits, as can be seen in one of the Danebury pits. Here the front half (with the exception of the legs) of a horse was deposited with two large flint nodules placed in the chest cavity and a broken whetstone lying next to the jaw. In the same pit two young pigs were deposited in a carefully arranged manner (Grant 1984a: 540, 1991: 110).

The objects that can be given special qualities are as can be seen here very varied. The basic assumption when using the artefact context as criterion for speciality must therefore be that any object not interpreted as ordinary refuse, or whose existence in the feature cannot be explained in any other "practical" way, can to some extent be an indication of speciality.

3.3.2 Feature type

A deposit's location in or near various types of features such as pits, ditches, wells or houses has often been used as an indication of ritual activity. The idea that ritual action could be associated with special types of features seems logical and is most likely inspired by both ethnographic and historical sources.

One typical example of this indicator are deposits associated with houses either in connection with the construction of a building, to protect it during its time of use, or after its abandonment. In many areas ritual deposits have been linked to specific parts of houses, especially roof bearing post holes, walls, entrances, pits below the floor and hearths (see for example Haarnagel 1979: 223-230; Carlie 1992: 58-58; Carlie 2004: 112, 190; Larsson & Lenntorp 2004: 24)

Other features that have often been related to rituals are pits, and to some degree ditches. These features have sometimes, predominately in Iron Age England, been considered an important part of the local belief system (Wait 1985: 152; Cunliffe 1995: 86-88). The idea that pits have a special meaning in connection with ritual activity is however not exclusive to England or the Iron Age. Special deposits are identified in pits in virtually all time periods and all over the world, and the use of pits and shafts for offerings can also be found occasionally in written sources. In *Vatnsdœla saga* special offering pits belonging to a man named Torolf are mentioned, where humans and animals were thought to have been sacrificed.

> *"Torolf stole other people's property and was a notorious thief; he also had offering pits where it was thought that he sacrificed humans and animals."*
> (Vatnsdœla saga, chapter 41, translation by the author from Näsström 2001)

Other categories of features sometimes associated with ritual activity are shafts and wells. Such features containing different forms of unusual or highly structured deposits have, especially in Great Britain long been considered special. Particularly the shafts and wells dated to the Iron Age and Roman Period have been given a ritual meaning that is accepted by most researchers (Wait 1985: 152; Merrifield 1987: 40-43; Wilson 1992: 341).

The general tendency to deposit unusual things or combinations of objects in the ground, regardless of whether it is in pits, shafts or wells, has been subject to a lot of speculation regarding the basic ideas behind the behaviour. One recurring explanation for this seemingly pervasive habit is that the deposits should be interpreted as offerings to the powers of the underworld (Wait 1985: 152; Merrifield 1987: 41). As with all explanations of this kind it is difficult to be very specific, at least if there are few or no written sources available. Deposits in the earth seem to be very important in virtually all societies, but we must not forget to be critical. There are problems that must be taken into consideration when using this indicator to identify special and ritual deposits, and when trying to explain its meaning and importance.

The first of these problems has to do with the actual series of events of the ritual activities. Even if a deposit is the result of such an activity it is not certain that the assemblage we find was the main focus. The deposit can of course be an offering placed in the feature as the final part of the ritual, but it can just as well be the remains of a ritual, unceremoniously dumped in a suitable hole in the ground after the actual event. If that is the case the exact type of feature used would be of less importance. Even if the deposit is actually placed in the ground for a specific purpose, for example as an offering, the previous function of this feature might very well be of no significance whatsoever!

The second problem is the taphonomic fact that objects, especially organic material such as bone, are best preserved if they are buried in the ground. In most circumstances being buried is actually the first condition that organic material will be preserved at all. This does of course pose difficult questions regarding the assumption that ritual deposits were primarily placed in the ground, since they are actually the only ones we can ever hope to find.

To conclude, the type of feature in which a deposit was found can be an indicator of ritual activity, but only if the actual placement of the objects in the feature was an integrated part of the ritual act. If we on the other hand find the objects in an unassuming feature, we might still be observing the remains of a ritual, but one that was originally performed in another place and where the final place of disposal was of little concern.

3.3.3 Stratigraphical position

Stratigraphy is used in archaeology to determine relative chronology as well as to understand depositional processes within a feature. The stratigraphical position of a deposit within a feature can also be of interest when searching for unusual deposits. It is possible that ritual deposits were regularly placed at specific levels of features. The relevance has been shown for building offerings in post holes where the placement can indicate whether objects were placed before the erecting of the pole, during its use, or deposited in the fill after the house was demolished or moved. This can be exemplified with the already mentioned Brogård site in Southern Sweden where a complete ceramic vessel was placed on the bottom of a post hole and then most likely crushed during the rising of the roof-bearing post (Carlie 1992: 59).

Another example of the importance of stratigraphy in this respect comes from J.D. Hill's study of patterns of deposits in Iron Age England. He concludes that different types of bone assemblages (including such that could be classified as special) were in this area deposited in specific vertical parts of the pit fills (Hill 1995: 50-53).

The stratigraphical position of special deposits must always be studied with a critical mind. The fact that many of the unusual Iron Age pit-deposits from Southern England are found in the bottom parts of the features has been strongly criticised from a taphonomic perspective by the archaeozoologist Bob Wilson, stating that the bottom parts of the pits more frequently have better preservation conditions for archaeological finds. This is due to the fact that the bottom parts of the features are both better protected from later disturbances, for example ploughing, and often contain a moist environment suitable for the preservation of organic material such as bones (Wilson 1992: 342-343).

Wilson's critique is certainly important but does not make stratigraphic position irrelevant as a method. There might still exist real differences between where special material was deposited in the features, but we have to be aware of the taphonomic history of the material when we draw such conclusions (compare with Hill 1995: 50).

3.3.4 Spatial context

As with the stratigraphical position within a feature, the placement of a deposit in relationship with other deposits or features in a settlement can be used as an indicator of speciality. The situation where a deposits' inclusion within, or in proximity to a house is used to signify speciality has already been mentioned. Most often these situations are interpreted as building offerings connected with the building or demolishing of a house (see for example Lauwerier 2002: 67). In other cases the deposits have been seen as guardians of the house, as was the case with the two dogs found in a small house from the Roman Period at Lundsgaard in Denmark, or the two dog mandibles deposited on either side of the entrance of a small bronze age house in Apalle, Sweden (Carlie 2004: 112, 131).

Sometimes a find is associated with a building even if it was actually not positioned in the house itself. This is the case for a "Head and feet" deposit in a Migration Period pit from Sorte Muld on Bornholm, Denmark. Here the cranium and four lower legs of a horse were deposited a few meters from the house in question (Klindt-Jensen 1957: 83-88). It is important to note that Klindt-Jensen does not directly discuss any connection between the pit and the house, but this has instead been made later by other researchers (Carlie 2004: 124-126).

There are also examples of spatial relationships that do not include associations with buildings in order to be identified as special or ritual deposits, for example systems of pits positioned in certain patterns or areas of a settlement. These however seem less frequent than the deposits connected with houses, possibly because association between other features such as different pits are more difficult to identify during an excavation, or because such relationships seem less convincing. Some studies that at least partly include the idea that features with special content were systematically placed on certain locations within sites have been produced, with varying results (see for example Wait 1985: 138-146; Lauwerier et al. 1999: 180-183; Gerritsen 2003: 150-167; Hedman 2003; Therkorn 2004; Groot 2007: 144-145).

On a larger landscape scale the importance and variation of the placement of ritual areas are more commonly stated. The idea that rituals were performed at different places in the landscape is widely accepted and can be exemplified with the many bog offerings in Northern Europe during large parts of the prehistorical and historical periods (see among many others Hagberg 1967; Nilsson 1995; Stjernquist 1997; Ersgård 2002; Jørgensen 2003). Another example are the Sami offering sites known from more recent periods in Northern Scandinavia that were usually placed beside large rocks or other prominent features in the landscape (see for example Rydving & Kristoffersson 1993).

It is important to note that when spatial context is studied on site level, the dating of the deposits are particularly important. When we observe such relationships we assume that the features in question existed at the same time. Often we lack detailed datings and cannot say for certain that for example a deposit in a pit close to a house was actually placed there while the building was in use. Furthermore it is vital to investigate the normal spatial distribution of animal bones on a site as well. It has been shown that different activities such as butchery and cooking often took place at designated locations within a settlement, something that could affect the disposal pattern of ordinary refuse. Other processes such as cleaning of activity areas and the moving of bone fragments by animals might also affect the general spatial pattern of bones within a site (Wilson 1996). As always the entire material and the context it is deposited in must be thoroughly assessed before speciality can be acknowledged.

3.3.5 General find context

The exact manner and position in which the animal bones and any associated artefacts were deposited, if this for some reason differs from what is normally the case, are usually regarded as indicators for ritual activity. If the general context seems odd enough it is one of the most accepted signs of speciality, even for the more sceptical researchers. Bob Wilson for example states that an odd, and preferably formal (non accidental) placement of bones can be a convincing way to identify special deposits. He also gives an example from the Romano-British site of Smithfield in Oxfordshire where a cow burial had been made with the correct anatomical order of the bones replicated, even though they had not been deposited until after the animal had been completely butchered (Wilson 1992: 345, 348).

This category can sometimes be identified already during excavation, and is therefore an important indicator since it often leads to a more thorough excavation of the specific feature, and a heightened interest for other similar features at a site.

Two examples of this category have already been given in the form of the anatomically "reconstructed" cow from Smithfield and the tradition to deposit ceramic vessels upside down, often with something inside. This tradition seems to have been widespread and an illustrative find is the five ceramic pots, the four smaller ones placed underneath the largest pot, dated to the 5th century AD that was found in Hundestrup mose, Denmark (figure 3.3) (Carlie 2004: 65; Becker 1972: 16-17).

Another example that could qualify for this category is a pit from Danebury which the complete skeletons of a horse and a dog were deposited, whereas the head of the horse was removed from the rest of the body and placed at the edge of the pit near the dog skeleton (Grant 1984a: 536).

A final interesting example is the Iron Age deposit from the *terp* Englum in Northern Netherlands. Eight human skulls without the lower jaws had been placed in a semi-circle next to two articulated cattle limbs, some sheep/goat bones and three ceramic vessels with holes in their bases (see figure 3.4). This whole find complex was situated in a layer of manure underneath some ash that indicated burning on top of the covered find (Nieuwhof 2008: 193-213).

As with many complex deposits the general find context is not the only category of speciality that applies in this specific case, on the contrary the deposit from Englum actually includes most of the categories discussed in this chapter:

- Species selection (humans and a dominance of cattle among the animal bones).
- Element distribution (the human skulls without mandibles and the articulated limbs of cattle).
- Sexual selection (at least six of the eight human individuals were female).
- Age structure (no very old or very young individuals were represented).
- Artefact context (the specially modified ceramic vessels).

Figure 3.3. Five pots dated to the 5th century AD found in Hundestrup Mose in Denmark. The four smaller pots where found underneath the upside down turned bigger one (after Becker 1972).

Figure 3.4. The ritual find complex from the *terp* Englum (Nieuwhof 2008).

- Stratigraphical position (in manure and partially underneath an ash lens).
- Spatial context (on a platform of the *terp* that was under construction).

Even if many researchers view the general find context as a very convincing indicator, there are a couple of serious problems with using it as a criterion. As has been said already it is not a formal criterion and can as such be freely interpreted. This is not an unusual situation in archaeology, but it does often result in scepticism towards the interpretation, at least if the specific find context is not repeated, has no strong symbolic content, or does not correspond with historical sources.

One part of what makes us interpret a find context as unusual is, as Wilson points out, the structured way the deposit was built. Included in this structure is however also often a symbolic interpretation. We think that the deposit was structured for a specific reason, for example to convey some form of idea. But as always our contemporary views on symbolism are reflected in how we interpret archaeological finds.

3.4 Analogical reasoning and the use of written sources

3.4.1 Introduction

As was stated earlier in the sections 3.1.4 and 3.1.5 the use of purely archaeozoological and archaeological methods can only identify deposits that in one way or another can be called special. When the nature and meaning behind such deposits are to be discussed we have to use complementary tools. These tools come in the form of analogies of different kinds and, more rarely, as direct historical sources (if we are working within an historical period).

I have already stated in chapter 1 that the focus of my thesis lies on the methods used to identify human behaviour behind deposits, and not on the specific analogies and written sources that can be used to interpret and understand these deposits on a more profound level. The reason behind this focus is not that I consider this step less important, or much more problematic to achieve. Instead this focus was chosen because the explanatory step of an analysis is more specific in nature than the initial identification. The available written texts are different for every region and time period and the analogies used would most likely vary accordingly.

Despite this I will attempt to interpret and explain the special deposits found in the different materials I have chosen as my case studies, even though these interpretations might not have much cross cultural significance, and would hence be of little use outside the selected region and time frame. The discussion will nevertheless stand as example for how such an interpretation might be formulated. Because of this, and because of the necessity to understand the problems involved in this step of an analysis even if no general methods can be formulated, I will here discuss the usage of some of these tools in a broader sense. The discussion will mainly include different forms of analogical reasoning, and the use of written sources in general.

3.4.2 The types and qualities of analogies

Analogy is the transferring of information from one particular to another particular and is assumed to be reliable if there exists a relevant relation of some kind between the two. Analogy is one of the most fundamental concepts within archaeology, and within most fields of science.

There are indeed many different methods beside analogy available for the modern archaeologist that can be used in the interpretation of a site or an artefact, but they are only supporting an originally formulated analogy, as Ian Hodder points out with his theoretical axe-example in the book *The Present Past*:

> *"To support the axe interpretation the archaeologist might examine the edge of the artifact with a microscope to show that it has traces of wear from cutting; he might conduct experiments to show that it is possible to cut down a tree with such an object; and he could conduct pollen studies of the environment of the prehistoric site where the object was found to show that trees had been cut down in the vicinity. Yet all these subsidiary studies are developed to support or weaken the initial analogy."*
> (Hodder 1982b: 11)

One of the most important aspects of the use of analogies in archaeology is their quality, e.g. their suitability for a certain situation or place. According to Hodder there exist two main types of analogies, formal and relational. Formal analogy states that if two objects or situations have some common properties, they probably also have other similarities. Such analogies are according to Hodder weak since the connection might be accidental and since no qualitative aspect is really applied to the connection. Relational analogy on the other hand seeks natural or cultural links between the different particulars of the analogy. For the analogy to be strong the various links have to be of importance for the specific subject, and not merely accidental. In reality analogies are seldom either strictly formal or relational, but vary on a scale between the two extremes. It has many times been convincingly argued that the more relational an analogy is, the more useful and reliable it is in archaeology (Hodder 1982b: 16; Wylie 2002: 147-149).

With that said it is not enough just using a relational analogy to guarantee good quality. An analogy (regardless of type) can never be verified as being correct or incorrect, but it can however be strengthened or weakened. According to Hodder there are three main ways to determine the strength of an analogy (Hodder 1982b: 22):

- Relevance (how are the similarities in the analogy actually of interest in the specific case).
- Generality (how much cross-cultural viability does the analogy seem to have).
- Goodness-of-fit (how many are the similarities between the particulars).

Finally another way to strengthen an analogy is by showing that the aspects that differ between the particulars involved are of limited concern for the specific subject that is to be investigated (Wylie 2002: 149-150). This method is also supported by Hodder although not included in the three main points above:

"The aim is to throw light on what is not known (the uncertain likeness) by demonstrating the strength and relevance of the positive analogy, while showing that the negative likenesses are of limited consequence."
(Hodder 1982b:25-26, my emphasis)

3.4.3 Analogies on different levels of interpretation

When using the word analogy I have adopted a very wide definition of the term. The view on how widely analogies are used when we make interpretations varies and some researchers would most likely want a more narrow definition and formal use of the concept (compare for example with Binford 1967; Tringham 1978). I do however totally agree with those who suggest that analogies are used on all levels of interpretation, and as such permeate virtually all stages of archaeological research, both on a conscious as well as on a subconscious level (Hodder 1982b: 11-12; Van Gijn & Raemaekers 1999: 43-44; Torsetnes 2004: 111).

The subconscious use of analogies is actually a serious problem when making interpretations of human behaviour. We as researchers possess an endless number of scraps of information from different sources. They can come from books we have read, films we have seen, lectures we have attended and many other types of personal experiences. Our view of these scraps of information is further filtered by our political attitudes and social preconceptions. All these things influence the way we view and interpret the world around us, and consequently also the past societies we try to understand, as has often been suggested in the theoretical debate of the last thirty years (see for example Hodder 1982b: 196-209, 1986: 187-188; Shanks & Tilley 1989; Renfrew 1989: 38-39; Trigger 1989: 379-382).

If we accept this as well as the earlier statement that analogical thinking applies to all levels of interpretation, it is natural that unconscious analogies, especially from contemporary situations, are very common within archaeology. This has also been repeatedly pointed out within for example gender archaeology, where contemporary cultural values clearly have influenced the views on gender roles and family structures throughout the history of archaeology (Conkey & Spector 1984: 5-14; Gilchrist 1994: 4).

To avoid or at least lessen this problem it is vital that we as far as possible are aware of which analogies we use when we make interpretations and what their sources are. If we do not use conscious analogies but instead rely on "common sense" (which, however useful, are mainly formed by our personal experiences), the risk is great that we actually subconsciously use a contemporary analogy that might be of little relevance for the specific situation.

3.4.4 The use of historical sources in archaeology

Traditionally the historical sources are the domain of historians while archaeologists deal with material culture and human interaction with the environment. In reality however written sources have been used by archaeologists since the very beginning of the discipline (see for example Worsaae 1886: 112-113; Montelius 1888: 117-121; Näsman 1988: 125-126). Historical sources are important within archaeology in three different ways.

First they are used in historical and protohistorical periods to give a general understanding of the technological, cultural and political situation in the geographical area that is studied, something that is vital for understanding more complex structures in a society.

Secondly, when available the written sources can give direct information about human behaviour, which in turn can be used to interpret archaeological finds. Such situations are however rare for protohistorical and early historical periods due to the limited number of available written sources. Basically the archaeologist is very lucky if a text that deals with the specific archaeological problem in question existed and has survived.

The third way an historical text can be used by an archaeologist, and the most common manner for protohistorical and early historical periods, is as an analogy. A text might not deal with exactly the same subject, but may show some resemblance to it. Maybe the text is from a later period or deals with another, but culturally similar area, making it possible to formulate an historical analogy, similar to how ethnographic or contemporary observations are used (see for example Näsman 1988).

3.4.5 Available sources for pre-Christian Northern Europe

The period and area of my research, roughly 400-1000 AD in Northern Europe, is really on the border between protohistory and history. Strictly speaking we are dealing with an historical period since some locally produced texts from the time exists. They are however very few in number, mainly a few runic inscriptions and coins with text, and rarely deal with the subject of explicit ritual behaviour. Most of the sources mentioned in this thesis are therefore either written by outsiders, or are part of an oral tradition written down centuries after the events are supposed to have taken place.

The available written texts can be divided into rather distinct groups that will be shortly presented here. I have been focusing on texts that to some extent deal with ritual customs, and that are either produced in, or relevant for the area where the case studies originate, that is the present day Northern Netherlands, Northern Germany and Scandinavia.

Runic inscriptions

The only contemporary pre-Christian texts that exist in Northern Europe are runic inscriptions. Runes are a Germanic alphabet probably inspired by Latin that was utilized at least since the third century AD and was still widely used in Scandinavia during the whole of the Middle Ages, in the later periods parallel with the Latin alphabet, and in remote parts of Sweden at least as late as the 19th century. Runes where mainly used to carve on wood, birch bark, bone, metal objects or chiselled in stone. Some five thousand Roman Period and Medieval runic inscriptions have been found in Scandinavia and a few hundred in the rest of Northern Europe (Sawyer & Sawyer 1993: 10; Page 1995; Looijenga 1997: 156, 173, 193). Unfortunately most of these inscriptions are very short, sometimes only indicating the maker or owner of an object. The longest pre-Christian rune inscription, Rökstenen in Östergötland, Sweden, contains only some 760 letters, which indicates the limited amount of text the runic inscriptions contain, even though such a number of letters chiselled in stone certainly is an impressive achievement in itself.

As the only true indigenous historic documents from the Early Medieval Period in the area, the runic inscriptions are very important for the study of the period. A number of runic inscriptions are of a religious or magical nature and give valuable insights into the ideas about the supernatural (Steinsland 2005: 40). The runes were in themselves of ritual importance and the very carving of them could be a vital part of magical acts, as is shown several times in *Egil Skallagrimson's Saga* where Egil uses runes both to avoid getting poisoned as well as to cure a woman from illness, and in the already quoted curse on king Erik and queen Gunhilda (see section 3.2.2).

The mythological explanation for the origin of runes in Scandinavia also shows a strong relation to sacrifice. According to the *Edda* the knowledge of runes was acquired by Odin through him hanging himself in the world tree. This ultimate sacrifice of himself unto himself was needed to gain the hidden knowledge that he then could pass on to humanity.

> *"I know that I hung, on a wind-rocked tree, nine whole nights, with a spear wounded, and to Odin offered, myself to myself; on that tree, of which no one knows from what root it springs.*
>
> *Bread no one gave me, nor a horn of drink, downward I peered, to runes applied myself, wailing learnt them, then fell down thence."*
> (Hávamál, 139-140, translation by Thorpe, B. 1906)

Early Scandinavian Christian manuscripts

The large body of written texts dealing with the pre-Christian period in Scandinavia are the early medieval manuscripts written by Christian scholars mainly on Iceland, between the 12th and 14th centuries. These texts are usually named sagas after the Old Norse word saga, meaning "to say" or "what is said". Several types of sagas can be distinguished among the well over 40 preserved texts. The different types are however not completely agreed upon by different researchers, and the categorising of several sagas is ambiguous. The categorisation used here is the one presented by Steinsland in her book *Norrøn religion. Myter, riter, samfunn* (Steinsland 2005: 59-61). She divides the sagas into three groups. The Royal sagas mainly deal with the history and the Christianisation of the kingdom of Norway and contain a lot of personal and political information about the upper classes of Western Scandinavia, but little about religion and the supernatural.

The sagas about the Icelanders (*Islendingasögur*) are also called the family sagas and mainly deal with the history of some influential families on Iceland. The sagas usually begin in Norway before the emigration to Iceland and follow the families through several generations. Since the people described mainly belonged to the upper class, the characters tend to travel a lot and most of Northern Europe is in this way touched upon in the sagas. These family sagas regularly deal with pagan religion and the supernatural, but usually only in passing or as plot concepts.

The ancient sagas (*Fornaldarsögur*) can best be characterised as a type of mythic-heroic stories that are supposed to have taken place far back in history. These sagas have little connection with the real world and supernatural motives regularly play a crucial role in the stories. The ancient sagas are the ones that were written down as the last of the three categories, often as late as during the 14th century.

Eddic and scaldic poetry

Intimately connected with the saga literature are the eddic and scaldic poetry. The art of poetry was highly regarded in early medieval Northern Europe and most of the main characters in the sagas are seen as authors of the poems in the texts. Our main knowledge of the pre-Christian mythology in Scandinavia, and in extension the whole of Germanic Europe comes from the eddic poetry.

What makes the poetry different from the sagas and especially interesting is the dating of these texts. Old Norse poetry was written with very complicated verse forms that are thought to hinder more comprehensive changes to have been implemented over time, thus making them more authentic than the late date of them being written down (the 12-14th centuries) would indicate. The problem with the poetry is that in order to produce these complicated verse forms a very elaborate language with countless synonyms, so called *kenningar*, was used that requires extensive knowledge to be able to interpret its meaning (Steinsland 2005: 47-48).

The prose Edda

Something of an oddity here is the prose *Edda*, or the younger *Edda*, that was written by the Icelandic priest and historian Snorri Sturluson around 1220 AD. The book describes many of the same motives as the older poetic *Edda*, but in an accessible prose form. It was actually written as a text-book teaching how to produce the old forms of poetry. It is a very interesting text unexpectedly free from degrading remarks about the pagan subject, but of course influenced by Snorri's interpretation of the old poems, as well as by his aims towards the writing of a poetry-guide (Steinsland 2005: 52-54).

Classical authors that wrote about Northern Europe

The earliest written sources with information on Northern Europe come from classical authors that describe the area and its population either from personal experience or, more commonly, based on second hand sources. Several of these texts, such as books by the three Greek geographers Pytheas, Xenophon and Philemon have been lost and are only known because they are mentioned by other authors (Grane 2003: 129-130).

The surviving sources are in most cases Roman and include the works of Caesar, Livius, Augustus, Strabo, Velleius Paterculus and Jordanes and mainly deal with the Germanic areas close to the Roman border. Areas further north have however partly been described by the authors Pomponius Mela, Plinius (the older), Tacitus and Ptolemaios (Clunie Ross 2002: 13-14; Grane 2003: 135-139).

Of these Publius Cornelius Tacitus is especially interesting, mainly because of his in-depth descriptions of the Germanic people in the book Germania. In contrast to other available authors he does not only provide geographical or political observations but also describes many aspects of the Germanic society such as history, customs, laws and religion. Especially the description of the religious practices has been repeatedly used by archaeologists. The problems with Tacitus' *Germania*, that he personally never travelled in the areas he described and that there is a clear political agenda behind the writing of the book, have often been overshadowed by the fact that the book (together with some parts of one of his other books, *Agricola*) is the only existing written source concerning these aspects of Northern European society before the Middle Ages.

Arabic travellers that described Rūs in modern day Russia

The Arabic travellers that met Scandinavians, or their descendants, in towns and trading sites along the rivers of present day Russia, Belarus and Ukrain near the end of the first millennium is a very interesting source despite its context being outside Northern Europe. The most famous of these travellers is Ibn Fadlān who has already been quoted several times in this chapter. Ibn Fadlān was part of the deputation sent by Caliph al-Muqtadir from Baghdad to the King of the Volga Bulghārs in the year 921 AD. He wrote about his experiences and his text is an invaluable source of information about the culture of the different ethnic groups that inhabited the area at the time.

Of most interest for western researchers is his description of a meeting with a people he call Rūsiyyah. These people, for short called Rūs, have in the Scandinavian research tradition been identified as traders from present day Sweden, that in accordance with archaeological finds and other written sources (mainly the *Nestor Chronicle* and several runic inscriptions), were active in what is present day Western Russia, Ukraine and Belarus (see for example Jansson 1991:214-216; Stalsberg 1991: 51 Näsström 2001: 13; Clunie Ross 2002: 14; Steinsland 2005: 63). This identification is not agreed upon by most Russian scholars who usually identify them as Slavs. This subject is a delicate matter involving both national pride and historical animosity from official Russia (and earlier the Soviet Union) towards the idea that Germanic groups wielded power in the area and were involved in the foundation of the Russian nation. The debate has been long-lasting and even if the so-called Normanist theory, that the Rūs were of Scandinavian decent, now seems to be the strongest, a somewhat balanced view must be called for. The Rūs as described by Ibn Fadlān undoubtedly bear many resemblances to Scandinavians as we know them from the archaeological material and other written sources, but they were presumably also quite influenced by, if not yet totally integrated into the Slavic culture (Montgomery 2000).

Ibn Fadlān's observations of the Rūs are interesting for several reasons. He was a direct eye-witness, a learned man and obviously an interested observer who does not seem to have had an immediate political agenda affecting his observations. Finally he was a Muslim living well before the first crusade, making him neither directly hostile towards westerners, nor having the Christian antipathy towards the pagans he encountered. His cultural background as a Muslim of high social class however influenced his views to a degree, something that for example can be seen in his disgust towards the Rūs' bad hygiene.

The most interesting aspect of Ibn Fadlān's text is however not his reliability as an observer but that he observes and closely describes several ritual acts performed by the Rūs. Even though some doubt can be cast upon whether he really was an eye-witness during the complete procedures, he was certainly on site observing first hand most of what he describes. These direct observations of pre-Christian rituals of at least partly Scandinavian origin are unique and have undoubtedly influenced many religious historians and archaeologists regarding their views on the Old Norse, and in a wider sense Germanic religion.

Ibn Fadlān is the most extensive Arabic observer of the Rūs, but not the only one. Several other Arabic authors describe these people although not as detailed as

Ibn Fadlān and not always as eye-witnesses. These other writers are less well known in Western Europe, mainly due to the limited number of translations of their work. Unfortunately they often seem to copy from similar sources and the origin of their information is thus difficult to find out. One of the more interesting of these texts is *Kitāb al-alāq an-nafīsa* that was written by the Persian geographer Abū Alī Ahmad ibn Umar ibn Rusta. He writes about the Rūsīya that lived in the town of Novgorod during the 10th century. Among other things, he describes offerings made by the Rūs and mentions that they had a certain person, a truth-speaker, who interpreted the effects of these offerings. It has earlier been assumed that Ibn Rusta visited Novgorod and was an eye-witness in par with Ibn Fadlān, but similarities between his writings and other Arabic texts however make this uncertain. It is at least not possible to rule out that Ibn Rusta copied some or all of his information regarding the Rūs from earlier written sources (Göckenjan & Zimonyi 2001:34-35).

Texts dealing with Christian missionaries in Northern Europe

During the second half of the first millennium the Christian faith expanded slowly towards Northern Europe from the Mediterranean. New areas were converted, sometimes voluntarily and sometimes after violent episodes. At the end of the 7th century Christianity reached what is present day Northern Netherlands and Germany. The conversion of the Frisians and Saxons in this area was a long and complicated political and religious process involving many important individuals. This process has mainly been described in a number of biographies (*vita*) written by men of the church about deceased predecessors or colleagues active as missionaries. Texts such as for example *Vita Willibrordi*, *Vita Bonifatii*, the letters of Bonifatius, *Vita Vulframni* and *Vita Liudgeri* are important documents for understanding the political situation and the process of transformation in the research area during the 7th, 8th and early 9th centuries AD. It would also seem natural that these biographies described the pagan traditions encountered by the missionaries, but unfortunately this is not the case. Most of the biographies mention pagan temples, pagan godly idols and a few heathen customs, but further details about these things are few (Halbertsma 1984: 24-25).

Two of these few examples come from the biography written by Alcuin about Willibrord, one of the earliest missionaries in Northern Europe, who was active at the very end of the 7th century. After a failed attempt to convert the Frisians and their king Radbod to Christianity, Willibrord travelled to Denmark and king Ongendus to try to spread the Christian faith there instead. Failing here as well, Willibrord left and travelled back southward. On this trip he came to an island situated between the realms of Denmark and Frisia that was called Fositeland (probably Helgoland) after the god worshiped there. Fosite was a god of laws and trials that appears in Scandinavian mythological texts under the name of Forsete, the son of Balder (Steinsland 2005: 218). Several temples dedicated to Fosite were to be found on Fositeland, as well as a holy spring and sacred cattle grazing freely. Willibrord desecrated these sites and was for this taken to the king, but was spared punishment. At a later occasion Willibrord travelled to Walichrum in the present day Dutch province of Zeeland and there he found an idol of an ancient god. He smashed the idol but was then assaulted by the custodian of the holy site who slashed the missionary on the head with a sword. According to the text the blow had miraculously no effect.

The biography of Wulfram, another missionary in Frisia whose chronological position in the process of Christianisation is unclear, presents more examples of pagan customs, although these possibly are quite exaggerated. In Frisia Wulfram was faced with the ritual sacrifice of humans, especially children, by hanging, partitioning or drowning. According to the text the victims were chosen during pagan festivals by drawing lots and then immediately killed.

After the conversion of the people of Frisia and Saxony, the church continued the process of Christianisation further north in the still pagan Scandinavia. The power of the Frankish realm was by then dwindling and the missionaries could no longer count on the support of a strong military power (compare with Olsen 1992:152-153; Heidinga 1997: 24-25).

One of the biographies most used by Scandinavian archaeologists about Christian missionaries is *Vita Ansgari*. Ansgar was called "the Apostle of Scandinavia" and was one of the first missionaries mentioned in the written sources to be active in the area. *Vita Ansgari* was written by Rimbert shortly after Ansgar's death 865 AD. Ansgar was a German monk who travelled to Denmark around 820 to preach. According to the text, Ansgar established two churches in Denmark but was also sent to Sweden, where he established a church in the trading town of Birka (Steinsland 2005: 63, 425-426).

The book, being mainly about Ansgar's missionary activity, touches upon the subject of the pagan faith several times, especially in the chapters about his travels in Sweden. Unfortunately the pagan ritual traditions are not described in detail. The existence of several pagan gods and the fact that people were supposed to bring regular sacrifices to the idols of the gods are mentioned on several places.

> *"They gathered together then, a second time and, as they possessed no power of resistance and had no hope of securing refuge, they exhorted one another to make vows and to offer greater sacrifices to their own gods. Herigar, the faithful servant of the Lord, was angry with them and said, 'Your vows and sacrifices to idols are accursed by God. How long will ye serve devils and injure and impoverish yourselves by your useless vows. You have made many offerings and more vows and have given a*

> *hundred pounds of silver. What benefit has it been to you? See, your enemies are coming to destroy all that you have.'"*
> (Vita Anskarii, chapter XIX, translation by Robinson 1921)

The existence of a temple to a pagan god, the deified former king Erik, is also mentioned in the text.

> *"For they had resolved to have a temple in honour of the late king, and had begun to render votive offerings and sacrifices to him as to a god."*
> (Vita Anskarii, chapter XXVI, translation by Robinson 1921)

Adam of Bremen's *Gesta Hammaburgensis Ecclesiae Pontificum* (Deeds of Bishops of the Hamburg Church), written in the second half of the 11th century, slightly later than the Ansgar biography, is another important document dealing with the then still partly pagan Scandinavia. Adam was a Christian chronicler working at the Church of Bremen, at the time the archbishopric of Scandinavia. Since the Church of Bremen was behind the intensive missionary activities towards Scandinavia, it was a natural place for the writing of an historical and geographical survey of these lands. Just as with Ansgar, Adam of Bremen travelled personally in Scandinavia, at least in the southern parts, and spent some time at the Danish king Sven Estridsen's court. Adam probably also used many earlier texts and contemporary information from other missionaries in his writings.

At the time of Adam of Bremen, Scandinavia had already partly been converted to Christianity. Especially the Danish and Norwegian areas could by then be considered mainly Christian, even though some pagans most likely still lived there. Much of present day Sweden was however not yet converted and one of the most interesting, and therefore most quoted, parts of Adam's book is the colourful description of the bloody sacrificial feasts at Old Uppsala in Central Sweden. It is however highly unlikely that Adam of Bremen actually attended these ceremonies and this part of the text is therefore based on secondary information (Steinsland 2005: 298). Despite this, Adam of Bremen's accounts are generally considered to be quite reliable, since he seems to have been an educated and interested observer, although he did show obvious disgust towards the pagan customs he described (Steinsland 2005: 295-299).

Early medieval laws

The earliest Frankish laws written for the people of present day Northern Netherlands and Northern Germany give a few insights into the old pagan traditions in these at the time newly converted areas. The two main laws preserved are the *Lex Frisionum* dealing with the Frisian people and *Lex Saxonum* regarding the Saxons east of Frisia. Both these laws were written down some time around 800 AD with the Frisian law being slightly older. Both texts are mainly built upon Frankish laws, but may also contain older traditional laws as was often the situation when Christian laws were introduced into new areas (compare with Sawyer & Sawyer 1993: 17-21).

The two law texts are rather short but a couple of interesting remarks regarding heathen rituals exist. Maybe the most interesting part is that the laws seem to accept the existence of pagan temples. In the Saxon law it is stated that the Christian churches are to be held in higher regard than pagan sites, while the Frisian law states that it is punishable by death to desecrate a temple, regardless of faith. The laws furthermore mention the existence of both animal and human sacrifices in connection with the pagan religion.

In Scandinavia a number of provincial laws were issued during the first centuries following the conversion to Christendom and of these laws 18 have been preserved. The laws are a slightly chaotic collection of texts covering different parts of Scandinavia that were written down from the late 11th century onwards. The individual laws are also themselves rather unorganized, sometimes for example presenting contradictory rules. The laws are usually considered to be a mixture of old traditional laws and influence from outside, a similar situation as for the Frisian and Saxon laws (Sawyer & Sawyer 1993: 17-21; Steinsland 2005: 64).

Some paragraphs of the Scandinavian laws are interesting with regard to pagan ritual tradition. Several rituals such as the old burial customs, the use of witchcraft and sacrificing to the old gods are forbidden in some of the laws (Raudevere 2002: 49; Steinsland 2005: 440-442). There are however indications, for example in *Islendingabok*, that the earliest laws actually allowed some pre-Christian rituals such as sacrifice, as long as they were not performed in public (Näsström 2001: 300-301).

Old English poetry

There are several Old English poems such as *Widsith* and *Deor* that at least partly deal with Northern Europe, but give no information regarding pre-Christian ritual customs. A bit more information is given in *Beowulf*, the by far longest and most well known Old English poem, not the least because of its literary qualities. *Beowulf* is an 8th century poem that describes events supposed to have taken place in present day Denmark and Southern Sweden and has for a long time been a focus of interest in the historically related sciences. The only surviving manuscript is dated to around 1000 AD but it is assumed to be partly drawn from earlier oral tradition.

The poem tells the story of the Geatish hero Beowulf and his three battles with the monster Grendel, with Grendel's mother and with an unnamed dragon. The story is said to have taken place in the late 5th century but is clearly coloured by Christianity despite solely dealing with characters that should be pagan. *Beowulf* is an interesting piece of literature that even though it was written long after the events should have taken place, still might tell us something about the situation in Northern Europe before the Christianisation (Bazelmans 1999: 108-110). The poem for example puts much emphasis on the, partly ritual, custom of giving and receiving gifts and how this tradition played an important political role in the area during the pre-Christian period. The Christian context of the poem however prevents more details about the pre-Christian religion and ritual behaviour from being dealt with (Bazelmans 1999: 75).

3.4.6 General and specific problems with using written sources

Source criticism regarding the use of written documents plays a major part in any historical research and if we as archaeologists are to use written sources in our research this part cannot be neglected (Steinsland 2005: 36-38; see also Bazelmans 1991; Sawyer & Sawyer 1993: 1-26; Norr 1998: 9-27).

Unfortunately archaeological analyses do sometimes use written texts with little source criticism. As is often the case with interdisciplinary studies this can partly be blamed on the difficulty of applying methods from one research field in another. An archaeologist must certainly be allowed to use written sources without first becoming an historian, but he or she still has to apply source criticism to the best of his or her ability. It is also important to form a personal view on the sources that are to be used, and not only follow the often changing trends within history and religious history. It is for example interesting to see the dramatic changes in perceived usefulness among historians of the Icelandic early medieval manuscripts over time. From being used almost totally uncritically in the late 19th and early 20th centuries as sources to Viking age Scandinavia, these texts were after the Second World War almost totally rejected as sources for anything except possibly the 13th and 14th centuries in which they were written. In the last decades the pendulum has once again turned back towards them being suitable as sources to the earlier periods, albeit much more critically than a hundred years ago (Steinsland 2005: 37-38).

When it comes to the question of how to use the written sources it is also important to remember that an archaeologist has a different perspective when reading and interpreting a text than a historian or religious scholar. The archaeologist is mainly interested in the material culture, something that can be a positive thing and a possibility to extract something new from the written texts.

The written sources that were discussed in this chapter are quite varied and therefore present different source critical problems, some very specific for a text or group of texts, while others more or less general. One major problem not really connected to the texts themselves is the general shortage of available sources from, or dealing with, the region and time period. This does not affect the reliability of the individual texts, but it results in problems with cross-referencing different sources and hence checking their reliability. This situation is impossible to solve, at least not without the miraculous discovery of many more written sources. Six of the other problems that will be briefly discussed here are however at least possible to deal with.

Errors of copying

One of the problems specific for many of the documents described in this chapter is that the texts only exist as copies, or that it is uncertain if the texts we have are the originals. The runic inscriptions are certainly original and the majority of the Icelandic sagas and most of the continental Church-biographies are also original documents.

An exception among the Church-biographies and a good example of this problem is *Vita Ansgarii*. This text is thought to have been written by Rimbert the archbishop of Bremen, before Ansgar's death in 888 AD, but the oldest surviving manuscript is dated to the 10th century. Later additions have been made to the text, something that for example is shown with the inclusion of Greenlanders as one of the neighbouring races of the Swedes and Danes in chapter XIII. This island was most likely not settled, nor even known, by Europeans before the end of the 10th century.

The problem also exists for some of the original documents to which additions were made, not necessarily by the original author, as is for example the case with the marginal notes in Adam of Bremen's Deeds of Bishops of the Hamburg Church.

Errors in translation and linguistic interpretation

Even if the original document is available a problem of translation and interpretation of the text still remains. All of the texts discussed in this chapter were originally written in Latin, Arabic, Old Norse or Old English, languages not widely known today, with the exception of Arabic. It would be a great advantage for the researcher to be able to read these documents in original, but since such a language arsenal is seldom combined with professional education in archaeology most texts used within this field are translations into modern languages. The translation will undoubtedly have resulted in information loss, and it is highly advisable to at least read several different translations if the original language is inaccessible to the reader.

One example of the problem is demonstrated by the different translations of Adam of Bremen's Deeds of Bishops of the Hamburg Church. In the passage already partly quoted above about the sacrifice at the pagan temple in Old Uppsala my English translation says that Adam was informed by a 72 year old Christian that he had once seen bodies hanging in the sacred grove (Adam of Bremen 2002). The translations into German and Norwegian on the other hand claim that the Christian saw 72 bodies hanging in the trees (Adam of Bremen 1983; 1993). Most likely the English translation is mistaken because while the Latin sentence is undoubtedly confusing, a marginal note in the original document clearly tells us that it was 72 living creatures that were sacrificed at the occasion.

Even a person able to read the original texts is however faced with translation problems since the concepts dealt with sometimes are unknown to the modern reader, and since the individual author's use of the language differs. This problem is especially obvious in the translation of the Old Norse poetry and its complicated chains of analogical phrases (Näsström 2001: 14-15).

Time gap and transformation

A problem in the same category as copying and translation is the time gap between the event and the writing of the text, and the subsequent transformation of the content, that is relevant for some of the texts. Especially the Icelandic sagas and *Beowulf* are at least partly based on oral tradition and were written down hundreds of years after the events described are supposed to have taken place. In such cases both conscious and involuntary transformations of the text are to be expected.

It has long been assumed that oral transferring involves huge losses of information, and since the time gap for the above mentioned texts is extensive, it has been assumed that little of the original information is unaltered. If this is true these texts would, as is claimed by sceptical scholars, be of very little use except possibly for telling something about the time and context in which they were written down (see for example Steinsland 2005: 36).

This problem might however be exaggerated and ethnological studies of oral information transfer regarding rural murders and assaults in Norway from the 17th until the early 20th century show that much information can correctly be passed on over centuries before it was written down. The study showed that "action-supporting" information such as the outcome and motive behind an act was best preserved. It was also shown that time was not the most important factor in the reliability of the information transfer, but instead the aspects that were emphasised in the society during the time period in question. Very specific information could remain unaltered for centuries if it was of interest for the society. As an example information about whether a criminal repented his or her deeds could survive long in a deeply religious society, but disappear rather quickly at other places or in other time periods (Norr 1998: 17-18).

Another interesting thing to note is that the time gap actually might have made it possible for Christian authors to describe some things that had been impossible if the story would have been contemporary. This is especially relevant for the description of pagan motives in the Icelandic sagas in comparison with those in the continental missionary biographies. These motives would have been difficult for the Icelandic authors that were often men of the church, to deal with in such a positive or at least neutral way as some of the sagas do. Several manners did however exist that would lessen the problem, one being to put the story in an earlier time period, while others were the placement of the act in the peripheries of the world or describing it using Christian terminology. The usefulness of putting the story in an early time period is for example clearly seen in the Ancient sagas (*Fornaldarsögur*) that could quite freely describe pagan motives since the stories were supposed to have taken place in a mythical time long ago (Clunies Ross 2002: 20-23).

The agenda and perspective of the author

One of the most difficult problems with using written sources, regardless of authenticity and age is to determine and evaluate the influence of the agenda and perspective of the author, be it from his or her personal opinion, social background or political pressure. Even a very superficial reading of the above presented written sources shows clear signs of how the author's personal position affects the views on things. The early Christian missionaries are portrayed in a favourable light, as are their benefactors within the Frankish empire, while the pagans are dangerous at best, and stupid or cruel at worst.

Tacitus on the other hand is sometimes, despite the obvious barbarism, portraying the Germanic people in a positive way in the book *Germania*. The discussion regarding the author's actual agenda behind the text shows that this could partly be explained as a criticism of aspects of the Roman Empire, and hence a political statement (Bazelmans 1991: 100-101; Widht 1997: 12).

As a contrast Ibn Fadlān does not seem to have had any political intentions behind his description of the Rūs, but his cultural background as an upper class citizen in the Muslim Caliphate clearly affects his views (compare with Montgomery 2000: 7-8).

Ibn Fadlān is not the only author belonging to the wealthy and powerful part of society. Being literate in itself meant that a person usually came from the upper classes. Some of the Christian authors might of course not have had personal wealth, but did nevertheless wield substantial power. Subsequently most of the texts deal with the upper classes as their subject. The Icelandic sagas might describe their main characters as farmers, but they were indeed no small scale farmers but wealthy landowners often with dozens of retainers and the capacity to equip ships of war from their excess income. Ordinary people and slaves do appear in these stories, but seldom have very important parts to play.

Another group in society seldom described more than in passing are women. The authors were always men (with some rune inscriptions being the sole exception) and the society was dominated by men. The gender perspective is thus definitely skewed, even if some sources are worse than others when it comes to omitting women. The Icelandic sagas are in this respect slightly more open-minded with several portraits of strong female characters, both of positive demeanour such as Aud in *Gisle Sursson's*

Saga, and of more questionable character as Hallgerd in *Njál's Saga*. The women seldom play the most prominent parts in the sagas, but are nevertheless often the driving force behind the events in them, frequently pushing their male relatives into action to uphold the honour of the family during conflicts.

All these varying perspectives undoubtedly affect the manner in which we should read the texts. Propaganda, idealised description and topics that are unmentionable and have hence been changed, disguised or totally omitted are facts we just have to deal with.

The reliability of the author's sources

The authors of only a few of the texts mentioned here actually visited the places described or saw the events first hand. Even in the cases that the author actually was at the site, as for example Ibn Fadlān, all the things described were most likely not observed in person (compare with Montgomery 2000: 9). In most of the other written sources the author based the text on hearsay, earlier writings or informers that might have been eye-witnesses, but that also might have received their information from yet another source. In other cases, as was discussed above, the knowledge behind the text is mainly based on oral tradition.

It is obvious that information transferred through several persons, all with their own personal perspective and reasons for describing something in a certain way results in a loss of information even before the author has written it down. This does however not necessarily make the text totally unreliable. As was shown for the orally transferred narratives above, important information seems to have been conveyed correctly in many cases, but the trustworthiness of the authors sources must still definitely be taken into consideration when the general reliability of a text is determined.

The literary genre

The final major problem of interpreting the written sources is the conventions of the genre in which the text was written. In early historical research many of the texts discussed in this chapter were treated in the same way as a scientific text. Most researchers eventually became aware of the biases that all texts contain regardless of whether they were written in a scientific context or for other reasons. Scientific texts were nevertheless still regarded by most researchers to be more useful for historical purposes and hence the old documents that showed "scientific" tendencies, such as early ethnographic observations, were deemed more trustworthy. It was not until the early 20th century that this notion was challenged and it was shown that most of the early written sources should be viewed as belonging to a literary genre and not as modern ethnographical or historical documents. As an example it has been argued that even if Tacitus wrote *Germania* primarily as a description of the Germanic lands and people, he still worked within a literary genre based on earlier Greek and Roman writings, even though he as an independent author could deviate somewhat from this framework (Bazelmans 1991: 93-95; Widht 1997: 11).

This fact is really not very surprising since all texts are produced within a certain genre, even the style of contemporary scientific books can vary between different disciplines, universities, countries and the intended audience. When reading older written sources this has to be taken into consideration. To some extent all texts try to give the readers what they want, and to understand what the readers of the specific literary genre were expecting is a vital part in literary source criticism.

3.4.7 Addressing the problems

After this overview of the problems of the early written sources dealing with Northern Europe it might seem useless, or at least very difficult, to use this kind of information in connection with archaeological research. I do however not consider this to be the case. It is true that the written sources from the Roman Period and Early Middle Ages dealing with Northern Europe are few and seldom directly inform us about the structures and procedures behind special and ritual deposits. It is also true that using the texts involves many problems, but so does the use of any source of information, including the material culture. The problems involved are only somewhat different.

It would definitely be strange to ignore such a rich source of information that the written sources provide. As a comparison the elimination of a complete category of archaeological material because it is difficult to interpret would most likely be considered preposterous.

What is even more important is that it is impossible to disregard the written sources even if we choose not to use them in our archaeological research. Regardless of our intentions the sources from this period are few and therefore frequently quoted. Even such basic things as the existence of the Frankish Empire, Frisia, Saxony or the early Scandinavian state formations, the political tension between these entities or the manner in which Christianity spread through the area are mainly known to us from written sources. They are irrevocably an integrated part of our preconceptions about the period and will affect our way of thinking about and interpreting the archaeological finds even if we avoid using them explicitly. In such a situation being familiar with the sources and hence being able to critically view them is the best available option.

The first step towards a critical usage of the written sources is being aware of the problems presented in this chapter. The next step is to learn as much as possible about the written sources and especially about the context they were written in. Only by knowing the individual situation and circumstances of a written text can we hope to address the problems of interpretation. If it is at all possible, the researcher should study the original text, but otherwise analyses made by researchers that studied the original text should be used as complements to the different available translations.

3.5 Getting it all together

This chapter demonstrated that a large number of methods exists that can be used to identify special deposits. Several ways for using analogies, primarily involving historical sources, for interpreting such deposits from the Northern European Early Middle Ages have also been presented.

These presented methods are unfortunately often vaguely formulated and when used on their own seldom produce an entirely convincing result. Moreover most of the methods described here only work in certain circumstances and even then have to be adapted to the specific situation of the research area or site. Generally they also defy the application of statistics, either since they the criteria are too imprecise and the samples are small. Because of this it is clear that a more flexible approach has to be sought.

The role played by rituals within human society, be it religious or otherwise, is so complex that one aspect of the archaeological material will seldom give us enough support for an interpretation. We will thus most likely neither find one method that works for identifying all ritual deposits or one analogy that explains them, even within a single society and time period. Instead we must be prepared to use the whole array of methods and as wide a range of analogies as can be found, and apply them to the archaeological materials. We will then have to find the methods and analogies that work in a specific situation, but still, as is virtually always the case within human science, the result will not be a clear answer but only, if we are lucky, a convincing interpretation.

It is here worth pointing out that I do not argue for an "anything goes" type of approach. Correctness and a strive towards objectiveness are a part of the process, however ultimately unobtainable. There will always be several alternative explanations available for the sceptical reviewer. This is impossible to avoid and the point is that the most likely explanation should be chosen, until another convincing interpretation is presented, maybe based on new data or old data seen from a new perspective.

4. Dongjum

4.1 The excavation

The excavation of the Dongjum *terp* was part of the ambitious Frisia project that unfortunately was ended prematurely, as was mentioned in chapter 1. For this reason the excavation of the *terp* Heringa in Dongjum has only recently been thoroughly published in the master thesis *Uitwerking steilkantonderzoek van de terp Dongjum* (Azier 2010). In addition to this text the information about the excavation techniques and archaeological results in this chapter are based on the field drawings, find lists and day reports from the excavation, as well as on the articles *Zoden aan de dijk. Kleinschalige dijkbouw in de late prehistorie en protohistorie van noordelijk Westergo* by Jos Bazelmans *et al.* published in 1999 and *Die Wurten von Dongjum-Heringa, Peins-Oost und Wijnaldum-Tjitsma: kleinmaßstäblicher Deichbau in ur- und frühgeschichtlicher Zeit des nördlichen Westergo* by Jos Bazelmans published in 2005.

The archaeological investigation at Dongjum-Heringa was a *steilkantonderzoek*, a type of excavation in which an already existing section of the *terp*, created by earlier commercial digging of *terp*-soil, is cleaned up and drawn (see figure 4.1). The technique was at the time a novelty but has since then been used at several other *terp*-sites such as Englum, Peins-Oost, Wierum, Anjum and Achlum (Nieuwhof 2006: 2; Dijkstra *et al.* 2008: 308, Nicolay 2010: 7-8).

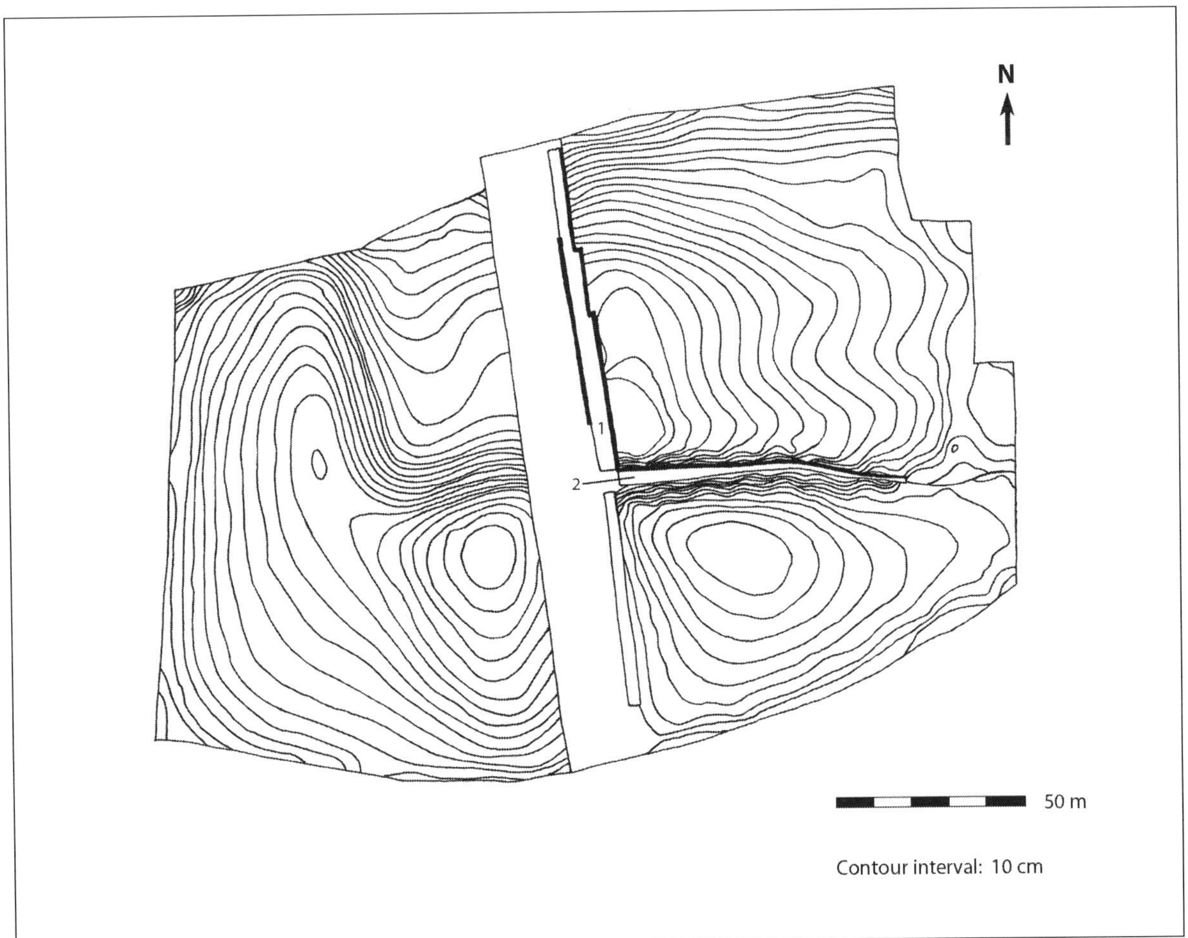

Figure 4.1. The excavation trenches at Dongjum (1 and 2) with the sections marked with bold lines (after Bazelmans *et al.* 1999).

The excavation method used in Dongjum is well suited for answering questions regarding the stratigraphy and chronology of a site, especially considering the limited time and manpower required for such an excavation (Bazelmans *et al.* 1999: 17). The technique unfortunately provides very little information regarding spatial relationships within the site since only a narrow slice of the settlement is examined. This makes it difficult to use the placement of finds within the site to identify special deposits. Because a majority of the settlement is left unexcavated or already dug away this type of investigation produces a limited amount of find material and runs a substantial risk of missing many archaeological features.

4.2 The archaeological results

The *terp* Heringa in Dongjum was first inhabited in the 2nd century AD when a single at least 15 meter long house-podium was raised. During the 3rd century this initial *terp* was expanded at least five times and eventually reached a diameter of more than 30 meters (Bazelmans *et al.* 1999: 15, 23-25, Nicolay 2010: 9). At the end of the 3rd century the *terp* was apparently abandoned, an event that the settlement shares with most *terps* in Friesland and Groningen (see section 1.5.1). In the early 5th century the *terp* was resettled, but a 25-30 cm thick homogeneous brown grey layer covering most of the excavated area indicates that this part of the *terp* was for some time mainly used as arable farmland. This layer presents a clear divider between the Roman and the Early Medieval building phases, enabling a reliable dating of structures and finds to one of these periods.

Until the 11th or 12th centuries the excavated parts of the *terp* were again intensively habited. The *terp* was expanded both horizontally and vertically and the sections show that a number of sunken huts and platforms for sod-walled houses were built (Nicolay 2010: 9).

The find material from Dongjum is mainly made up of pottery and animal bones. Imported *terra sigillata* is present in the Roman Period but domestic pottery dominate the material. In the 11th or possibly 12th centuries the habitation gradually moved towards the south east and eventually the *terp* was abandoned (Bazelmans *et al.* 1999: 26-27).

4.3 The archaeozoological material

The animal bone material from the Dongjum excavation dated to the selected time period (ca 400-900 AD) consists of 842 bone fragments weighing together 17,628 grams. Of this material 681 fragments weighing 17,007 grams could be identified to species.

The bone material from Dongjum is on the whole well preserved with a small degree of modification by taphonomic processes. This is also shown by the medium weight per fragment of 20.9 grams for the material, a comparatively high figure, even though this figure to some extent is due to the fact that sieving was not used during the excavation (Nicolay 2009, personal comment). 57 bone fragments in the material show signs of heating

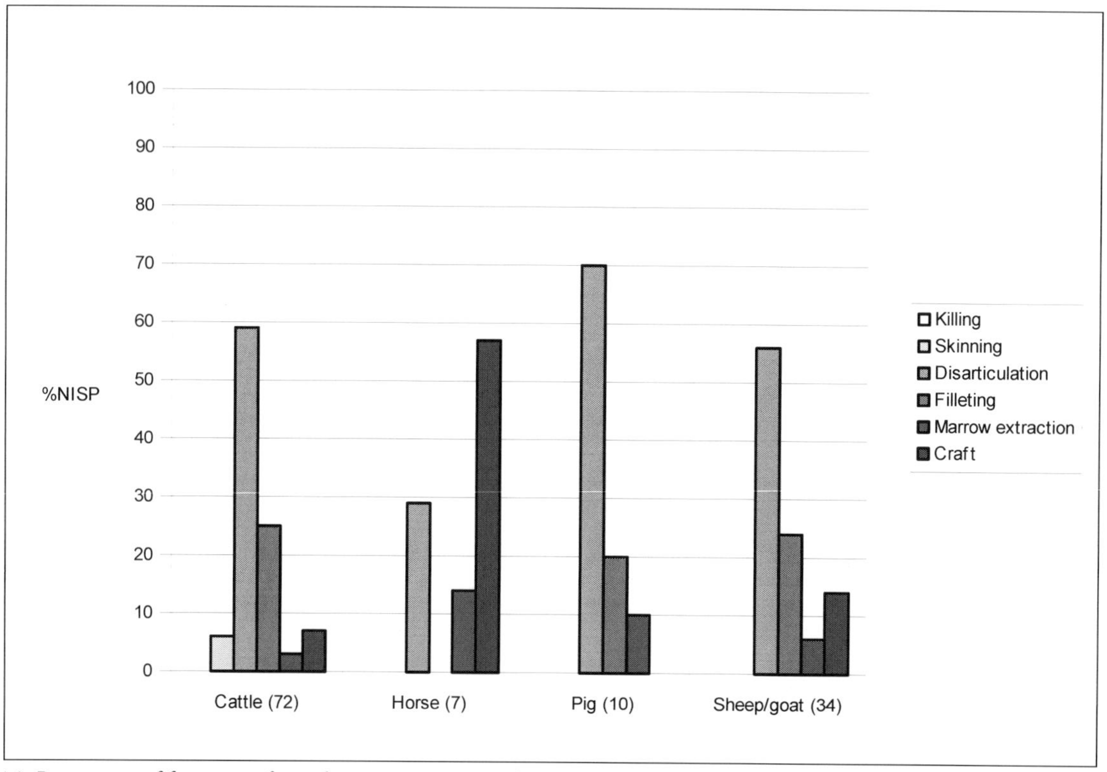

Figure 4.2. Percentage of fragments from the major species in the Dongjum material showing marks belonging to the different animal processing categories (the total number of fragments is presented within parenthesis).

and of these 20 fragments had been affected by very high temperatures (more than 675 degrees Celsius). 56 bone fragments show signs of animal gnawing whereof seven fragments can be considered heavily affected. In addition some 163 butchery marks were recorded on 131 identified bone fragments. Of these, 123 marks could be placed into one of the five animal processing categories presented in chapter 2. The general pattern of animal processing in the material is the same for cattle, sheep/goat and pig, with a majority of the marks (around 60 percent) indicating disarticulation and almost half as many indicating filleting (figure 4.2). The horse bones show a large percentage of marks from bone craft while the pig bones totally lack this category.

In figure 4.3, 4.4 and 4.5 the statistics for weathering, trampling and fracture freshness for the material are presented as numbers of identified fragments per stage or index step. The diagrams show a well preserved material with only few traces of weathering and trampling. The fracture freshness index diagram shows that most fractures were indeterminable (FFI 3) and that fresh and dry fractures are about equally represented.

The NISP-, weight- and MNI-numbers for the identified material are presented in table 4.1.

In figure 4.6 and 4.7 the proportions of the major species from Dongjum are presented as NISP- and weight-percentages. The material is dominated by cattle and sheep/goat, of which in turn cattle is showing a higher representation both in number of fragments and, more expectedly, in weight. A smaller part of the material has been identified as pig and horse while the category "Other" includes bones from cat, dog and red deer as well as domesticated and wild bird species. The bones in this group are all small fragments, something that explains the smaller percentage of this category when weight is measured. Fish bones are completely missing from the material, something that can at least partly be explained by the fact that sieving was not used.

The distribution shown in figures 4.6 and 4.7 does not demonstrate any deviations from what could be expected for a *terp*-material. In figure 4.8 the NISP of the major domestic species from six comparative bone materials from early medieval *terp*-settlements are shown together with the corresponding numbers for Dongjum and Oldehoofsterkerkhof (see chapter 5). As can be seen

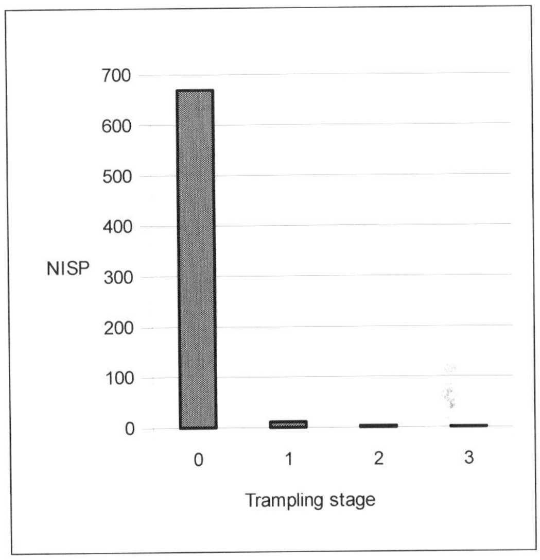

Figure 4.4. Trampling stage diagram for the Dongjum material (0: no marks; 1: few, but clear marks; 2: several and clear marks; 3: most of the surface is covered with marks).

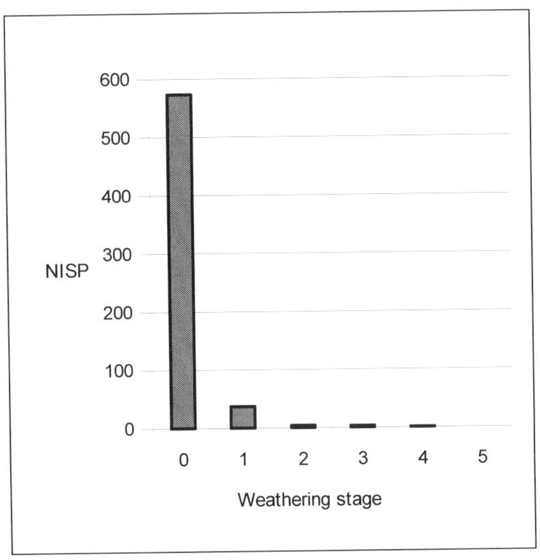

Figure 4.3. Weathering stage diagram for the Dongjum material (0: no cracking or flaking; 1: cracking parallel to fibre structure; 2: flaking of outer surface; 3: fibrous texture; 4: coarsely fibrous and rough surface, inner cavities visible; 5: bone falling apart).

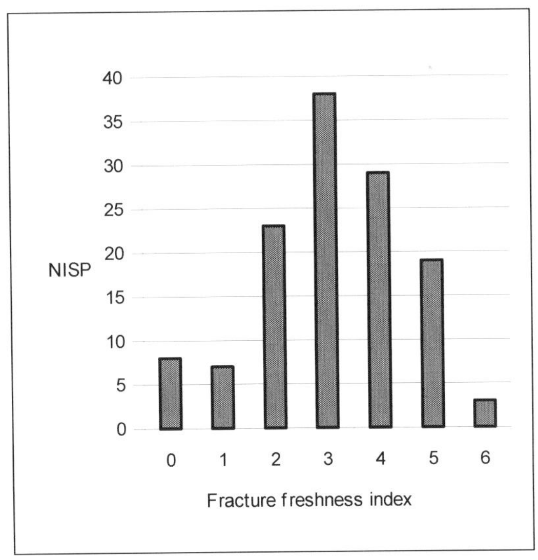

Figure 4.5. Fracture freshness index diagram for the Dongjum material (0-2: fresh fracture; 4-6: dry fracture).

Table 4.1. NISP, weight and MNI for the identified material from the selected time period.

Species	NISP	Weight (g)	MNI
Cattle (*Bos taurus*)	302	12838.6	11
Sheep/goat (*Ovis aries/Capra hircus*)	228	2112.6	11
Pig (*Sus scrofa domesticus*)	37	580.9	7
Horse (*Equus caballus*)	20	1238.3	4
Dog (*Canis familiaris*)	3	38.8	3
Cat (*Felis catus*)	51	44.2	2
Red deer (*Cervus elaphus*)	1	4.0	1
Tame mallard (*Anas platyrhynchos domesticus*)	2	10.8	3
Tame/wild mallard (*Anas platyrhynchos domesticus/ Anas platyrhynchos*)	6	4.5	2
Domestic chicken (*Gallus gallus domestica*)	5	8.1	3
Greylag goose (*Anser anser*)	2	29.1	1
Bean goose (*Anser fabalis*)	1	3.2	1
White-fronted goose (*Anser albifrons*)	3	34.5	2
Brent goose (*Branta bernicla*)	3	6.7	1
Unspecified goose (Anser sp.)	2	36.5	3
Wild mallard (*Anas platyrhynchos*)	4	8.5	3
Wigeon (*Anas penelope*)	1	0.8	1
Common teal (*Anas crecca*)	1	0.3	1
Common teal/garganey (*Anas crecca/Anas querquedula*)	1	0.4	1
Pintail (*Anas acuta*)	2	1.9	1
Goldeneye (*Bucephala clangula*)	1	0.9	1
Long-tailed duck (*Clangula hyemalis*)	1	0.8	1
Red-breasted merganser (*Mergus serrator*)	1	0.7	1
Bar-tailed godwit (*Limosa lapponica*)	1	0.8	1
Lesser black-backed gull (*Larus fuscus*)	1	0.7	1
Common gull (*Larus canus*)	1	0.4	1
Total:	**681**	**17007.0**	

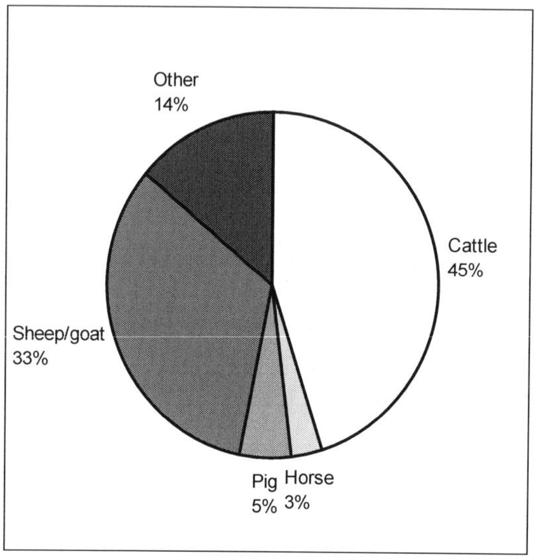

Figure 4.6. NISP-percentage diagram for the Dongjum material.

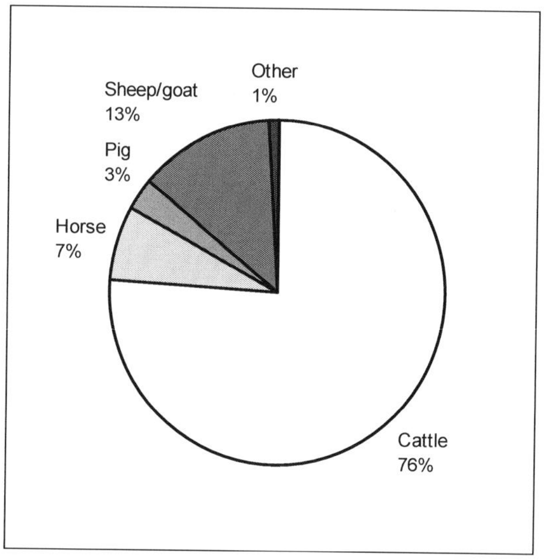

Figure 4.7. Weight-percentage diagram for the Dongjum material.

cattle and sheep/goat are the two dominating species in all these materials, while pig only makes up a small part of the number of fragments. The relation between cattle and sheep/goat as well as between pig and horse does not however show any clear pattern. Most likely the distribution of these largely depends on the landscape in which the *terp* was situated, as well as economy and traditions. One of the few aspects that do stand out is the high proportion of horse in Feddersen Wierde (and to a lesser degree in Wierum).

In figures 4.9 and 4.10 the element distribution of the bones from the major species in the Dongjum material are presented as percentages of fragments from each anatomical body group. As can be seen in these two figures the element distribution for the major species does not deviate very much from the expected percentages given in chapter 2 (see figure 2.3 and 2.4). There are less vertebrae, ribs and carpal bones found in Dongjum than in the ideal and expected distributions, but these bones are either less likely to be recovered (small carpal and tarsal bones) or

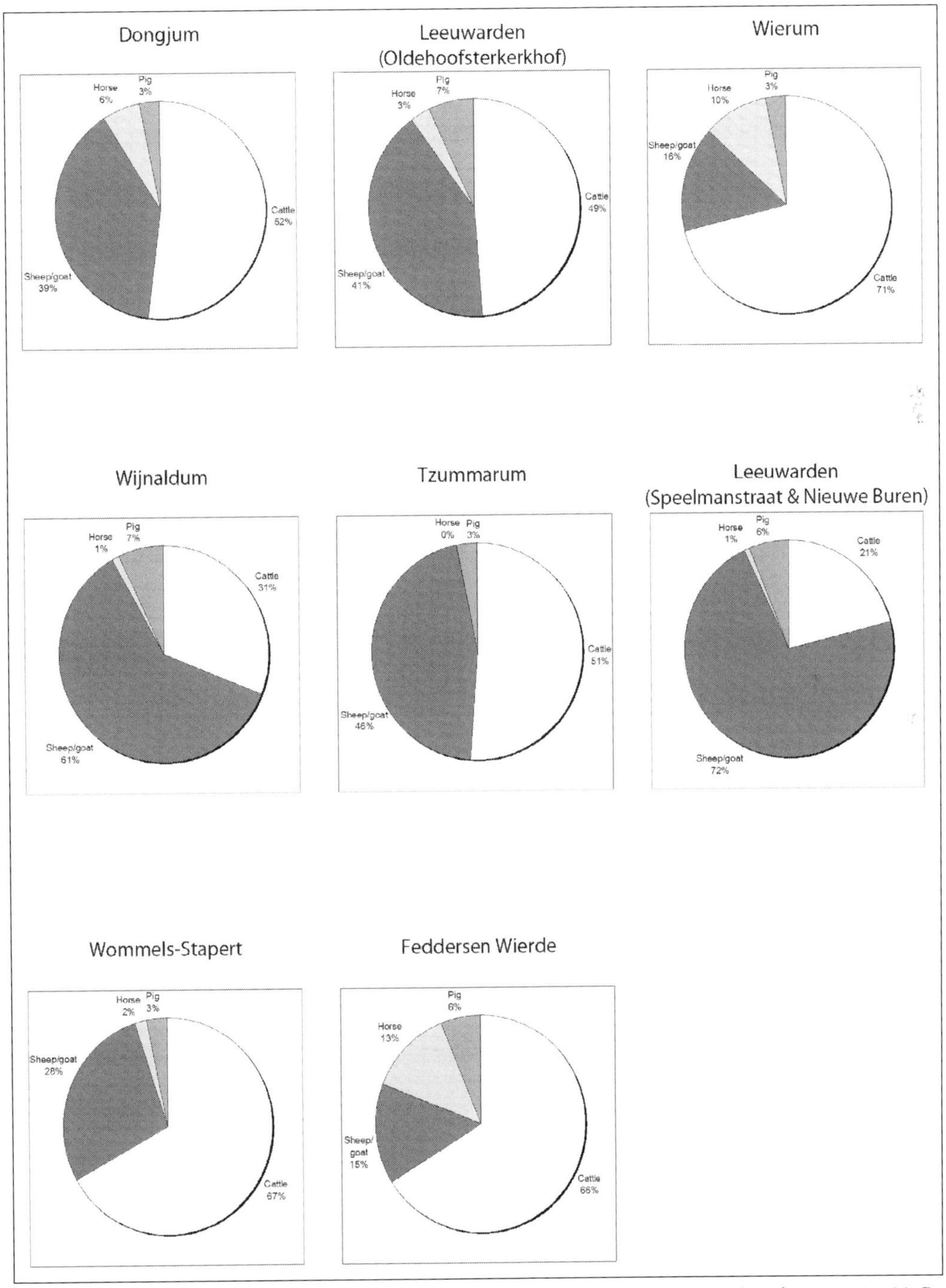

Figure 4.8. NISP diagrams for the major domestic species from Early Medieval *terp*-settlements (Reichstein 1991: 28; Prummel 2006: 45).

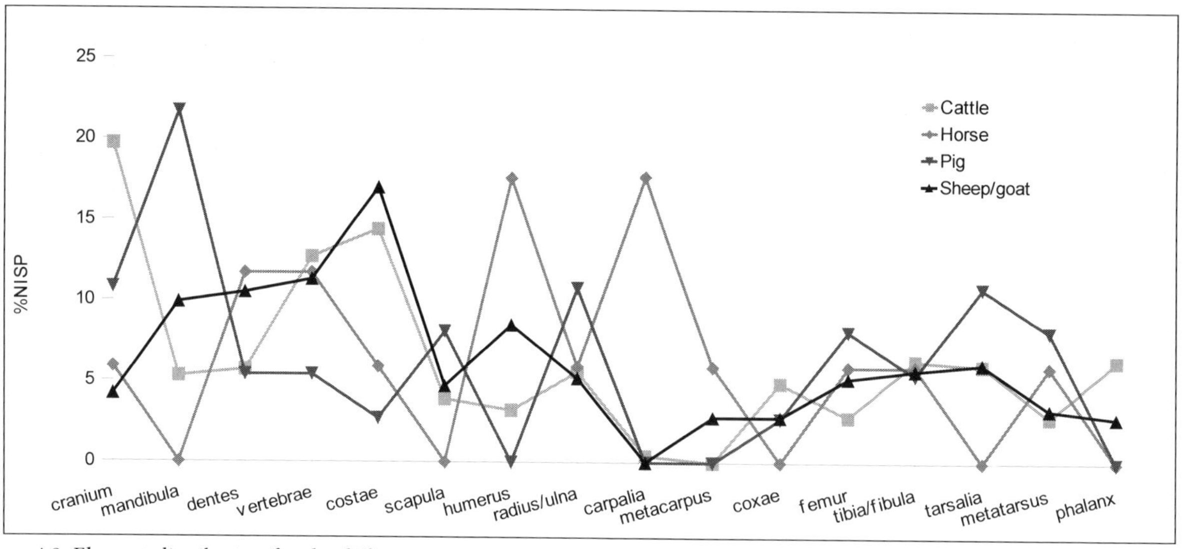

Figure 4.9. Element distribution for the different major species in the Dongjum material.

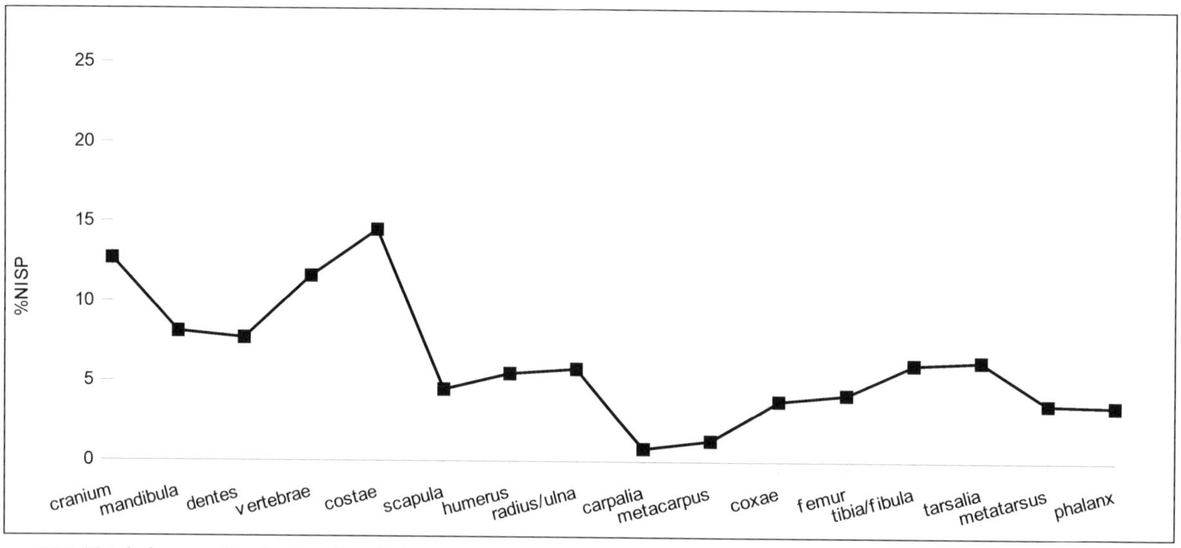

Figure 4.10. Total element distribution for all the major species in the Dongjum material.

difficult to identify to species if not complete (vertebrae and ribs). The overrepresentation of cranial and mandible fragments can be explained with the opposite argument, since these are bone elements that easily fall apart into pieces that still are possible to identify.

Only five bone fragments in the Dongjum material, all from cattle could be identified to sex giving a distribution of three male and two female animals.

The vast majority of the animals were fully-grown. Only less than 20 percent of the fragments of the major species originate from young individuals (see figure 4.11). The percentage of young individuals is lowest for horse, as could be expected since these would primarily have been used as riding and draft-animals and not for food, and highest for sheep/goat and pig. One possible explanation for the low representation of cattle killed at a young age is that at least some of the cattle were used for milk production and as draft-animals and were for that reason slaughtered later. Pigs and sheep/goats were slaughtered at younger ages than cattle for their meat. This would in that case indicate that the production of wool and milk played a small role in the keeping of the sheep/goats. It is also interesting to note that only a small number of very old animals were identified in the material.

The measurements of the bones from Dongjum produced one unusual result in the form of an almost complete very large cattle atlas from a pit (feature 1003) dated to the 5-7th centuries AD (figure 4.12). The bone was much larger than any of the other cattle atlases from the Dongjum material. Since cattle from the *terp*-area in this time period were generally small (see for example Reichstein 1991: 48-51) the atlas was suspected of either coming from an imported animal or even possibly from an aurochs (*Bos primigenius*). While the bone is very large compared to other cattle atlases from early medieval Northern Europe, it is still slightly smaller than most aurochs (see Appendix II). It is nonetheless possible that the bone is from a female aurochs.

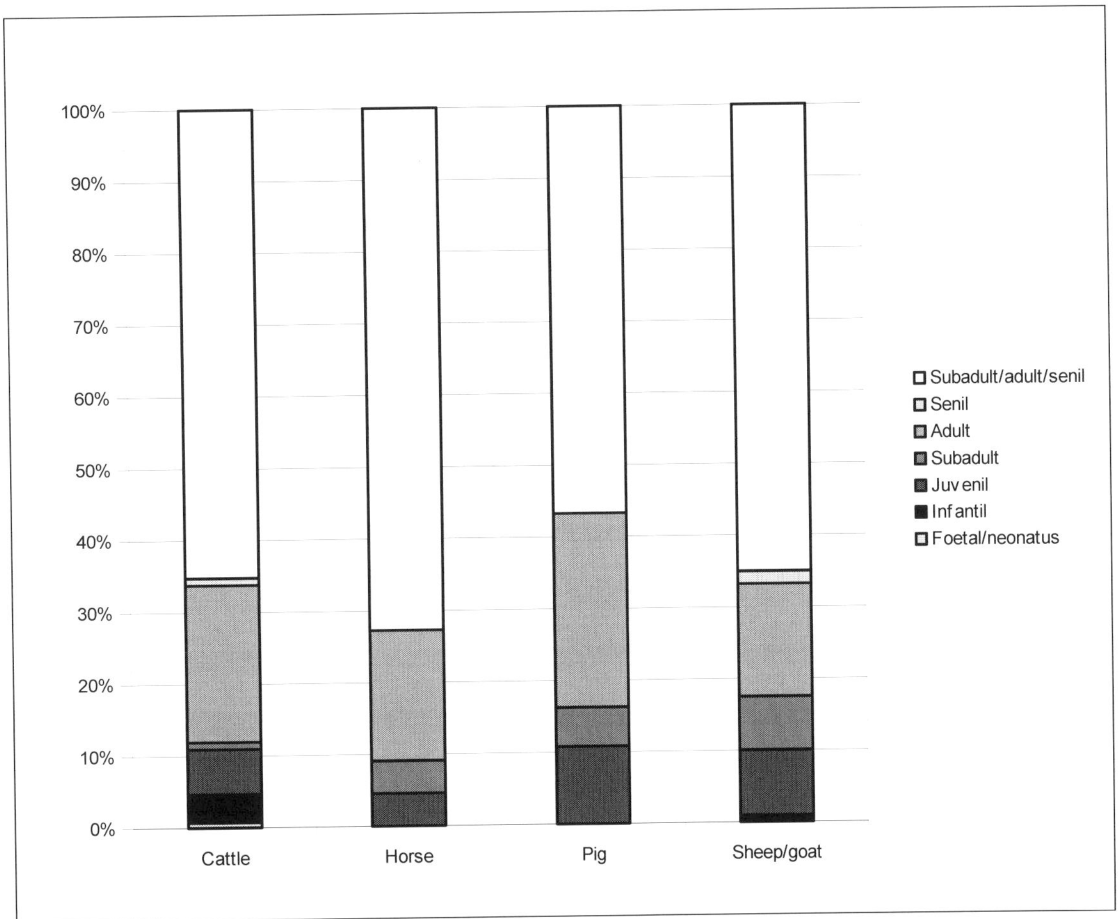

Figure 4.11. Percentage of fragments belonging to different age groups for the major species in the Dongjum material.

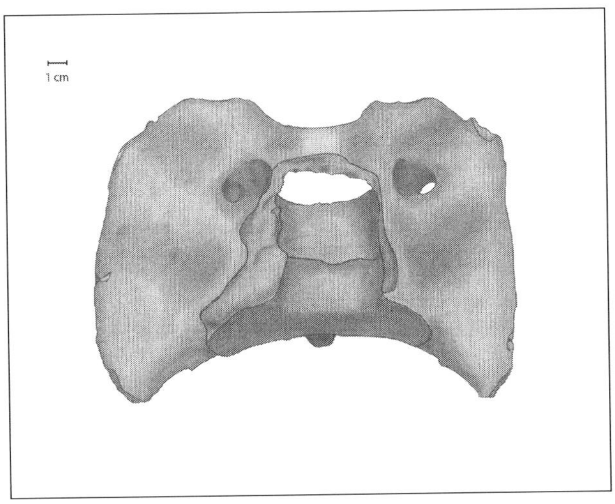

Figure 4.12. The very large cattle atlas from feature 1003.

4.5 The special deposits

4.5.1 Introduction

In the bone material from Dongjum only three deposits from the selected time period have been interpreted as special. These deposits will be described in this section together with the reasons for their selection. In table 4.2 the basic data for the different deposits are presented.

4.5.2 Articulated body parts

The occurrence of deposits with articulated parts of animals are often considered to be special (see section 3.2.5). One such deposit was identified in the Dongjum material.

Table 4.2. Overview of the special deposits from Dongjum.

Feature	Trench	Level	Filling	Find nr	Feature type	Dating	Deposit type
1022	2	2	1	275	Ditch	VMEA	Articulated body parts (horse)
1023	2	2	1	209	Ditch	VMEA	Selected elements (cattle and sheep cranial fragments)
1088	1	2	1, 3	257, 259	Well	VMEC	Age structure (infant and juvenile cattle and sheep/goat)

Feature 1022 (left ankle of horse)

This feature is a part of a whole system of ditches (figure 4.13) which also includes feature 1023 that is described below.

The filling of this ditch was rather complicated but mainly contained brown-grey soil with high organic content (possibly manure). The ditch contained some pottery and a bone assemblage of 20 fragments weighing 463.6 grams. The bone material showed virtually no signs of weathering, trampling or burning and also had no visible processing marks. The only bones gnawed by animals were those belonging to the horse ankle described below. Except for the horse bones the material consisted of a few fragments of cattle and sheep/goat. These bones look quite ordinary with the sole exception of a single pelvis fragment from a cattle foetus.

The horse ankle mentioned above consists of the lower parts of the left tibia, the astragalus and the calcaneus of an adult horse. There are no processing marks on these bones but they do display plenty of animal gnawing. The bones fit together and the placement of the gnawing marks on the protruding parts of the joint and their absence from the articular surfaces strongly indicate that the ankle was articulated when it was placed in the ditch (figure 4.14).

It is possible that the horse ankle represents normal refuse since this part of the legs contain little meat. An articulated bone group placed in a feature without any other horse bones is despite this a characteristic that makes the deposit qualify as special.

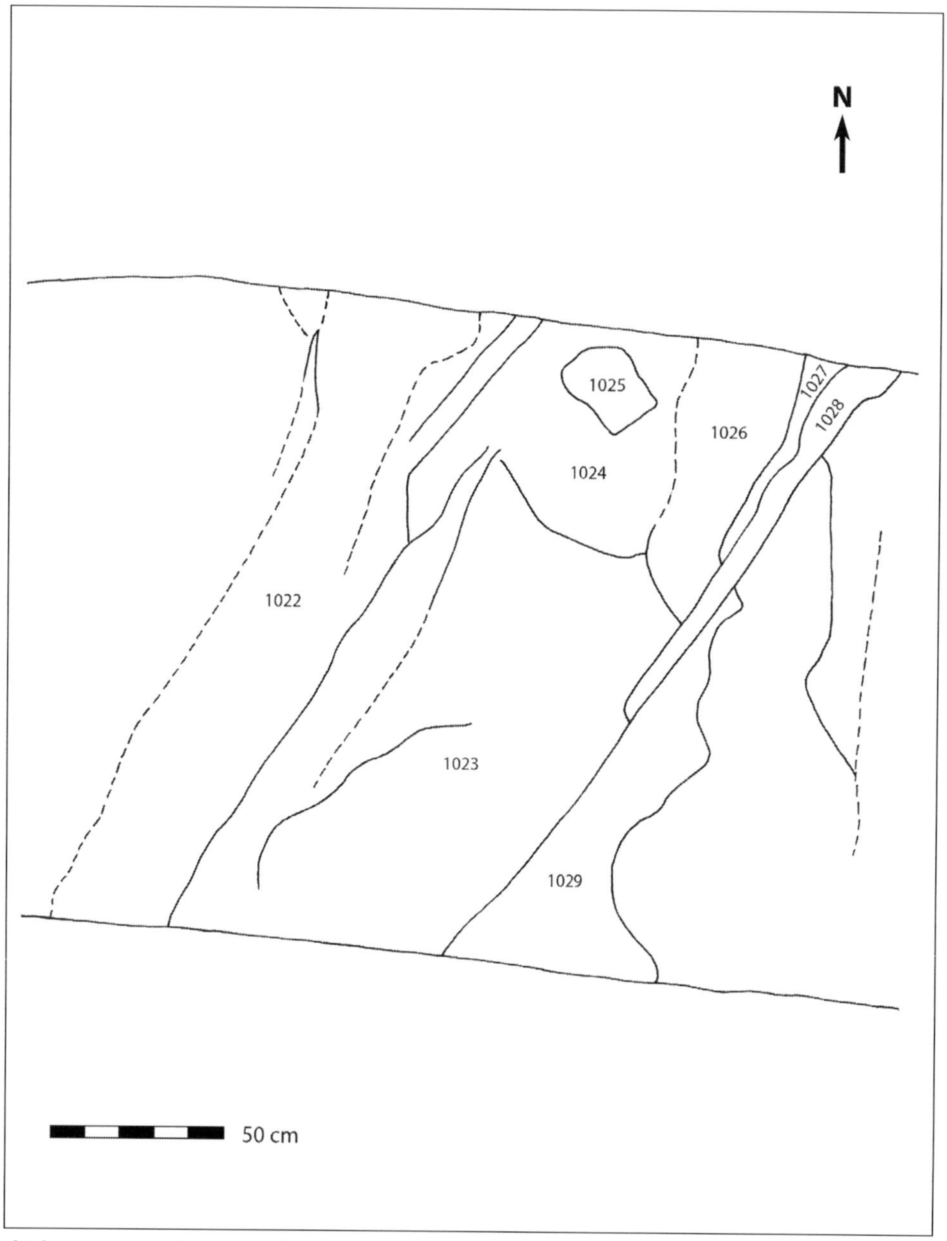

Figure 4.13. The ditch system with feature 1022 and 1023 (after field drawing).

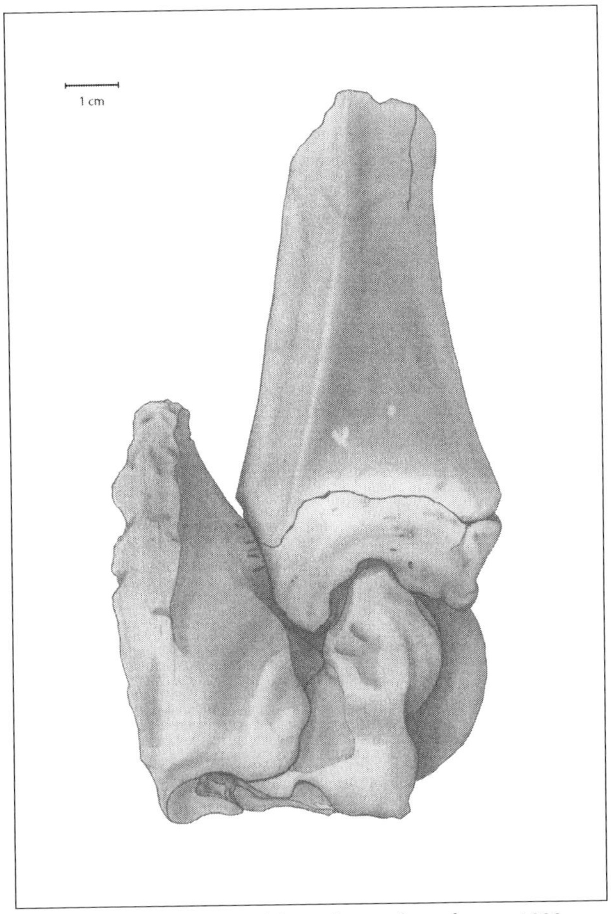

Figure 4.14. The articulated horse bones from feature 1022.

4.5.3 Selected elements

Assemblages that contain a strongly deviating distribution of bone elements from that of the rest of the bone assemblage could be an indication of speciality. One such deposit was identified in the Dongjum material.

Feature 1023 (cranial fragments of cattle and sheep/goat)

This feature belongs to the same system of ditches as feature 1022 above (see figure 4.13) and contained a similar complex filling mainly composed of brown-grey soil and manure. The ditch furthermore contained some pottery, wood and 40 bone fragments weighing 1324.3 grams. 39 fragments could be identified to species. The bone assemblage from the ditch does in most respects resemble the majority of the material from Dongjum. Cattle and sheep/goat are the dominating species while horse, pig and birds are represented by a few fragments. The taphonomic history of the deposit is also quite regular with few signs of light weathering, trampling, animal gnawing and marks from both skinning, disarticulation and marrow extraction. The age structure is normal with some young and some old individuals.

The aspect where this deposit stands out is the element distribution of the cattle and sheep/goat fragments. As can be seen in figure 4.15 the percentage of mandible fragments is substantially higher than what is normally the case for these species in the Dongjum material (see figure 4.9 and 4.10).

The tendency is even stronger in figure 4.16 where the percentages for only sheep/goat are shown.

Since the assemblage from the ditch is rather small the percentages in the diagrams do not represent large numbers of fragments. Five of the mandible bones are however large fragments that represent at least one cattle and three sheep/goat individuals (see figure 4.17). All sheep/goat mandibles come from adult animals.

The unusually high proportion of mandible fragments of cattle and sheep/goat qualifies this deposit as special.

4.5.4 Age structure

If the age structure of one or several species in a deposit differs distinctly from the rest of the material the deposit

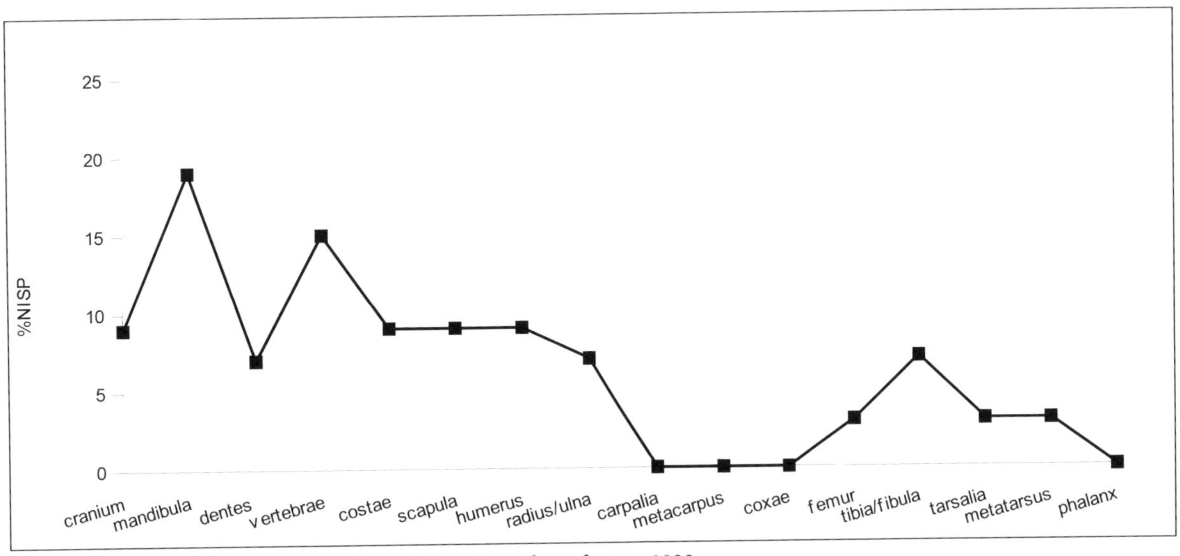

Figure 4.15. Element distribution for cattle and sheep/goat from feature 1023.

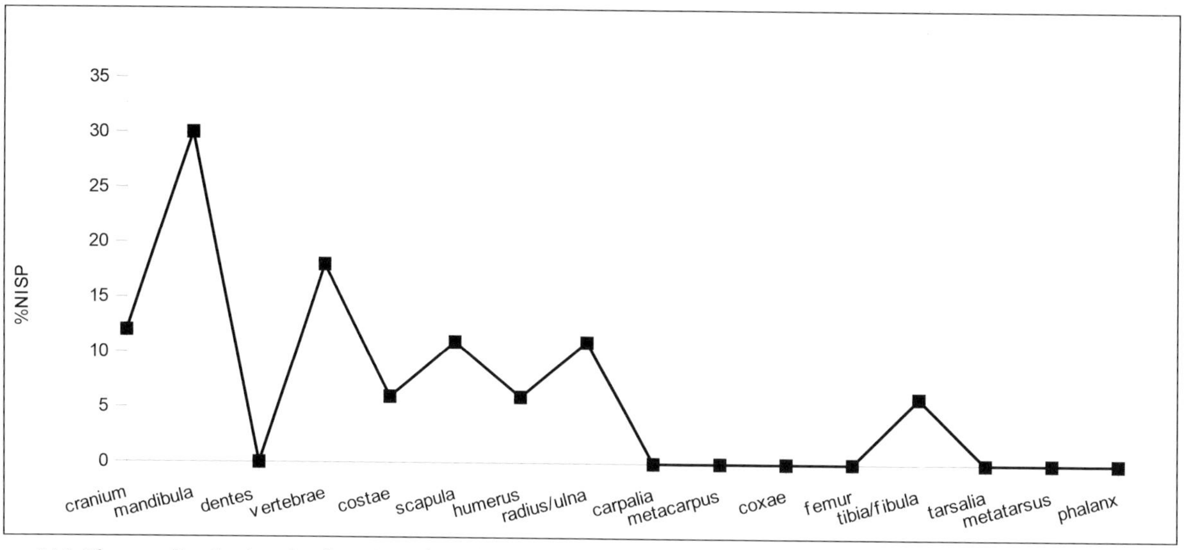

Figure 4.16. Element distribution for sheep/goat from feature 1023.

could be considered special. In the case of Dongjum one deposit shows such a pattern.

Feature 1088 (infant and juvenile cattle and sheep/goat)

This 2.2 meter deep beehive-shaped feature was interpreted as a well with three distinct fillings of which two contained bone material (figure 4.18).

The topmost filling consisted of light-brown sand with some ash and charcoal in it. Underneath this filling was a 6-8 cm thick clay filling. The bottommost filling was made up of several layers of darker soil with clay layers in-between. The topmost filling contained 23 fragments of animal bone weighing 160.3 g, as well as a grinding stone. 16 of these bone fragments weighing 146.1 g could be identified to species. The thin middle filling was empty while the bottommost one contained seven fragments of animal bone weighing 106.9 g, a small piece of slag (6 g) and a sharpening stone deposited at the very bottom of the feature (V254). Five bone fragments weighing 99 g from the bottom filling could be identified to species. The bone material from the feature was varied and did in

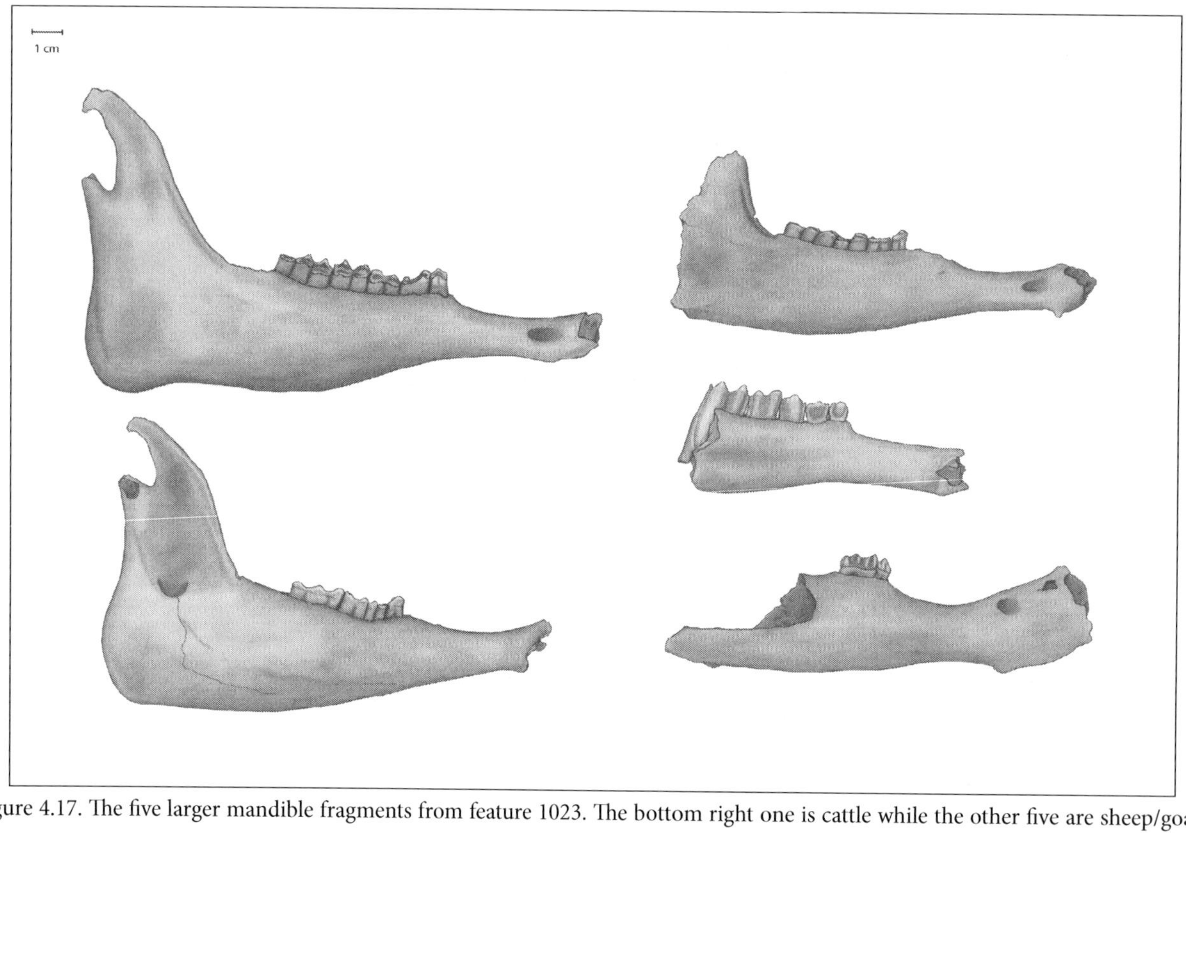

Figure 4.17. The five larger mandible fragments from feature 1023. The bottom right one is cattle while the other five are sheep/goat.

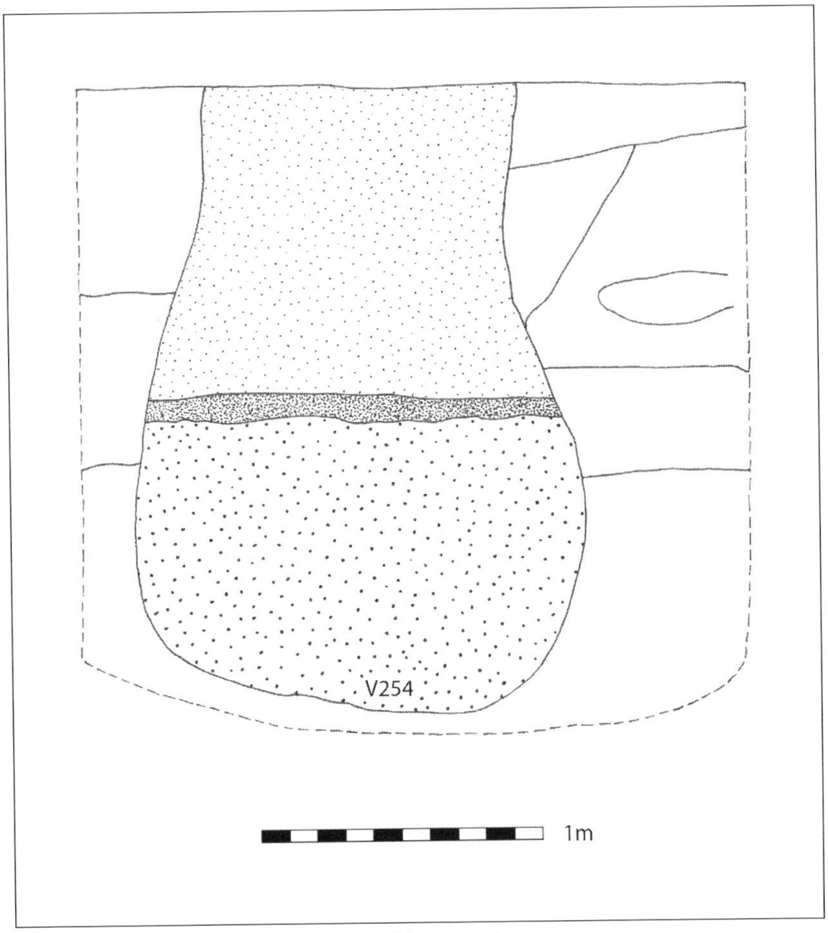

Figure 4.18. Section drawing of the well (feature 1088) (after field drawing).

most ways resemble the general pattern of the rest of the site. The assemblage from the well is dominated by sheep/goat and cattle and its taphonomic history indicates a rapidly deposited material, only very lightly weathered, not trampled and with only a couple of lightly gnawed bones. The butchery marks indicate that parts of the material have been disarticulated and filleted.

The only aspect that clearly differs from the rest of the Dongjum material is the age distribution and then especially for the topmost filling that has an unusually large number of bones from juvenile animals, as well as one fragment from an infant and one from a subadult individual. As many as 44 percent of the bone fragments from the topmost filling (38 percent from the whole feature) belongs to non-adult animals. For the material as a whole the corresponding figure is only 14 percent. This aspect together with the deposited grinding and sharpening stones, and the placement of the latter of these at the bottom of the feature is enough to qualify the deposit as special.

5. Leeuwarden

5.1 The excavation

The second selected bone material was excavated between winter 2004 and spring 2006 in the central part of the city of Leeuwarden in the north of the Netherlands. The construction of an underground car park in front of the church tower Oldehove would destroy parts of one of the *terps* that constitute the oldest settlement phases in the city and it had to be excavated. The archaeological project was carried out jointly by the company ADC ArchaeoProjecten and the Groningen Institute of Archaeology at the University of Groningen (Nicolay 2008e: 1).

The tight time schedule in connection with the project restricted the rescue excavation to two thirds of the actual construction area and forced it to be performed in a series of consecutive steps after which construction was immediately commenced. The excavation at Oldehoofsterkerkhof in Leeuwarden covered some 6000 m^2 (see figure 5.1) and was as such the largest *terp*-investigation since the excavation of the *terp* Tjitsma at Wijnaldum 1991-1993 (Dijkstra & Nicolay 2006). Originally the excavation was aimed towards the Early Medieval phases of the *terp*-settlement, but the well preserved Roman Period layers made the excavators

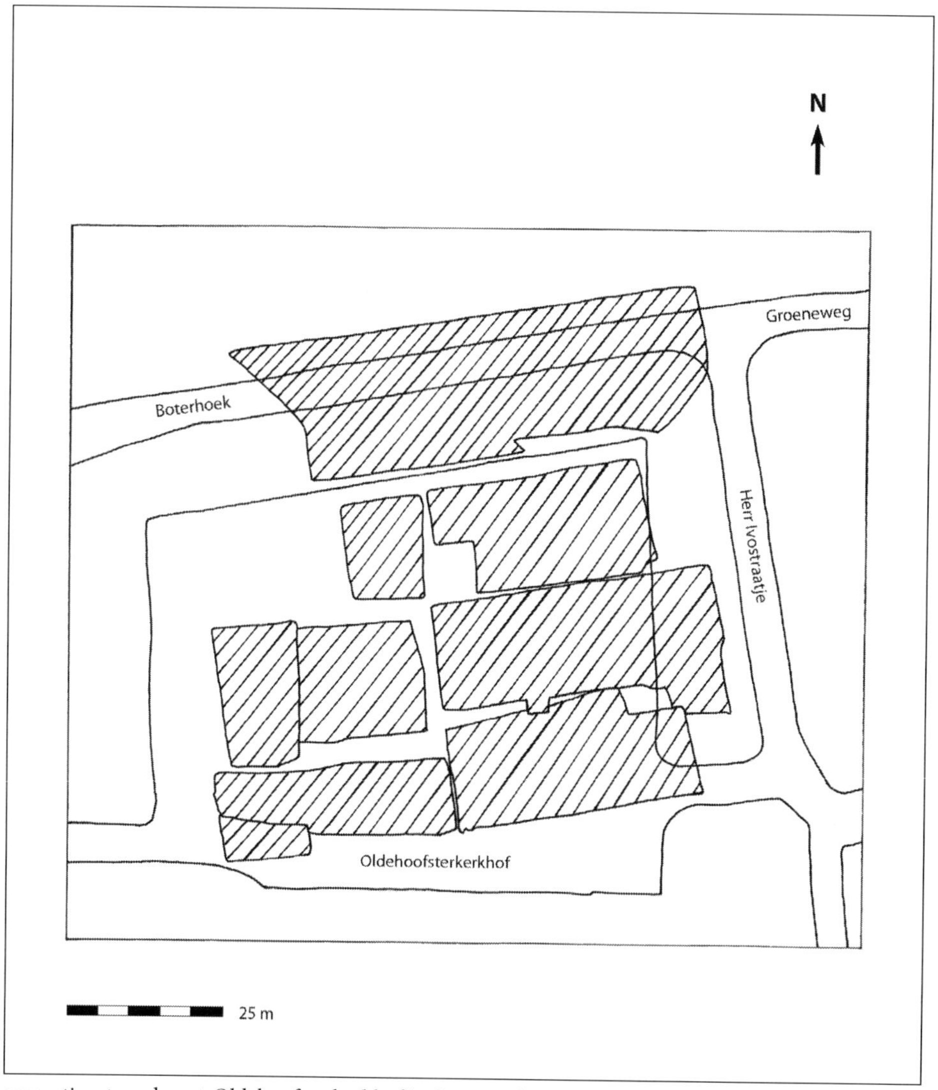

Figure 5.1. The excavation trenches at Oldehoofsterkerkhof in Leeuwarden (after Dijkstra & Nicolay 2006).

broaden the scope of the investigation to include this phase as well (Nicolay 2008e: 1).

5.2 The archaeological results

The excavated settlement remains in Leeuwarden date from the Early Roman Period up until the Early Middle Ages (1st-11th century). In the Roman Period the settlement constituted of a core podium with a single farm. Several well preserved house foundations from this period were excavated. Around 300 AD the site was abandoned as was the case with most *terp*-settlements in the Northern Netherlands, see section 1.5.1 (Dijkstra *et al.* 2008: 309, 317-318).

During the middle of the 5th century the *terp* was resettled, again following the pattern for most *terp*-settlements. New types of ceramics and fibulas indicate that the new settlers came from the northeast, most likely present day Northern Germany or Southern Denmark. During this period the *terp* grew both in height and in area and gradually developed into a larger settlement with two to four farms in the Merovingian and Carolingian periods (Dijkstra *et al.* 2008: 310, 317-318).

In the Early Medieval Period the *terp* had a round shape and was encircled with ditches. During this period the dwelling mound was raised with salt marsh sods at several occasions probably as a reaction to the formation of the Middelzee and the expansion of the Boorne tidal inlet that led to a higher water level in the area (Dijkstra *et al.* 2008: 314, 318-319).

In contrast to the Roman Period, large house plans could not be clearly seen in the medieval phase, only corners and parts of the walls, while several sunken huts dated to this feature were found (Dijkstra *et al.* 2008: 318-319).

Around 900 AD the *terp* was again raised with salt marsh sods, probably due to the further expansion of the Boorne tidal inlet. The settlement was abandoned in the 10th or 11th century and transformed into a religious site with the building of the St.-Vitus church and the accompanying cemetery (Dijkstra *et al.* 2008: 320).

The *terp*-settlement of Oldehoofsterkerkhof can generally be described as a normal agricultural community. It lacks high status finds directly connected with the ruling classes of society as well as indications of specialized craftsmen connected with the aristocracy that for example is evident in Wijnaldum (Dijkstra *et al.* 2008: 333, 337).

5.3 The archaeozoological material

The animal bone material from the Leeuwarden excavation that have been dated to the selected time period (ca 400-900 AD) consists of 2206 bone fragments weighing together 33,945 g. Of these 1257 fragments weighing 30,841 g could be identified to species.

The bone material is generally well preserved with a medium weight per fragment of 15.4 g. 110 bone fragments in the material show signs of heating, whereof 60 fragments were calcinated. 97 bone fragments show

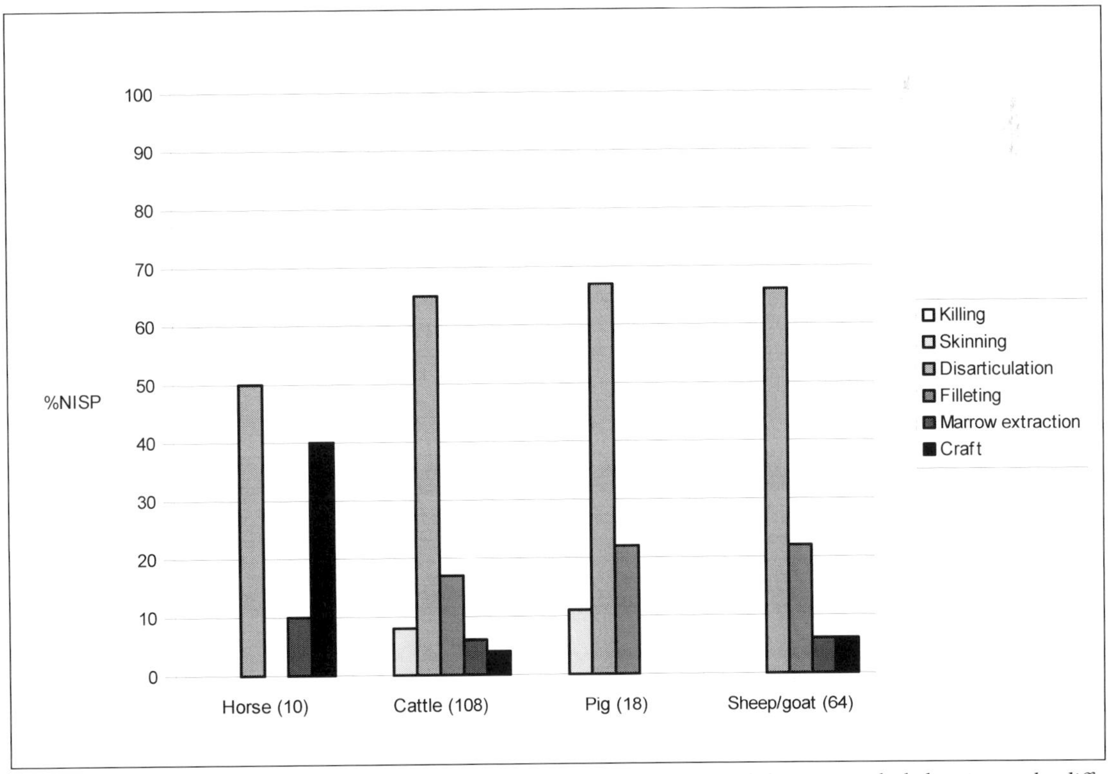

Figure 5.2. Percentage of fragments from the major species in the Leeuwarden material showing marks belonging to the different animal processing categories (the total number of fragments is presented within parenthesis).

animal gnawing marks and of these 26 fragments can be considered heavily affected. 367 butchery marks were recorded on 212 bone fragments. Of these, 200 marks could be placed in one of the five animal processing categories presented in chapter 2. The distribution of the processing marks can be seen in figure 5.2 and shows a strong resemblance with the distribution for the Dongjum material presented in chapter 4. The disarticulation marks dominate for all species except horse at roughly 65 percent, while the filleting is around 20 percent. As with Dongjum the horse fragments differ by showing far more traces of bone craft, while the other variations shown in the diagram are too small to be reliable.

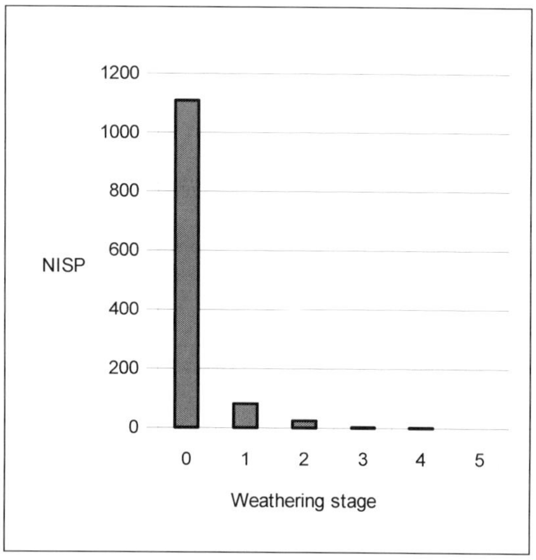

Figure 5.3. Weathering stage diagram for the Leeuwarden material (0: no cracking or flaking; 1: cracking parallel to fibre structure; 2: flaking of outer surface; 3: fibrous texture; 4: coarsely fibrous and rough surface, inner cavities visible; 5: bone falling apart).

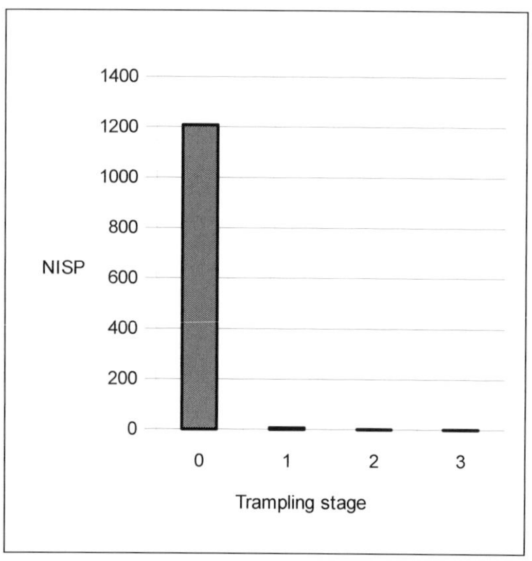

Figure 5.4. Trampling stage diagram for the Leeuwarden material (0: no marks; 1: few, but clear marks; 2: several and clear marks; 3: most of the surface is covered with marks).

In figure 5.3, 5.4 and 5.5 the statistics for weathering, trampling and fracture freshness for the material are presented as number of identified fragments per stage or index number.

The diagrams again display a similar picture as those for Dongjum, something that confirms the visual observation of the material as being well preserved.

The NISP, weight and MNI-numbers for the identified material from Leeuwarden (table 5.1) show the expected dominance of the major domestic species, but also 26 human bones, a substantial amount of bird bones as well as some fish. As a comparison human and fish bones are totally absent in the Dongjum material.

In figures 5.6 and 5.7 the proportions of the major species from Leeuwarden are presented as NISP- and weight-percentages. The figures resemble those for the Dongjum material with the exception of a larger portion of fragments not belonging to any of the four major species. This is mainly the result of the large number of bird and fish bones, while the slightly larger portion of weight from of species is the result of one whale bone and 26 human bones together weighing more than 1.2 kg compared to the remaining 1 kg of bones categorised as other species.

In figures 5.8 and 5.9 the element distribution of the bones from the major species in the Leeuwarden material is presented as a percentage of fragments from each anatomical body group.

The element distribution for the different major species (figure 5.8) in the material varies slightly from the corresponding numbers for Dongjum. The total numbers of fragments from horse and pig are however so small that these differences are quite uncertain. For cattle it seems as if the Leeuwarden material presents less cranial fragments but more loose teeth while the sheep/goat material has relatively few loose teeth.

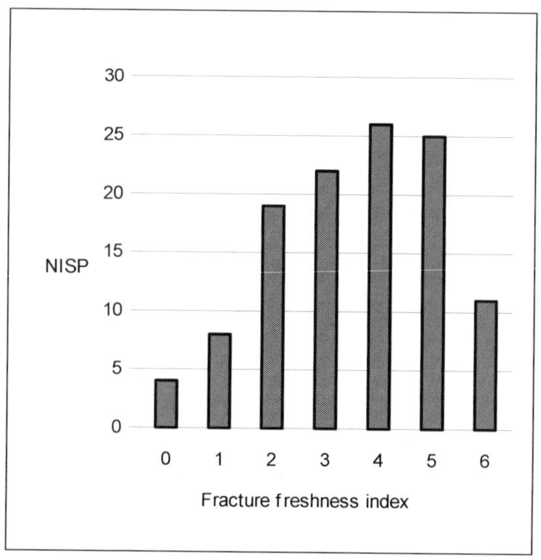

Figure 5.5. Fracture freshness index diagram for the Leeuwarden material (0-2: fresh fracture; 4-6: dry fracture).

Table 5.1. NISP, weight and MNI for the identified material from the selected time period.

Species	NISP	Weight (g)	MNI
Human (*Homo sapiens*)	26	918.5	2
Cattle (*Bos taurus*)	422	19432.5	16
Sheep/goat (*Ovis aries/Capra hircus*)	349	4406.4	10
Pig (*Sus scrofa domesticus*)	61	1361.6	3
Horse (*Equus caballus*)	27	3305.5	4
Dog (*Canis familiaris*)	3	14.0	1
Cat (*Felis catus*)	1	0.1	1
Red deer (*Cervus elaphus*)	6	482.6	2
Unspecified whale (Cetacea)	1	375.6	-
Domesticated goose (*Anser anser domesticus*)	7	29.0	7
Tame mallard (*Anas platyrhynchos domesticus*)	1	1.1	1
Domestic fowl (*Gallus gallus domesticus*)	5	31.4	2
Red-throated diver (*Gavia stellata*)	11	39.2	1
Grey heron (*Ardea cinerea*)	1	5.5	1
Mute swan (*Cygnus olor*)	1	7.2	1
Greylag goose (*Anser anser*)	109	180.3	11
White-fronted goose (*Anser albifrons*)	53	123.9	2
Brent goose (*Branta bernicla*)	1	1.4	1
Unspecified goose (*Anser* sp.)	34	68.6	-
Wild mallard (*Anas platyrhynchos*)	5	3.2	2
Gadwall (*Anas strepera*)	2	1.7	1
Wigeon (*Anas penelope*)	3	2.2	2
Common teal (*Anas crecca*)	35	10.8	7
Northern shoveller (*Anas clypeata*)	4	5.7	2
Goldeneye (*Bucephala clangula*)	1	0.1	1
Unspecified mergus (*Merganser* sp.)	1	1.3	-
Ringed plover (*Charadrius hiaticula*)	1	0.1	1
Eurasian golden plover (*Pluvialis apricaria*)	1	0.7	1
Dunlin (*Calidris alpina*)	3	0.3	1
Knot (*Calidris canutus*)	2	0.2	1
Unspecified wader (*Caladris* sp.)	3	2.0	-
Common redshank (*Tringa totanus*)	2	0.2	1
Unspecified shanks and tattlers (*Tringa* sp.)	1	0.4	-
Black-tailed godwit (*Limosa limosa*)	3	1.8	1
Bar-tailed godwit (Limosa lapponica)	1	0.4	1
Unspecified godwit (*Limosa* sp.)	2	1.2	-
Unspecified avocets and stilts (Recurvirostridae)	2	1.9	-
Cod (*Gadus morhua*)	1	4.9	1
Unspecified codfish (Gadidae)	2	0.5	-
Thin-lipped grey mullet (*Liza ramada*)	9	5.9	1
Righteye flounders (Pleuronectidae)	54	11.5	-
Total:	**1257**	**30841.4**	

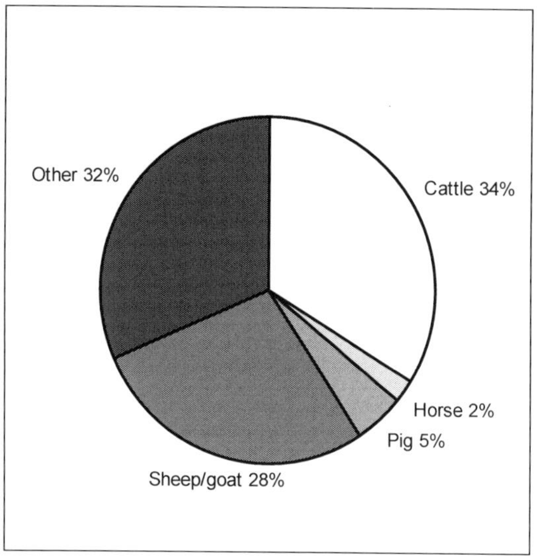

Figure 5.6. NISP-percentage diagram for the Leeuwarden material.

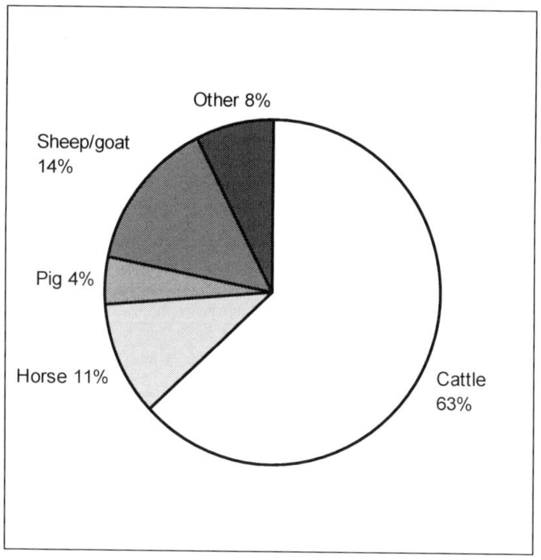

Figure 5.7. Weight-percentage diagram for the Leeuwarden material.

If we study the diagram for all major species together (figure 5.9) it is clear that there are more metacarpus fragments in the Leeuwarden material than in Dongjum (see figure 4.10). This can however partly be explained with the bone material from feature 90, a well that alone contained nine metacarpus bones, almost a fifth of all such fragments in the whole material (see further section 5.4.5 below).

27 bone fragments from the Leeuwarden material could be identified to sex, all belonging to either cattle, sheep/goat or pig (table 5.2).

There is, as can be seen a higher frequency of cows among the adult cattle and of boars among the adult pigs, while the two sexes are equally represented among sheep/goat. It is however important to consider the very small sample sizes, especially for pig and sheep/goat.

The age at death distribution in the Leeuwarden material is slightly different from that in Dongjum in that even less individuals were slaughtered young, only around 15 percent for cattle and below 10 percent for the rest (see figure 5.10). There are also more post-juvenile fragments that could not be determined precisely to age, hence the large portion of sub-adult/adult animals. These frequencies show that very few animals were slaughtered young for their meat. Instead most animals were kept at least until they were fully-grown. The almost total lack of senile individuals is however remarkable and indicates that even if few animals were slaughtered young, they were not kept very long past adult age as producers of milk, wool or as draught-animals.

5.4 The special deposits

There are 12 deposits in the Leeuwarden bone material dated to the selected time period that have been inter-

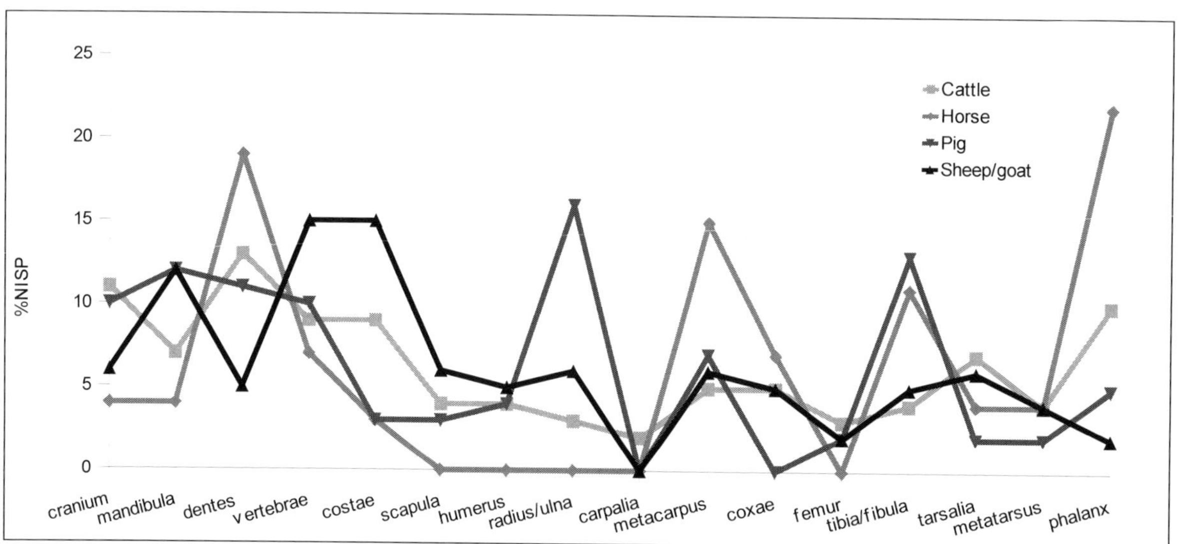

Figure 5.8. Element distribution for the different major species in the Leeuwarden material.

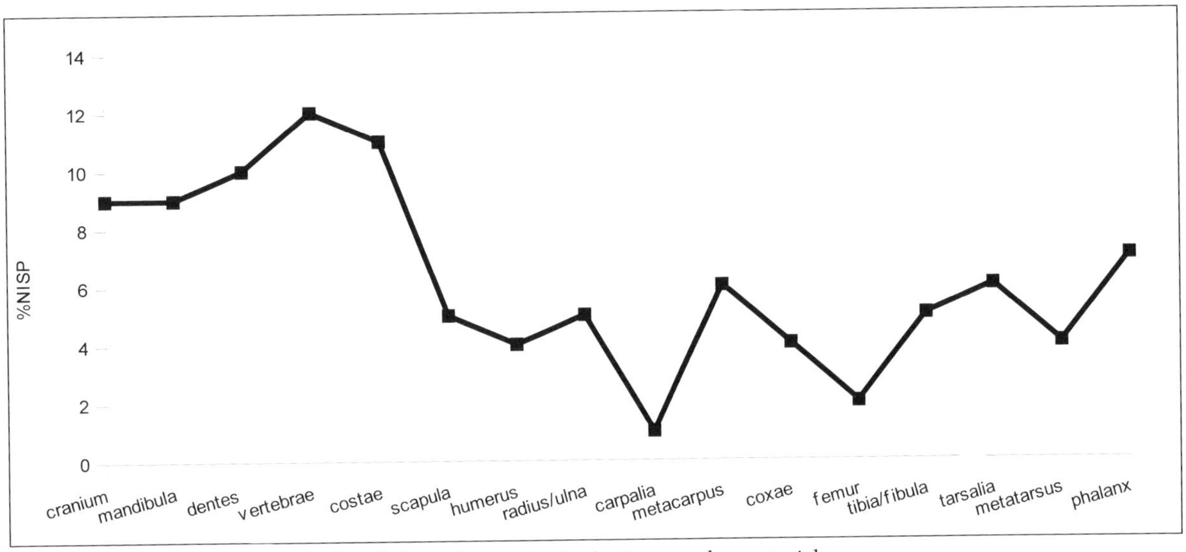

Figure 5.9. Total element distribution for all the major species in the Leeuwarden material.

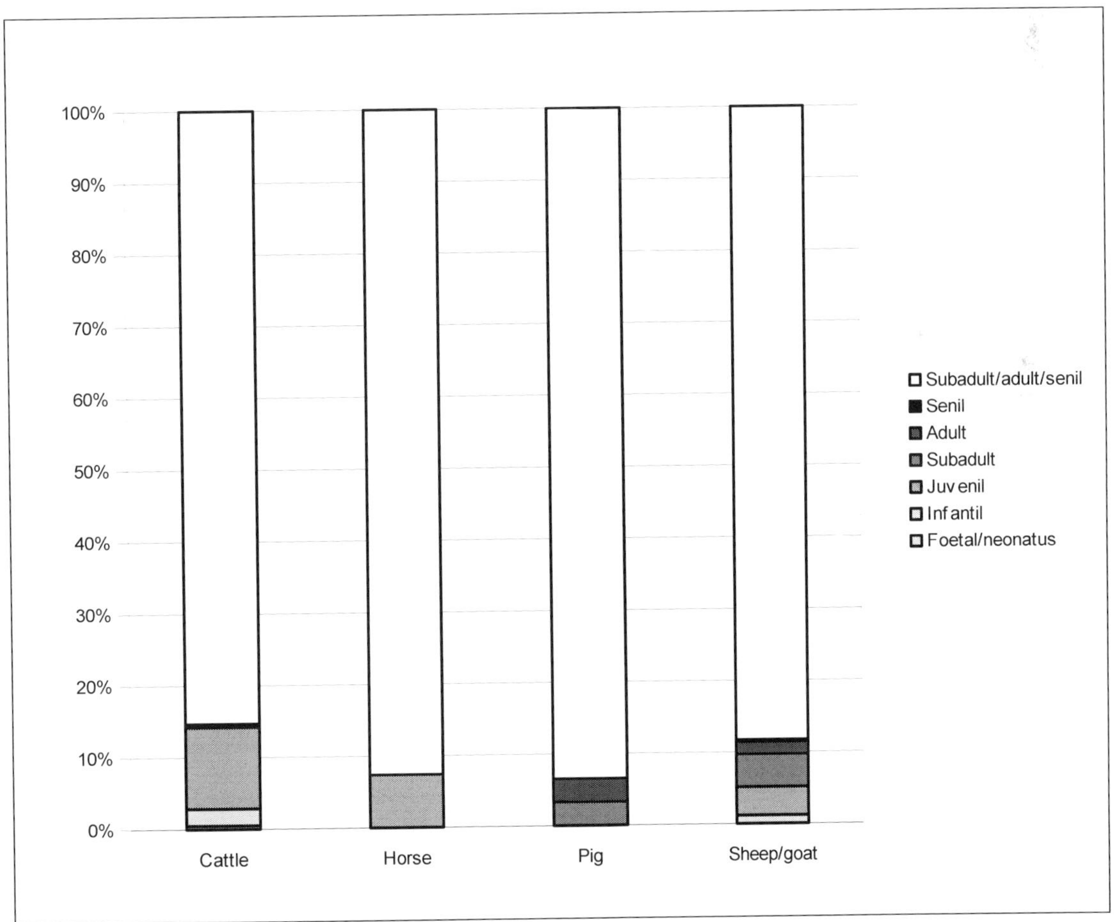

Figure 5.10 Percentage of fragments belonging to different age groups for the major species in the Early Medieval material from Leeuwarden.

Table 5.2. Sex-distribution for the Leeuwarden material.

Species	Male (fragments)	Female (fragments)
Cattle (*Bos taurus*)	4	12
Sheep/goat (*Ovis aries/Capra hircus*)	3	3
Pig (*Sus scrofa domesticus*)	4	1

preted as special. Each of these deposits will be described together with the reasons for these interpretations. The deposits have, as in Dongjum, been divided into general groups based on the primary category of speciality (table 5.3).

5.4.1 Articulated body parts

One recurring type of special deposit are articulated body parts from sheep/goat. Three such deposits were identified in the material.

Feature 146 (front part of juvenile/subadult sheep/goat)

This spotted light grey-brown layer contained one piece of Merovingian pottery and a number of animal bones coming from the front part of a juvenile/subadult (9-24 months old) sheep/goat. The bone elements included were the mandibles, parts of the cranium, four ribs and vertebrae number 3 to 11 (see figure 5.11). Two animal processing marks that indicate disarticulation were found, the chopped through vertebrae 8 and 10 (see figure 5.12).

Feature 246 (front part of juvenile/subadult sheep/goat)

This spotted grey-brown layer of sandy clay contained two pieces of early medieval pottery in addition to the front part of a juvenile/subadult (9-24 months old) sheep or goat. The elements include large parts of both mandibles, the atlas, two thoracic vertebrae (number 12 or 13 and number 18), one lumbar vertebra (number 21) and a single rib (see figure 5.13). In the feature was also a fragment from a sheep/goat tibia but the different colouring of this bone makes it uncertain if it belongs to the same individual. The bones in the feature showed no signs of being affected by taphonomic processes.

Table 5.3. Overview of the special deposits from Leeuwarden.

Feature	Trench	Level	Filling	Find nr	Feature type	Dating	Deposit type
14	146	2	1	571	Layer	VMEB	Articulated body parts (sheep/goat)
14	246	3	1	625	Layer	VMEC	Articulated body parts (sheep/goat)
14	55	1	1	353, 354, 373	Layer	VMEC	Articulated body parts(sheep)
15	127	101	1, 2	1388	Pit/ditch	VMEB	Selected elements (cattle and sheep astragali)
13	610, 611	4	1	467, 558	Sunken hut	VMEB	Selected species (bird bones, mainly wild geese and ducks)
14	828	5	1	770	Layer	VMEB	Skulls/crania (horse)
13	1591	8	1, 6	1161, 1162	Layer	VMEB	Skulls/crania (sheep)
21	90	3	7	2742	Well	VMEC	Human bones (crania and other large bones)
14	182	2	1	600	Ditch	VMEC	Human bones (vertebra)
13	555	3	1	426	Layer	VMEC	Human bones (two vertebrae)
13	85	1	1, 2	231, 232	Pit	VMEC	Human bones (fibula, tibia and calcaneus)
13	258	2	1	275	Pit	VMEC	Human bones (fibula, tibias, calcaneii, astragalus, tarsal bones, metatarsus bone, metapodial and phalanges)

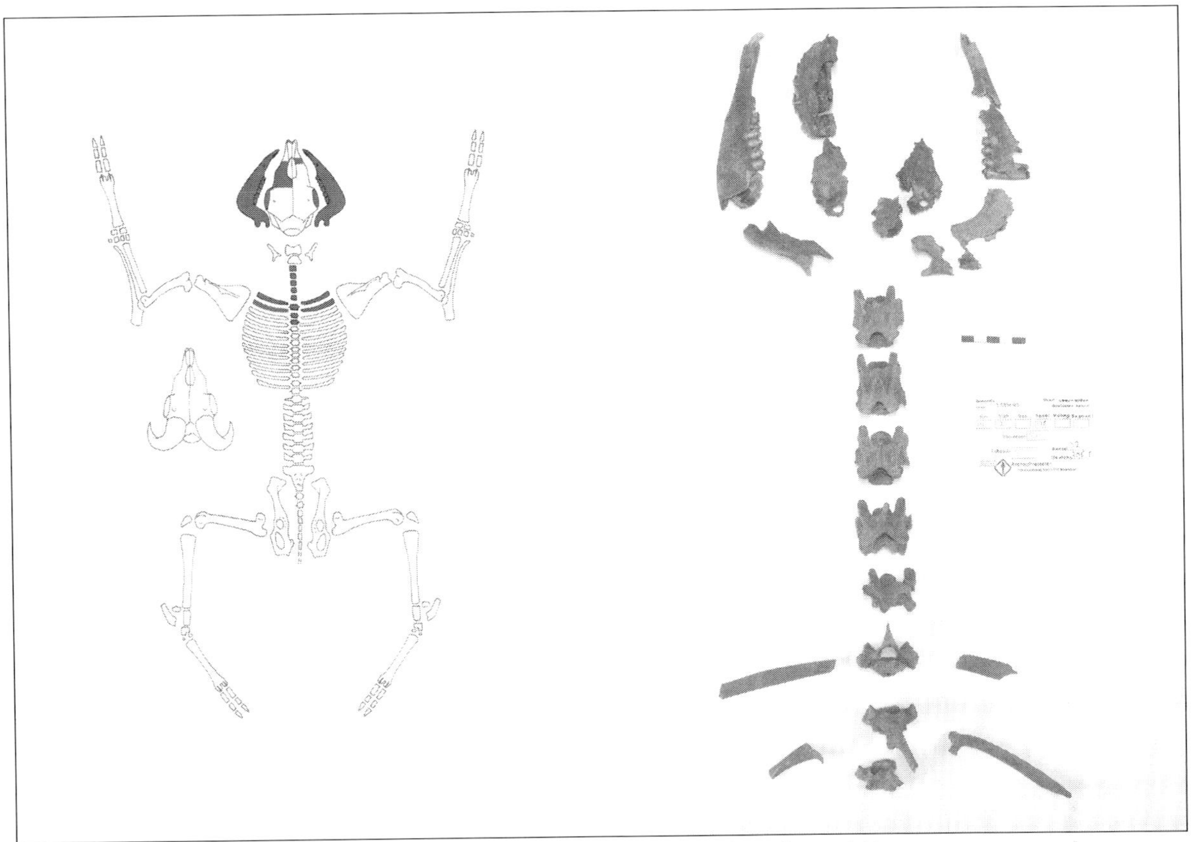

Figure 5.11. Schematic view (left) and photo (right) of the sheep/goat elements from feature 146.

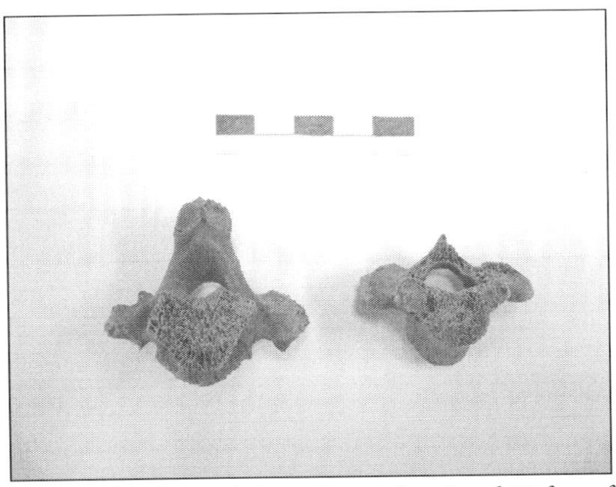

Figure 5.12. The chopped through vertebra 8 and 10 from feature 146.

Feature 55 (parts of two sheep, one subadult)

This partly ash and charcoal spotted dark grey and partly grey brown layer (see figure 5.14) contained three pieces of pottery in addition to a number of animal bones.

Except for two domestic goose bones and three cattle fragments (among which one was a cranial fragment from a foetus) the bones come from at least two sheep. For the majority of these bones their placement within the feature, the fragment's size, the colour of the bones, the age at death and how they actually fitted together were used to identify them as belonging to two different individuals. In addition three pieces of sternum, one rib slightly burned on its proximal end and one gnawed and weathered metacarpus could only be identified as sheep/goat and not be linked with certainty to one of the two animals. The adult (findnumber 354, ca 30 months or older) sheep in the feature had the largest number of bone elements with two humerus, two radius, one ulna, one metacarpus, four ribs, two cervical and two thoracic vertebrae (figure 5.15). None of these bones show signs of being affected by any taphonomic process.

The subadult (findnumber 373, ca 12-30 months) individual is represented by only a few fragments including one radius, one metacarpus, one cervical vertebra and possibly two smaller mandible fragments (see figure 5.16, the two small mandible fragments are not depicted).

The mandible fragments did most likely not belong to the same individual since the colour of the left one deviates from that of the other bones and since the right one was burned at a high temperature. The left mandible fragment shows a butchery mark indicating disarticulation. No other taphonomic traces were observed.

Despite the fact that none of the three bone assemblages discussed in this paragraph were observed in the field, much in these deposits indicates that we are dealing with articulated bone groups of sheep/goats. It is clear that the three deposits described in this section show a number of similarities with each other. All of them were found in concentrations in similar types of layers together with a small number of pottery shards. The bone material is dominated by the remains of mainly the front parts of

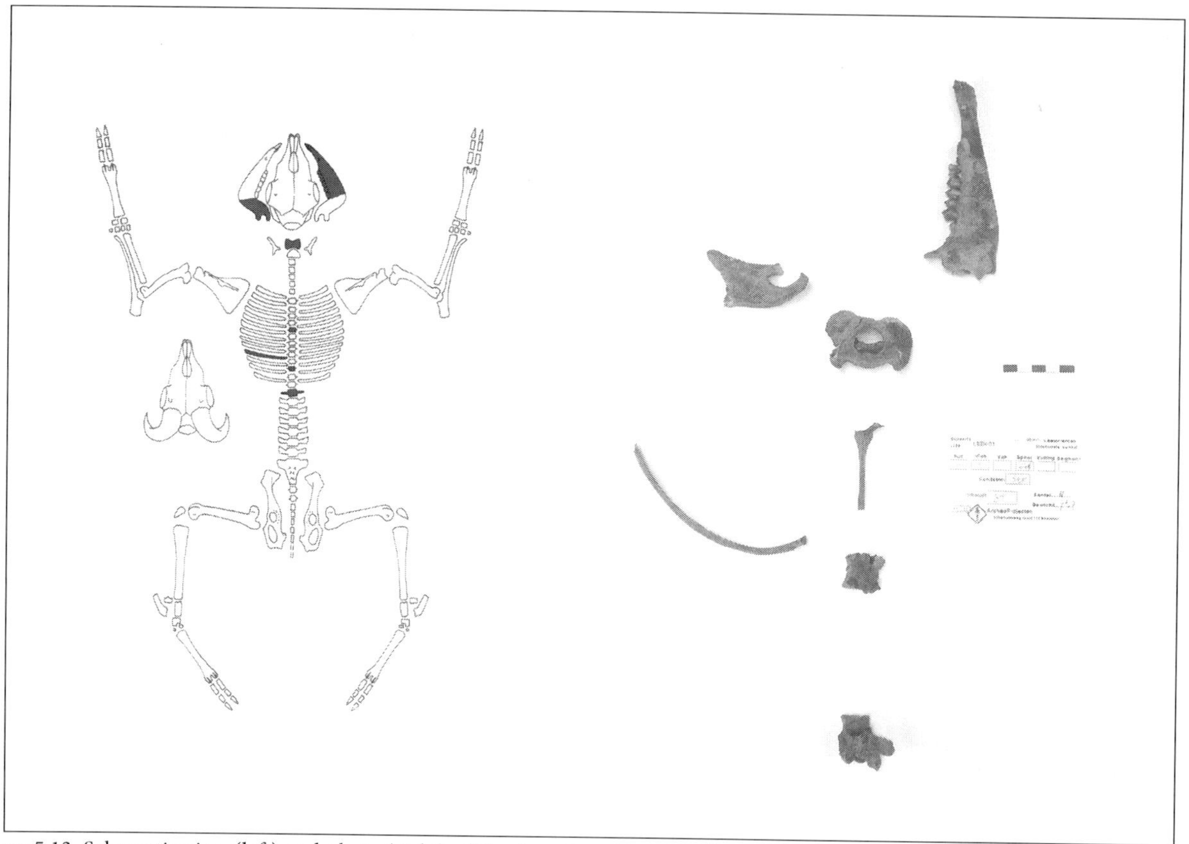

Figure 5.13. Schematic view (left) and photo (right) of the sheep/goat elements from feature 246.

sheep/goat bodies. Three of the animals were juvenile or subadult while the fourth was adult (but not necessarily very old). With the exception of one calcinated mandible fragment (that was probably not connected to the deposited sheep) the only traces on the bones are a couple of disarticulation marks indicating a very rough partitioning of at least one of the animals. The datings of the deposits place feature 146 in the Merovingian Period (VMEB-VMEC) and the other two either in the same time span or slightly later (VMEC).

The identical teeth eruption and wear stages for the animals in feature 146 and 246 indicate a similar age at death for these two individuals, despite the rather long general age period given. It is likely that the two animals were killed during the same time of the year (see section 2.5). The age at death is however too imprecise to determine when exactly during the year this should have occurred. It is also possible that the two animals in feature 55 were killed at the same time since their age is either overlapping with the first two (for the subadult animal) or older and thus could have been killed at just about any time of the year (for the adult animal).

5.4.2 Selected elements

One deposit where the distribution of bone elements differs from the rest of the bone material was identified in the material from Leeuwarden.

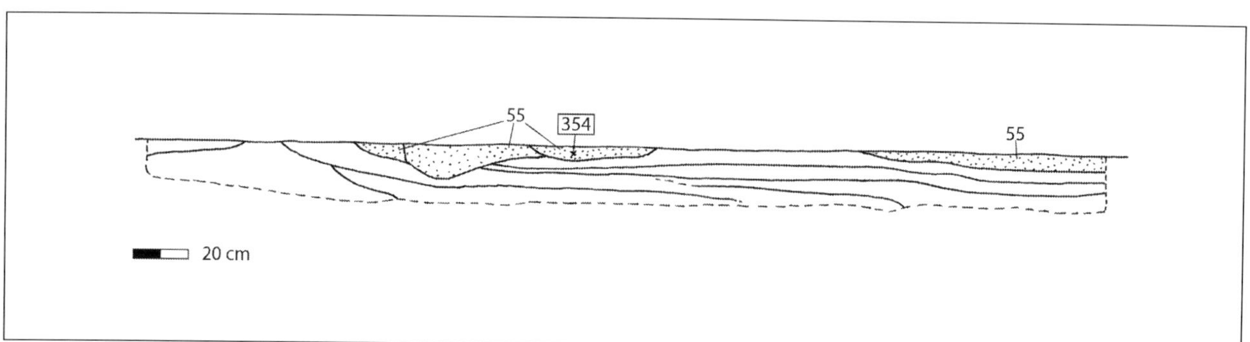

Figure 5.14. Section drawing of feature 55 with the bone deposit including the adult sheep having findnumber 354 (after field drawing by ADC ArcheoProjecten).

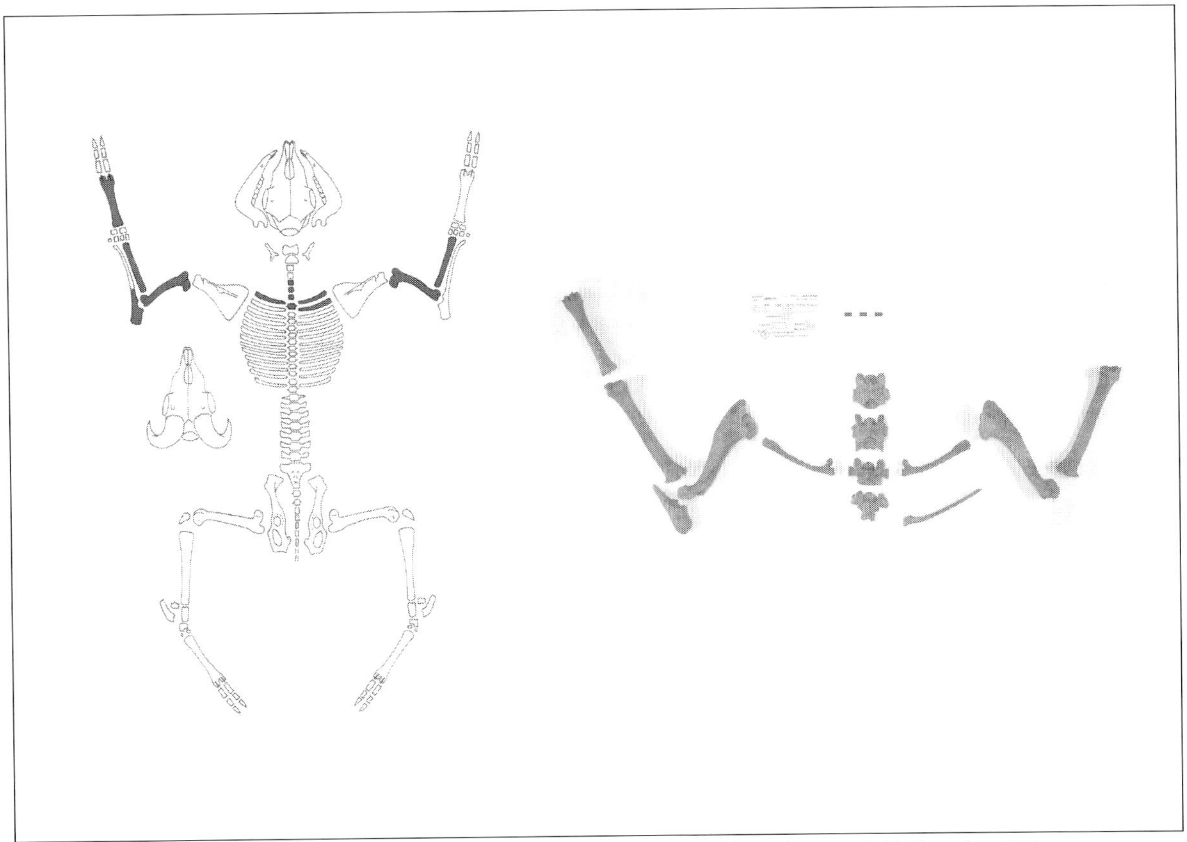

Figure 5.15. Schematic view (left) and photo (right) of the adult sheep elements from feature 55 (findnumber 354).

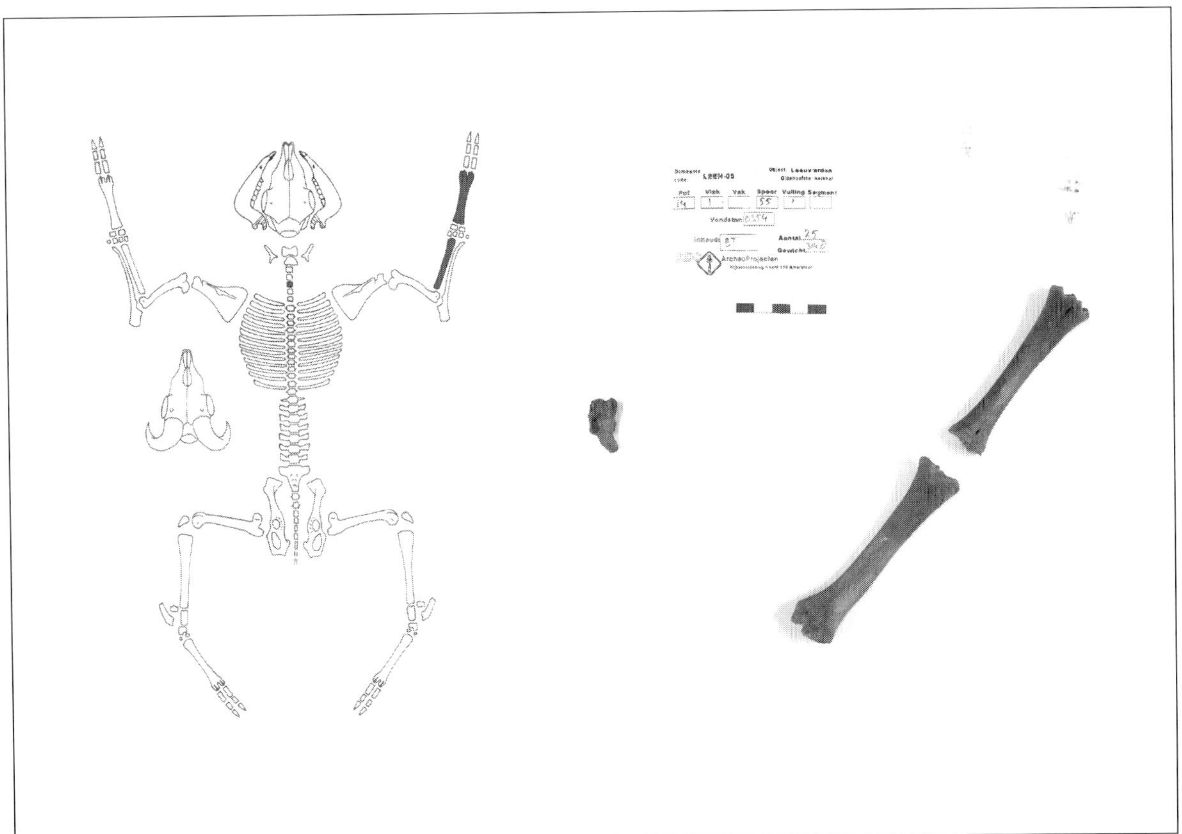

Figure 5.16. Schematic view (left) and photo (right) of the subadult sheep elements from feature 55 (find number 373, the two possible mandible fragments are not included in the photo).

Feature 127 (four left astragali from cattle and sheep)
This pit (or possibly ditch) dated to ca 525-725 AD (VMEB) contained two dark grey fillings and several soil lenses indicating periodical water coverage. The only archaeological material found was four complete astragali coming from the deeper of the two fillings. Three of the astragali were identified as cattle and the last one as sheep, all four belonging to the left side of the body. Two of the cattle astragali show signs of very light weathering and the last cattle astragalus shows traces of carving on its proximal end, possibly indicating the initial stage of bone craft.

Astragali are not an unusual element to find in archaeological assemblages, being compact and therefore often one of the best preserved bones of an animal. The element distribution in the material however shows that only seven percent of the cattle and sheep/goat fragments in the whole material belongs to the group tarsalia (see figure 5.8 above). The group tarsalia does however include a number of other tarsal bones besides the astragalus. That four astragali from the left side were found alone in a feature and show only superficial surface deterioration (thus not being the only remains of a once larger bone assemblage in the feature) is definitely out of the ordinary and enough to consider the deposit special. In this respect the traces of bone craft on one of the cattle astragali strengthens this view since it shows that the bones were used for some purpose after the butchery of the animals.

5.4.3 Selected species

Sometimes the actual types of species or the amount of fragments of a certain species in a deposit can be considered out of the ordinary. In the Leeuwarden material one such feature was identified.

Sunken hut 5 (feature 610 and 611, large number of bones from mainly wild geese and ducks)

This almost half a meter deep feature was first thought to be a pit but later reinterpreted as a sunken hut that was stratigraphically dated to ca 525-725 AD (VMEB). In the report this feature is designated as Sunken hut 5 (Nicolay 2008a: 67). In addition to a large number of animal bones, ten pieces of ceramics, a handful of comb fragments and an unidentified iron object were found in the hut. The feature was made up of several fillings of which two contained animal bones (filling 2 of feature 610 and filling 1 of feature 611) (see figure 5.17). These two fillings were quite similar both being very dark (610:2 is brown-black while 611:1 is grey-black) with a lot of charcoal, and have both been interpreted as ash layers. The two fillings were separated by cleaner layers, especially in the form of filling 1 of feature 610 where sods have been used to cover the older of the two bone carrying fillings (610:2). The two individual deposits were placed in the hut with some time in between, after it was no longer used as a building. The exact time that passed between the two episodes are very difficult to determine and could be everything between a few days and several years, although more likely somewhere in the middle of these two extremes.

The deepest, and therefore oldest, of the two bone carrying fillings (610:2) contained 404 bone fragments weighing 717.9 g. The non-bird bones consisted of only three unidentified fish bones, four cattle bones weighing 273 g (the high weight is mainly the result of one complete metatarsus) and six sheep/goat fragments weighing 37.8 g. As many as 383 of the remaining fragments (weighing 356.2 g) could be identified as bird bones. Of these 192 fragments weighing 309.8 g could be identified to species or group of species and are presented in table 5.4.

The material is dominated by greylag goose, white-fronted goose and common teal. These species also produce remarkably high MNI for a single deposit, and must thus have constituted quite an impressive amount of meat. It is interesting to note that all bird bones in the material were identified as coming from wild animals.

Also in the younger filling (611:1) a number of bird bones were found that could be identified to species, although fewer than in 610:2. This assemblage consisted of 27 fragments weighing 14.1 g (table 5.5).

Table 5.5 shows an assemblage containing more bird species but without any species that dominates the material. The greylag goose is still the most common bird, but even so is only represented by four fragments.

Only a few selected features at the Leeuwarden excavation were sieved and Sunken hut 5 was one of these. The sieved material from the youngest filling (611:1) only contained one identifiable fragment of jack snipe (*Lymnocryptes minimus*) weighing 0.1 g.

The sieved material from the older filling (610:2) contained 15 fragments weighing 11.5 g. Seven of these could be determined as common teal (*Anas crecca*) while seven were identified as white-fronted goose (*Anser albifrons*). Finally one fragment came from a starling (*Sturnus vulgaris*), a bird not very often found in the *terp*-region. With the exception of the starling the sieved fragments from this filling corresponded well with the hand collected bones where the common teal and white-fronted goose were the most common species after greylag goose, whose larger bones quite naturally would be rare in the sieved material. In addition to the starling, the only other change the sieved material from the older filling produced was the increase of the MNI for white-fronted goose from six to seven because of two additional left tarsometatarsus fragments.

The taphonomic situation for the two fillings is quite straightforward. A few of the non-bird bones from both fillings show slight weathering, superficial gnawing and some processing marks indicating normal mixed refuse. The bird bones do however show few traces of taphonomic processes. 19 of the bird fragments from the older filling (610:2) are black-burned (one only partially) while 17 are white-burned. Two greylag goose humeri from the same filling have butchery marks, but otherwise

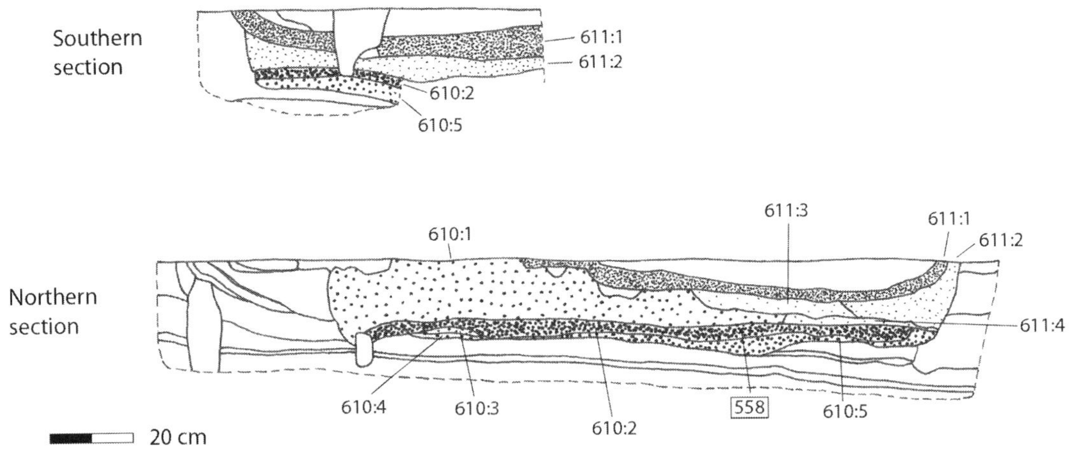

Figure 5.17. Section drawings of Sunken hut 5 with the two bone carrying fillings 610:2 and 611:1 (after field drawing by ADC ArcheoProjecten).

Table 5.4. NISP, weight and MNI for the identified bird bone fragments from filling 610:2.

Species	NISP	Weight (g)	MNI
Bar-tailed godwit (*Limosa lapponica*)	1	0.4	1
Common teal (*Anas crecca*)	31	9.4	6
Greylag goose (*Anser anser*)	105	179.1	10
White-fronted goose (*Anser albifrons*)	48	111.4	6
Wigeon (*Anas penelope*)	2	0.9	1
Wild mallard (*Anas platyrhynchos*)	2	1.1	1
Unspecified wader (Caladris sp.)	1	1.8	-
Unspecified goose (Anser sp.)	2	5.7	-
Total:	**192**	**309.8**	

Table 5.5. NISP, weight and MNI for the identified bird bone fragments from filling 611:1.

Species	NISP	Weight (g)	MNI
Black-tailed godwit (*Limosa limosa*)	3	1.8	1
Common redshank (*Tringa totanus*)	2	0.2	1
Common teal (*Anas crecca*)	2	0.5	1
Dunlin (*Calidris alpina*)	3	0.3	1
Goldeneye (*Bucephala clangula*)	1	0.1	1
Greylag goose (*Anser anser*)	4	1.2	1
Knot (*Calidris canutus*)	2	0.2	1
Ringed plover (*Charadrius hiaticula*)	1	0.1	1
Unspecified avocets and stilts (Recurvirostridae)	2	1.9	-
Unspecified godwit (Limosa sp.)	1	0.8	-
Unspecified goose (Anser sp.)	3	6.4	-
Unspecified shanks and tattlers (Tringa sp.)	1	0.4	-
Unspecified wader (Caladris sp.)	2	0.2	-
Total:	**27**	**14.1**	

the fragments show no traces of taphonomic processes. From the younger filling (611:1) 11 bird fragments are black-burned while seven are white-burned. Even if these figures are lower than those for the older filling, the percentage of heat affected fragments is actually much higher for the younger filling. 26 percent of all the bird fragments are burned, while the corresponding figure for the older filling is only 9 percent. The total lack of weathering, trampling and animal gnawing on the bird bones from both fillings indicates that they were deposited at two distinct occasions and not over a longer period of time, and that they were rapidly covered by a layer of soil.

The element distribution of the bird bones from the sunken hut shows a wide spread of body parts but with cranial fragments, ribs, pelvises and femurs underrepresented (see figure 5.18). The low number of ribs might however be explained with the difficulty of identifying these elements to species. Radius, ulna, and humerus are in contrast overrepresented, something that partly can be explained by the fact that these elements are easy to identify. The two fillings have a similar element distribution with the slightly more extreme numbers for the younger deposit (611:1) mainly being the result of this material being quite small.

The concluding remarks for this feature is that a very large number of bird bones from wild species were deposited in the, at the time, no longer used Sunken hut 5 at two distinct occasions. In the older filling large parts of at least 25 individual birds were deposited, while no less than 9 individual birds were laid down in the younger filling. Since these deposits, and especially the older one constitutes a focused selection of birds that will have produced a quantity of meat far beyond the normal daily consumption need of a settlement of this size, they qualify as special deposits.

5.4.4 Skulls/crania

Complete or almost complete skulls or crania (skulls lacking the mandibles) are often interpreted as special deposits. Whether this really is always the case does however depend on what was the normal butchery practice in the area during the specific time period. Skulls do have a practical value both in the form of meat (the tongue as well as other parts), marrow (mainly from the mandibles) and the brain, that can both be consumed and used in tanning (see for example Skinner 1913: 72; Driver & Massey 1957: 24; Klindt-Jensen 1957: 83-84; Roundtree 1998: 16; Outram 2001: 401). If these resources were used in a society, the normal procedure of butchery would include the disarticulation of the mandibles and/or the opening of the brain cavity. If such a butchery method was commonplace the existence of complete solitary skulls or crania could indeed indicate speciality.

20 percent of the cranial and mandible fragments (teeth not included) from the larger species in the Leeuwarden material showed animal processing marks indicating disarticulation or later animal processing stages. This is actually higher than for the rest of the material where 10-15 percent of the identified fragments show marks of disarticulation or later animal processing stages. Of the larger cranial fragments in the material (those weighing over 100 g) one was a complete horse cranium (feature 828), one was a complete sheep skull (feature 1591) and three were skulls cloven laterally. These numbers are not entirely conclusive but do at least show that disarticulation and filleting of skulls were commonplace and that the brain was often extracted.

Feature 828 (horse cranium)

This layer, dated to ca 525-725 AD (VMEB), consisted of grey brown sandy clay with turf patches. It contained a single complete horse cranium and a single unidentified (but quite large) rib fragment. The horse cranium

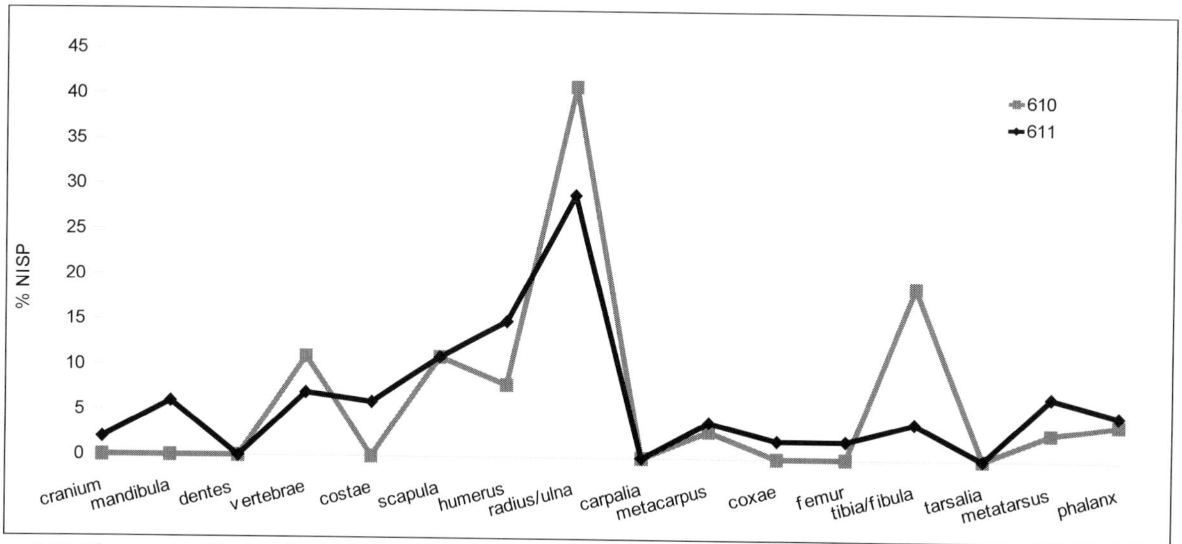

Figure 5.18. Element distribution for the bird bones from Sunken hut 5 (610 the older filling, 611 the younger filling).

weighed over 1200 g and lacked any marks or traces of taphonomic processes, something that together with the otherwise empty layer makes this deposit difficult to interpret, especially since the mandibles are missing. The above discussion showed that it was not an uncommon practice to open the brain cavity of butchered animals, but none of these crania came from horse. With the exception of loose teeth and a single mandible no other cranial fragments from horse were however identified in the specified period. If horse crania were not regularly processed, despite the removal of the mandibles and possibly the tongue, it is still not clear why the cranium should be neatly deposited alone in a layer (figure 5.19). All this qualifies the deposit as special.

Feature 1591 (sheep skull)

Feature 1591 was a roughly one meter wide and 80 cm deep ditch dated to 525-725 AD (VMEB) containing 4 pieces of ceramics, 11 animal bones and 1 human femur (see figure 5.20 and 5.21).

The ditch had six fillings of which number 2 probably is an older ditch fill while the remaining five belong to a second phase when the ditch was emptied and then filled again. All of the fillings except number 6 (situated almost at the bottom) seem to be consciously dumped into the pit and were made up of variants of grey clay or sandy clay and brownish clay (indicating some organic content). Of these fillings number 1 to 4 contained a handful of animal bones and an almost complete human femure.

The filling number 6, a slightly layered dark grey sandy clay filling contained a complete sheep skull (including a tongue bone) from a subadult individual (10-17 months) (see figure 5.22). The layering of this filling indicates natural sedimentation from water at the bottom of the ditch. This would mean that the sheep skull was placed in the ditch after it was recut and not later during its use or when it was finally filled with soil. The bottom of the ditch was probably always wet, since the skull lacks signs of weathering and animal gnawing.

The skull lacks processing marks, something that together with the fact that it was deposited completely with mandibles and tongue, show that it was not processed further after the killing and initial disarticulation of the animal. Since at least two longitudinally split sheep/goat crania were found in the Leeuwarden material, it cannot be considered normal butchery practice to leave the skulls unprocessed in this manner. This, together with the placement in the feature (and the moment of placement) indicates that the skull was deposited on purpose directly after the remaking of the ditch.

A complete human femur deposited in the ditch could also be considered an indication of speciality for this feature (see paragraph 5.4.5 below), but the uncertainty

Figure 5.19. The complete horse cranium from feature 828 (the damages seen on the rear parts of the cranium are modern) (photo by ADC ArcheoProjecten).

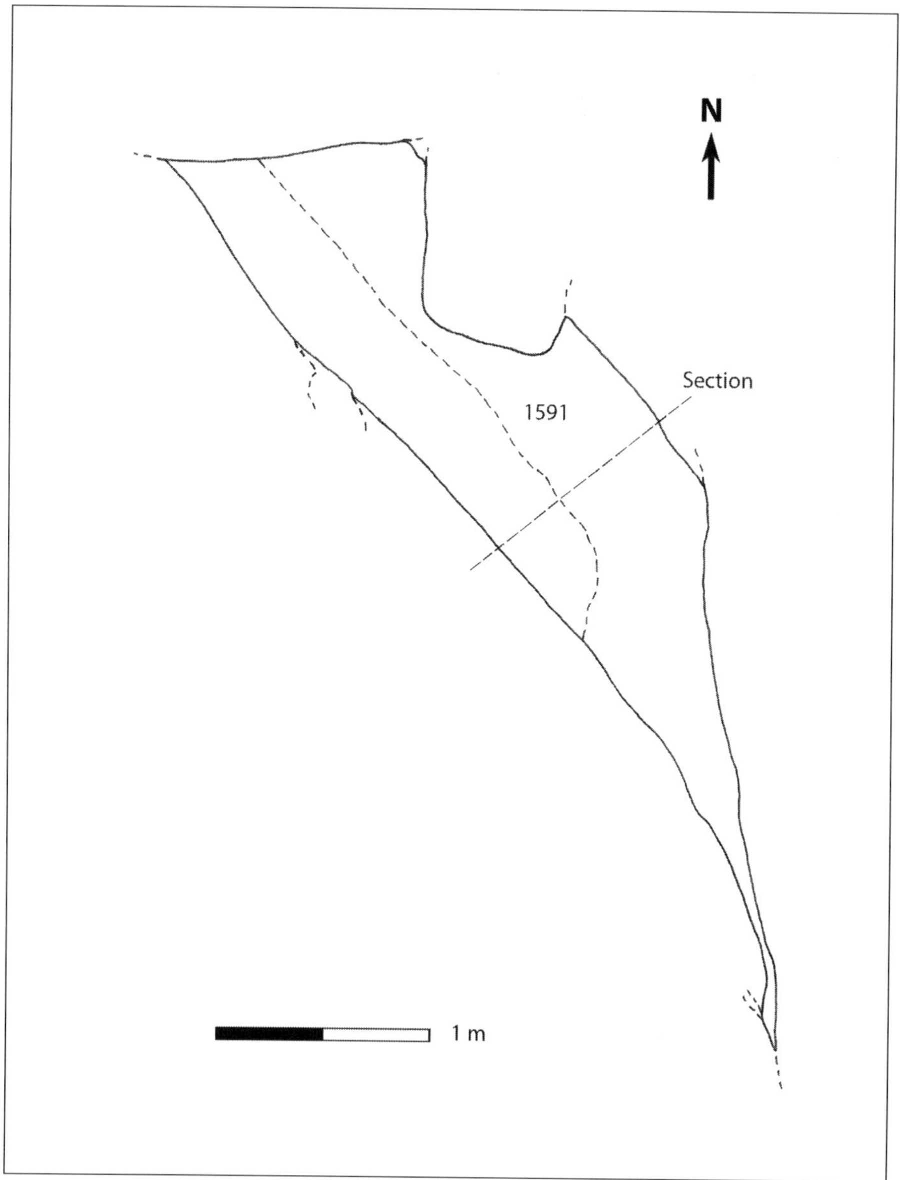

Figure 5.20. Plan drawing of feature 1591 with the section shown in figure 5.21 marked (after field drawing by ADC ArcheoProjecten).

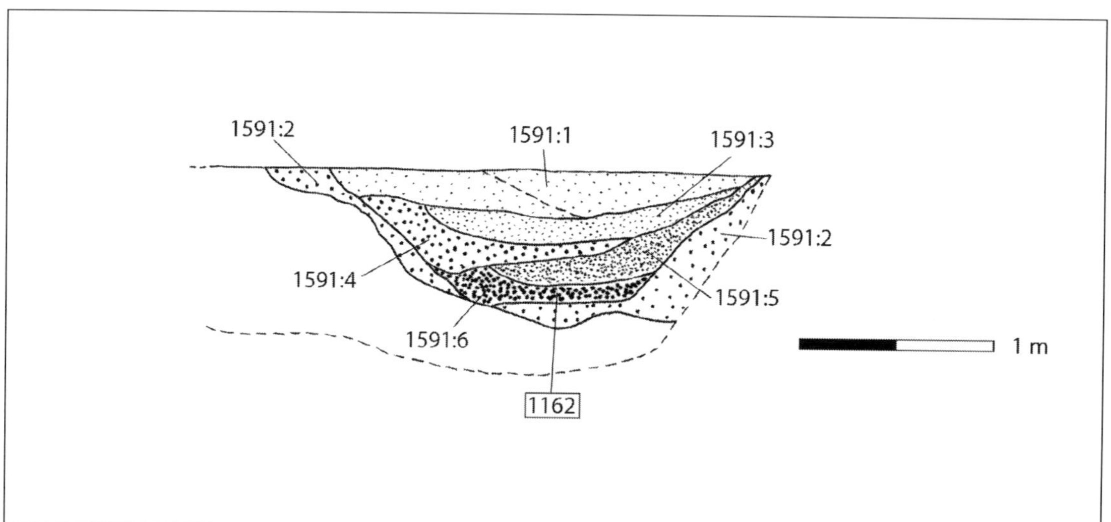

Figure 5.21. Section drawing of feature 1591 (with the sheep skull marked as find number 1162) (after field drawing by ADC ArcheoProjecten).

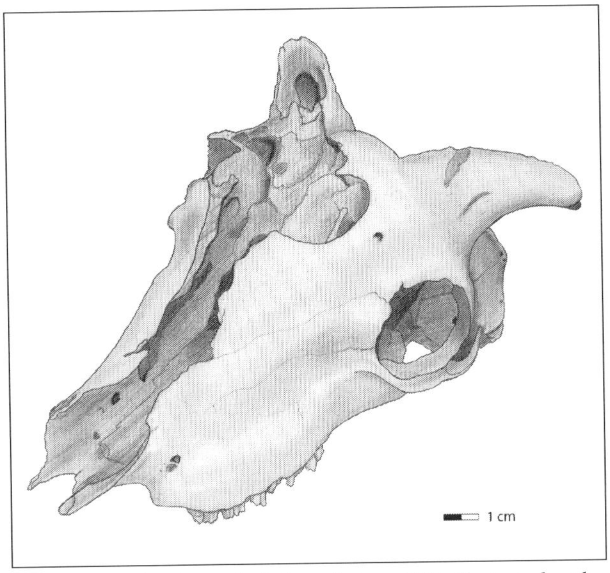

Figure 5.22. The sheep skull from feature 1591 illustrated without the accompanying mandibles and tongue bone (the damages on the top of the cranium are modern).

of its exact placement makes it difficult to draw any conclusions about it. It might very well have ended up in the ditch by coincident when this was finally filled with soil.

5.4.5 Human bones in non-grave contexts

Human bones usually receive a lot of attention when they are found during an archaeological excavation. According to most archaeological and theological definitions a human burial is inherently ritualistic (see for example Alekshin 1983; Pendleton 2001: 170; Crawford 2004) and consequently the (deliberate) inclusion of stray human bones in other deposits makes them almost automatically special. On sites with deep and complex cultural layer sequences the question of whether the human bones were deliberately placed in a feature or not is however difficult to answer with certainty since human bones from earlier graves might accidentally end up in later deposits.

Intrusions of human bones from the later Christian cemetery that was situated on top of the settlement might at first glance be expected at the Oldehoofsterkerkhof. With the exception of downward movement of very small fragments (for example by burrowing animals) bones should however not end up in older features since the later graves were always dug through the earlier layers. A potential problem would instead be that a later grave was filled with older soil containing for example pottery that then produces a false older date for a much later feature. This stresses the importance of dating features both stratigraphically as well as with finds.

There are six features in the material from the Early Middle Ages that contain human bones. One of these, feature 1591, has already been described since it also contained a complete sheep skull, while the remaining five features are presented below.

Feature 90 (human skulls)

Feature 90 was an almost four meter deep well with a preserved wooden construction (figure 5.23). The feature was designated as Well 39 in the excavation rapport and has been dendrochronologically dated to 828 ± 6 AD (Nicolay 2008a: 78-80). The well contained as many as 19 fillings but the most interesting part of it is a deposit of three human skulls and other large human bones found in filling 7 at the bottom of the shaft inside the wooden construction. The human bones were placed directly underneath an upside down turned wooden trough made of ash (see figure 5.24).

The well was constructed during the very last period of habitation on the *terp* before it was transformed into a religious centre. Many Christian graves were found in the youngest layers of the *terp* and the human bones from these were put aside for reburial after the excavation. Unfortunately the human bones from Well 39 were initially also thought to come from a Christian context and therefore reburied. The mistake was not realised until later when the well was more accurately dated. This is very unfortunate since no further information can be gained from these human bones nor even their exact composition beside the fact that three skulls were included. It is however clear that the bones were deposited as parts of skeletons from at least three different individuals and not as complete bodies (Dijkstra *et al.* 2008: 339).

Animal bones were found in four of the fillings of the well, but these are all situated around the main shaft and thus outside the wooden construction. Because of this it is possible that a lot of this material came with the soil used to support the finished well and thus came from somewhere else on the *terp*, or was part of the very same soil that was removed during the initial digging of the shaft. These bone carrying layers have in any case no direct connection with the human bone deposit inside the well.

The animal bone assemblage from the four fillings (number 8, 13, 14 and 15) are made up of only 27 fragments that however weigh as much as 1666.8 g. The material is dominated by cattle and sheep/goat fragments but also include two dog metapodials, a pig metacarpus and a horse metacarpus. Regarding composition and taphonomy the assemblage gives a mixed picture. In many ways the material resembles the bone refuse found all over the site. A few bones are lightly weathered, two are trampled, two are water polished, six are gnawed and one is black burned. Five fragments show processing marks, one indicating disarticulation and one filleting.

This is basically a diverse, not to say chaotic taphonomic history for this assemblage. The species distribution with mostly cattle and sheep/goat fragments is also nothing out of the ordinary but the fragment size is unusually high with the average fragment weight being as much as 61.7 g. This can be explained by the only thing that seems odd, the element distribution in the form of a high proportion of complete or nearly complete metapodials

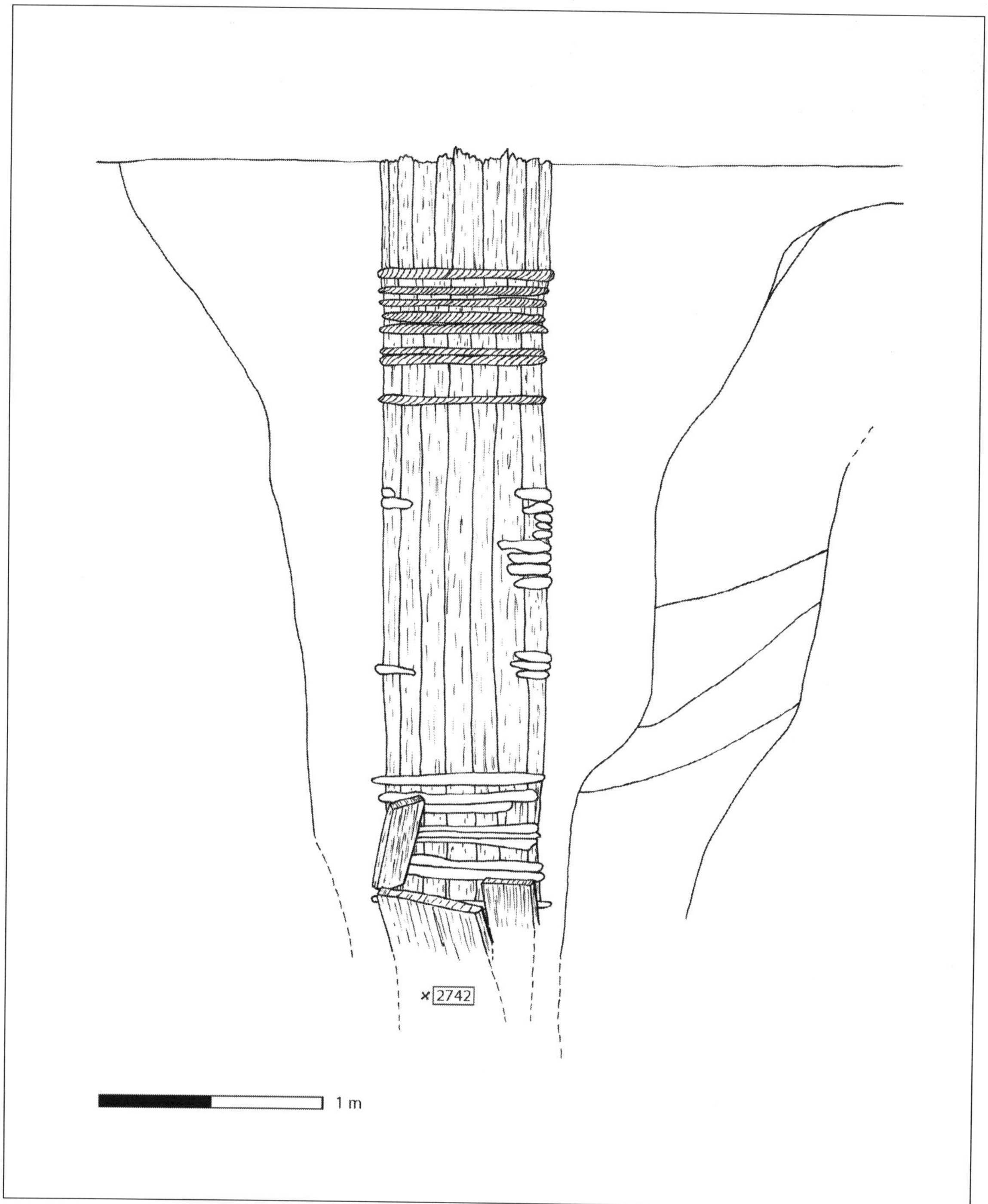

Figure 5.23. Section drawing of Well 39 (feature 90) with the human bone deposit and wooden trough from filling 7 marked as find number 2742 (after field drawing by ADC ArcheoProjecten).

(metacarpus and metatarsus bones). As much as 50 percent (12 out of 24) of the identified fragments in this deposit are metapodials and most of them (9 fragments) are metacarpus bones. For the whole material metapodials only constitute 10 percent of all bone fragments.

It is doubtful whether the animal bone assemblage actually is special since metapodial bones are among the elements that usually are best preserved in archaeological assemblages, possibly making their high proportion at least partly the result of taphonomic processes destroying less compact bone elements. The human bones and the wooden trough placed near the bottom of the well shaft is however a very convincing special deposit.

Feature 182 (human vertebra)

A single severely weathered human thoracic vertebra was identified in a small deposit of animal bones found in the top dark grey filling of a ditch dated to ca 725-900 AD (VMEC) (see figure 5.25).

The animal bones consist of cattle and sheep/ goat bones and a single pig bone with a quite normal.

Figure 5.24. The wooden trough (find number 2742) made of ash found upside down covering the human bones in Well 39, filling 7 (after Hänninen *et al.* 2008).

composition. The processing marks on these bones indicate disarticulation, filleting and further processing (either marrow fracturing or preparation for bone craft). This and some light weathering and animal gnawing indicate that the bones were mainly normal food refuse that most likely had been lying exposed for some time. Two pieces of ceramics were found in the feature as well.

A single human vertebra might have ended up in a ditch for a number of reasons and when no other special characteristics are present in the feature it is difficult to determine the cause of its placement. The severe weathering of the vertebra indicates exposure and possibly secondary deposition. The fact that a human bone is deposited outside a grave context does however make it special.

Feature 555 (two human vertebras)

In this small spotted grey layer with some humus content dated to 725-900 AD (VMEC) two lightly weathered human thoracic vertebrae were found together with parts of a cattle lumbar vertebra and a sheep/goat rib (see figure 5.26). The cattle bone was lightly gnawed and shows a disarticulation mark.

This deposit is very similar to feature 182 above and does for similar reason qualify as special. It is however also important to be aware that the two human vertebrae might have ended up in the layer by coincidence rather than on purpose, something that the light weathering on them also supports.

Feature 85 (human fibula, tibia and calcaneus)

This 33 cm deep round pit had two grey brown fillings with some charcoal and was dated to 725-900 AD (VMEC) (see figure 5.27).

The pit contained a for its small size large bone material consisting of 94 fragments weighing 831.4 g equally divided between the two similar looking fillings. The material is dominated by cattle and sheep/goat and

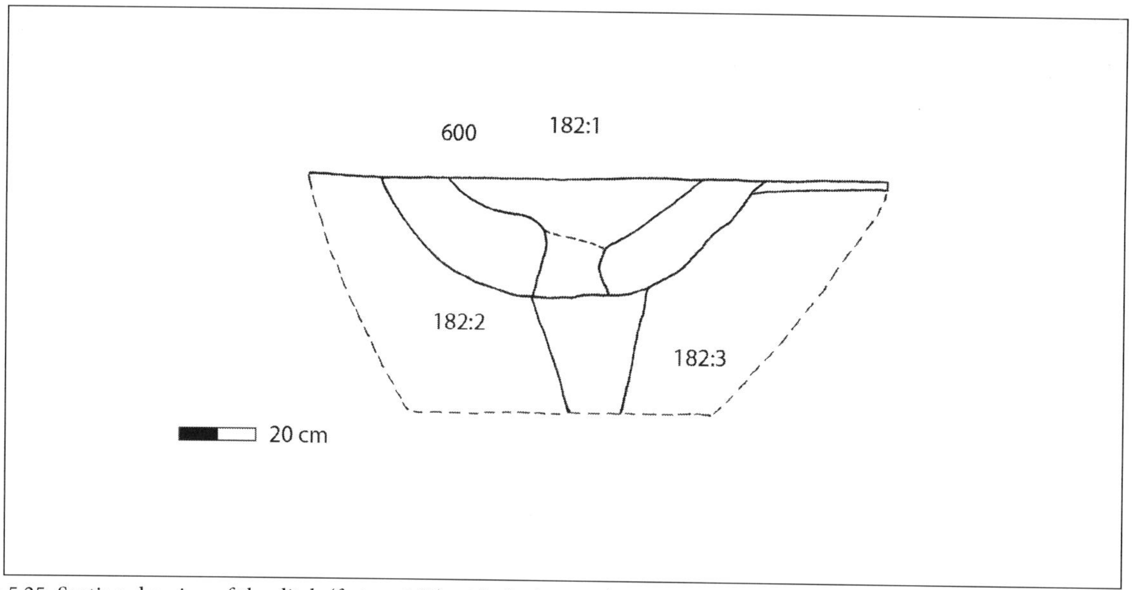

Figure 5.25. Section drawing of the ditch (feature 182) with the human bone marked as find number 600 (after field drawing by ADC ArcheoProjecten).

is complemented with a single pig bone, a single goose fragment and a number of fish bones. The animal bone assemblage seems to be refuse from late in the processing sequence with at least two bones being marrow fractured. A few fragments are white burnt and several bones show light weathering, supporting the picture of a material not immediately deposited in the ground. The bone element distribution does not deviate from the normal situation for the period, except that fewer cranial fragments than normal were present.

Of special interest is the find of four human bone fragments in the upper filling (85:1) of the feature. These include two fragments of the lower end of a left femur, the upper part of a left fibula, and the upper parts of a left tibia, all from the younger filling. The tibia shows very light abrasion on the shaft and both the tibia and the two femur fragments show light weathering. Otherwise no taphonomic processes were recorded.

The similar size, taphonomic history and fragmentation pattern suggest that the human bones represent the middle part of left leg with the kneecap missing (see figure 5.28).

It is impossible to tell if the human bones were placed in the pit in articulation or not. The small size of the upper pit filling as well as the high density of animal bones in the feature do however rule out an interpretation of the feature as a disturbed grave. Regardless of the state of articulation of the human bones their placement was almost certainly conscious. If the human bones were not in articulation the presence of three large bones from the same side by coincidence in such a small pit is unlikely. If the bones on the other hand were articulated the body part could hardly end up in the feature without being noticed. This reasoning definitely indicates that the deposit is special.

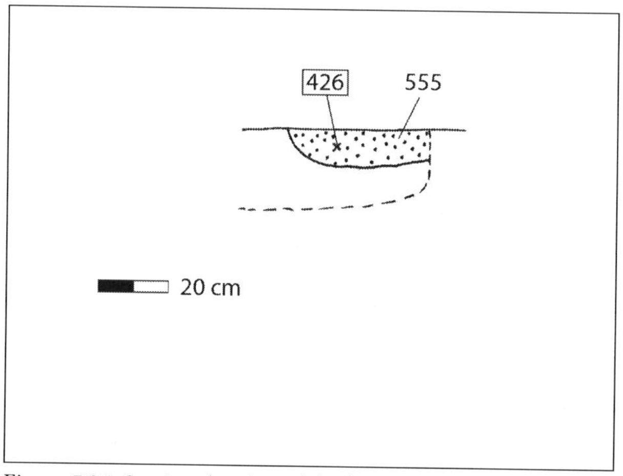

Figure 5.26. Section drawing of the layer (feature 555) with the human bone marked as find number 426 (after field drawing by ADC ArcheoProjecten).

Feature 258 (human fibula, tibia, calcaneii, tarsal bones, metatarsus bones, metapodial, astragalus and phalanges)

This only 6 cm deep irregularly shaped pit had a grey brown spotted filling (see figure 5.29).

The feature contained a bone material with 18 identified fragments weighing 193.1 g, all identified as human bones. The fragments come from the lower legs, both left and right, and include two parts of a right tibia, an almost complete right fibula, three metatarsus bones (plus one metapodial, most likely also a metatarsus), four tarsal bones, one astragalus, two calcanei and four phalanges (see figure 5.30).

With the exception of very light trampling on one of the tibia fragments, the bones lack taphonomic traces. The size, colour, element distribution and how the bones fit together strongly suggest that all fragments belong to the same individual. The incomplete fusing of the distal

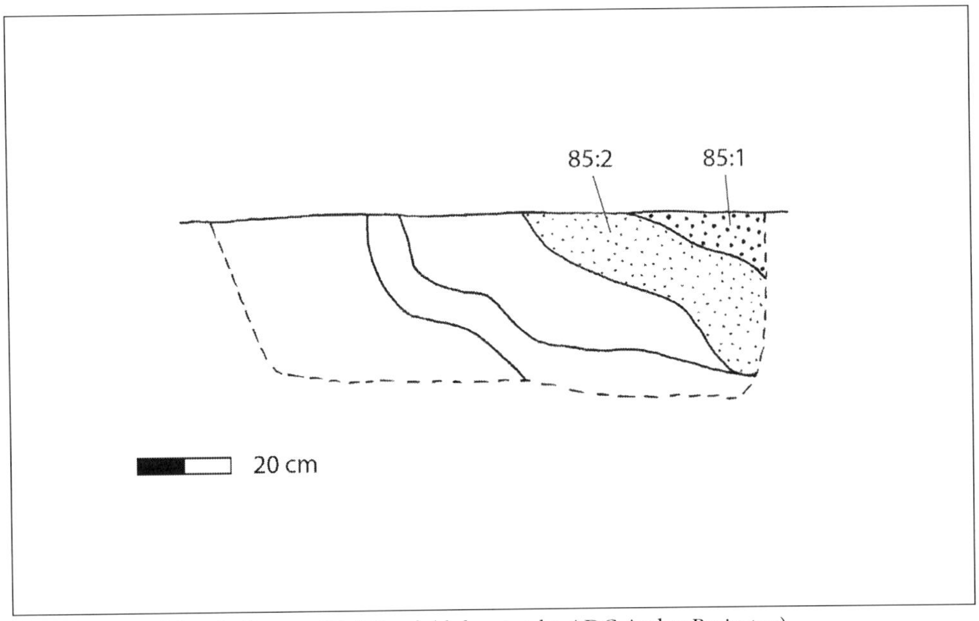

Figure 5.27. Section drawing of the pit (feature 85) (after field drawing by ADC ArcheoProjecten).

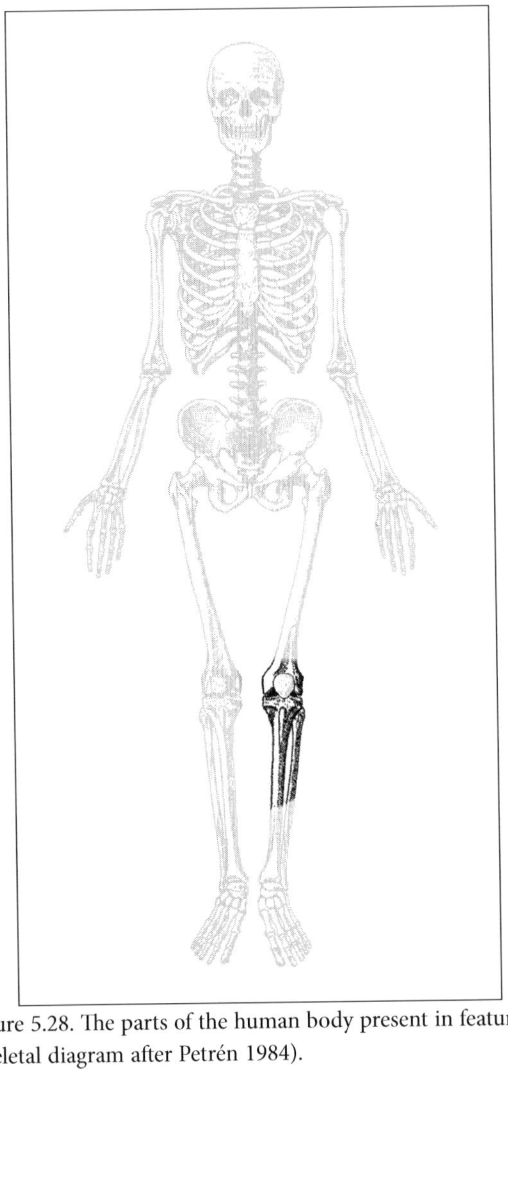

Figure 5.28. The parts of the human body present in feature 85 (skeletal diagram after Petrén 1984).

epiphysis on the fibula and the proximal epiphysis of the calcaneus indicate that the individual was around 17 years of age (Buikstra & Ubelaker 1994; Scheuer & Black 2004: 375, 403).

The placement of possibly articulated remains of a human in a very shallow pit without any other content could be interpreted as the remains of an almost completely dug away grave. Since the stratigraphy does not give any indication that this is the case and since the pit is very shallow even for being the bottom of a grave, it is however more likely that the human bones are the result of a conscious placement in a pit.

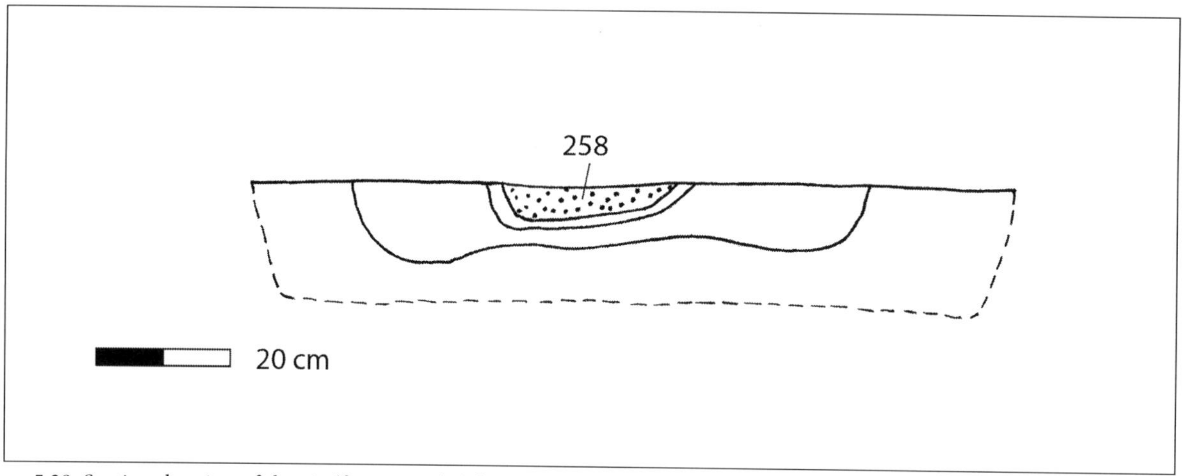

Figure 5.29. Section drawing of the pit (feature 258) (after field drawing by ADC ArcheoProjecten).

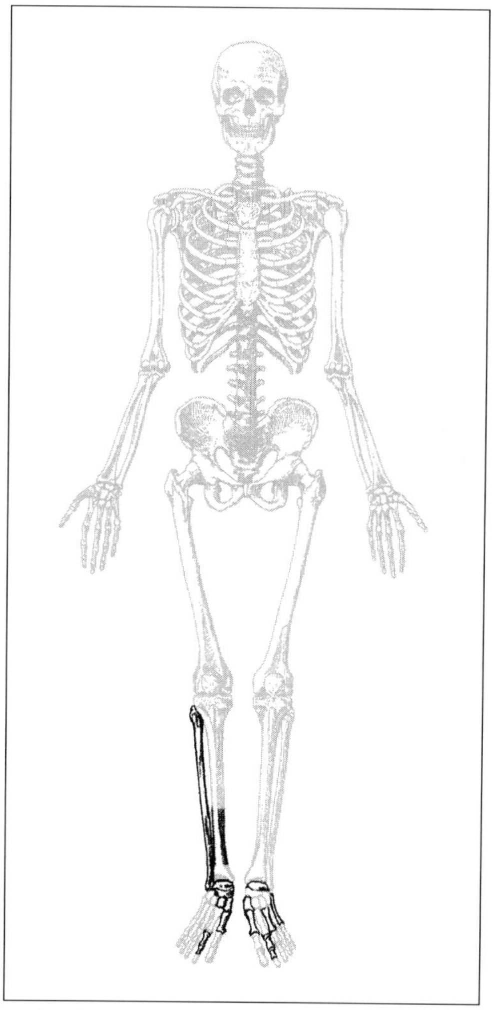

Figure 5.30. The parts of the human body present in feature 258. The exact placement of some of the elements is not certain (skeletal diagram after Petrén 1984).

6. Midlaren

The third bone material used in this thesis comes from a site located near the village of Midlaren in the north of the Netherlands. The excavation took place during 2003 and 2004 to enable an expansion of the holiday resort De Bloemert. The excavation was performed by the Groningen Institute of Archaeology at The University of Groningen in collaboration with the company ARC (Archaeological Research & Consultancy) (Raemaekers *et al.* 2008: 4-6).

6.1 The excavation

The archaeological investigation of De Bloemert covered 4.8 hectares and despite that the construction work had already commenced on the site when the excavation started, it was possible to integrate the archaeological project with the building activity in a successful way (Raemaekers *et al.* 2008:1). During the excavation finds were found dating from the Late Palaeolithic until modern times, but the site was permanently inhabited only from the Iron Age until the Early Middle Ages (Nicolay & Den Hengst 2008: 577).

6.2 The archaeological results

The permanent settlement in De Bloemert had its origin during the Middle or Late Iron Age, probably in the form of a single farm built by people already living in the region. The existence of several celtic fields from this period in the surrounding area indicates that other similar farms were present at the same time as De Bloemert. During the beginning of the Roman Period this scattered habitation seems to have been concentrated into two small villages, one at De Bloemert and one at Plankensloot, 1 km to the south. During the Middle Roman Period (2nd-3rd century AD) the settlement at De Bloemert consisted of two or three farmsteads and had long distance contacts both with the Roman areas to the south, with the *terp*-region to the north, and with Scandinavia (figure 6.1).

In addition to imported objects the site also shows signs of local craft production, including metalworking. In most aspects such as pottery forms and house layout (although not house construction techniques) the village showed strong connections with the coastal *terp*-area. In the Late Roman Period and Early Medieval Period the settlement decreased somewhat in size and was during its continued existence made up by one or two farms. Two graveyards from the same period connected to the village were also found. There is no indication of the village being abandoned at the end of the Roman Period as is mostly the case for the *terp*-settlements. After the incorporation of the area into the Frankish kingdom around 800 AD and the introduction of Christianity, the villages of Noordlaren and Zuidlaren developed and the habitation in De Bloemert gradually moved towards the present day location of the village of Midlaren. The excavated area was finally abandoned in the middle of the 11th century (Nicolay & Den Hengst 2008).

6.2.1 Archaeological features

A large number of archaeological features from the selected time period (ca 400-900 AD) were excavated at De Bloemert. Among the more interesting of these were 22 larger (at least 10 m long and more than 5 m wide) houses (Nicolay & Waterbolk 2008: 91-93). In addition to these, ten sunken huts from the period were excavated (a few of these might however belong to the Late Roman Period). Several of the sunken huts showed traces of metal working while some contained possible signs of glass working. None of the huts did however contain loom-weights, an otherwise common find in this type of feature, and the reason why weaving often is considered to have been performed in this type of building. A handful of smaller buildings interpreted as storages, sheds and granaries could also be dated to the Early Medieval Period (Nicolay 2008b: 127-128, 136-142).

During the Early Medieval Period the water supply for the settlement was provided by 22 wells and 14 waterholes. The wells included three different types, those constructed with a hollowed log (with or without additional structures of wood or stone), those lined with wickerwork and finally those with four main posts situated in a rectangle and covered on the outside with wooden planks. The waterholes have supplied drinking water for humans and animals but may also have been used in iron working and for retting of flax (Nicolay 2008c: 151-172).

Finally, hundreds of pits were found on the site, many of which could be dated to the selected time period. The pits were used for a variety of activities including cooking, baking, metal working, firing of pottery and refuse disposal (Nicolay 2008d: 182-190).

Figure 6.1. Reconstruction of the Midlaren settlement ca 200 AD (illustration by S. Boersma, GIA).

6.3 The archaeozoological material

The animal bone assemblage from Midlaren dated to the selected time period consists of 1740 bone fragments weighing together 1003 g (Prummel *et al.* 2008). Of these only 112 fragments weighing 576.7 g could be identified to species. In addition to this material a single cremation pit from the Carolingian Period contained almost 476,000 burned animal bone fragments weighing 12,160 g. This feature is, due to the enormous size of the material not included in the general description of the material that follows here, but is described further in section 6.4.

The bone material from Midlaren differs a great deal from the other three materials in this study in that it is poorly preserved and highly fragmented with a medium weight per fragment of only 0.6 g. The main reason for this is the location of the settlement in an area with sandy soils not suited for the preservation of organic material. The retrieved bone material is therefore either burned, as heat affected bone have a better survivability in these conditions, or comes from the lower parts of features such as pits and wells where the conditions were moist. Even when unburnt bone was found it showed signs of severe diagenesis and was in many cases very brittle and difficult to handle. Some of the unburnt bones were extracted as block lifts after being solidified with glue and are therefore not included in the bone weights given in this chapter. The condition of the bone material presents several challenges when the human actions behind specific deposits are to be discussed. The material was nevertheless included in the study since this type of material is quite common and these are problems that have to be addressed.

The taphonomic study of a material of this kind is naturally fraught with difficulties and most of the detailed taphonomic characteristics recorded for the other materials were difficult or impossible to document for the Midlaren assemblage, with signs of heating being the obvious exception. For this reason no diagrams will be presented for the different taphonomic categories. Of the 1740 fragments in the material as many as 1270, or 73 percent, showed signs of heating. Of these, 38 fragments were black burned while 1232 were white burned.

The NISP, weight and MNI's for the identified material from Midlaren are presented in table 6.1. From these three categories an image emerges of a small material with few species represented in predictable proportions with cattle being the most common animal, followed by sheep/goat and pig.

The small size of the material and the special taphonomical circumstances make it unfruitful to present diagrams for NISP- and weight-percentages as well as for element distributions. No fragments could be identified to sex. The age at death distribution in the Midlaren

Table 6.1. NISP, weight and MNI for the identified material from the selected time period.

Species	NISP	Weight (g)	MNI
Cattle (*Bos taurus*)	56	459.7	5
Sheep/goat (*Ovis aries/Capra hircus*)	34	45.5	2
Pig (*Sus scrofa domesticus*)	13	22.5	3
Horse (*Equus caballus*)	7	46.5	1
Dog (*Canis familiaris*)	1	2.4	1
Unspecified cyprinid (Cyprinidae)	1	0.1	-
Total:	**112**	**576.7**	

material is however based on enough fragments to be of interest (see figure 6.2).

As can be seen in the figure the material totally lacks young individuals, something that can probably also be explained by the poor preservation conditions especially for the small and less mineralised bones of young animals. They perish more quickly than the bones from adult individuals. The very high number of subadult pigs and the total absence of the same age group among sheep/goats is interesting. A possible interpretation is that sheep/goats were mainly kept for milk and wool production and therefore slaughtered late, while pigs only produce meat and were therefore often slaughtered as soon as they reached their full size.

6.4 The special deposits

Three different special deposits from the selected time period were identified in the bone material from De Bloemert. In table 6.2 the basic data for these deposits is presented.

6.4.1 Skulls/crania

Several deposits with seemingly complete cattle skulls or crania were found in pits and wells during the excavation in Midlaren. Two of these deposits could be dated to the Late Roman or Merovingian Periods (ROML-VMEA). The poor preservation made the deposits less obvious than they would have been with better preservation con-

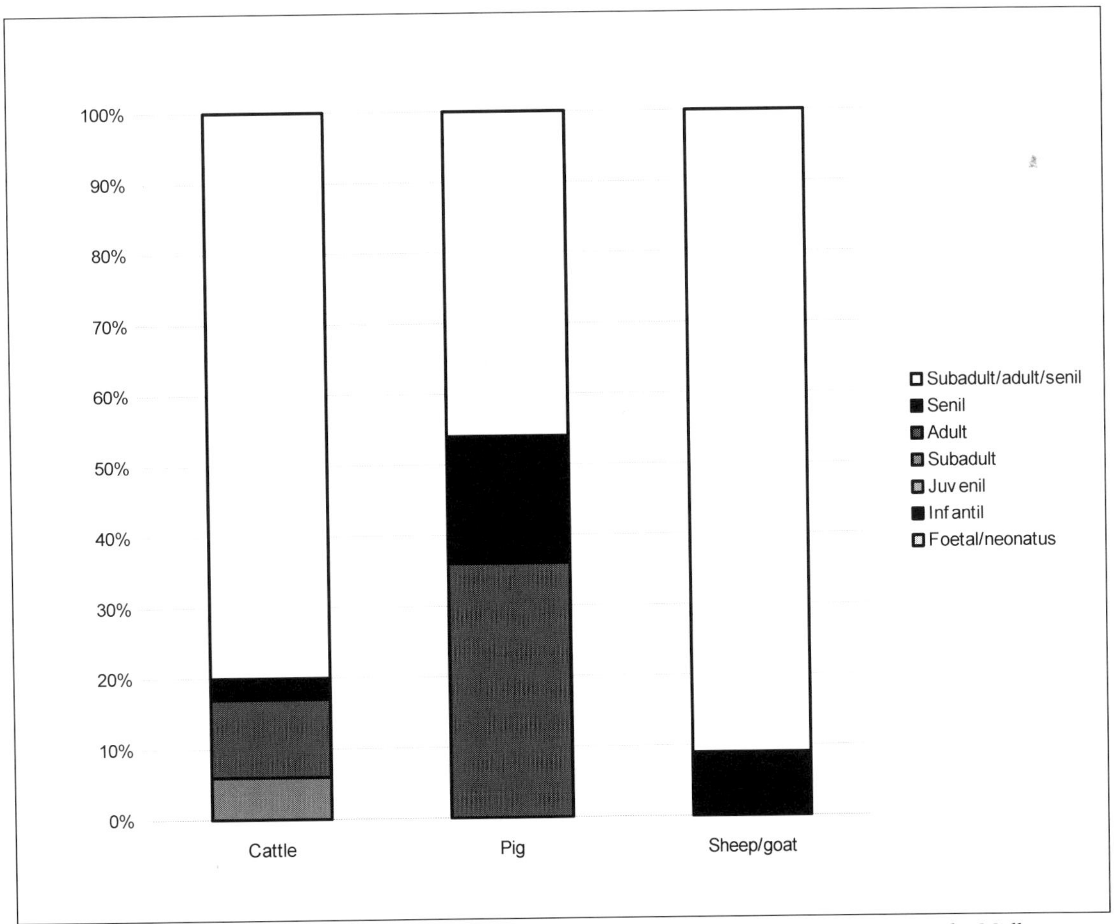

Figure 6.2. Percentage of fragments belonging to different age groups for the major species except horse in the Midlaren material.

Table 6.2. Overview of the special deposits from De Bloemert.

Feature	Trench	Level	Filling	Find nr	Feature type	Dating	Deposit type
21	51	1	1	785, 787	Pit	VMEA	Skulls/crania (cattle)
59	3	1	1, 6	1373, 1410	Well	ROML-VMEA	Skulls/crania (cattle)
45	22	1	1	1215, 1216	Pit	VMEC	Cremation pit (various animals)

ditions, and they would most likely have been impossible to detect during analysis if they had not been identified already in the field where several of them could be recovered as block lifts. If this had not been the case these brittle bones would almost certainly have been reduced to dust in the excavation process.

The deposits contained several examples of cattle skulls with the mandibles still attached to the cranium. This strongly suggests that the skulls were deposited in a complete state without further processing after the initial disarticulation from the rest of the body. The preservation conditions on the site do however also pose another difficulty in that a normal refuse material for comparison is lacking. If a deposit was not made in a deep feature and covered with soil it is unlikely that it would survive. It is therefore possible, although from an economic point of view less likely, that the normal butchery procedure in the area included the discarding of complete cattle skulls.

Feature 51 (cattle skulls)

This irregular shaped pit had a single dark brown filling of fine grained sand with pieces of charcoal and was dated to the very Late Roman Period or Early Middle Ages (ROMLB-VMEA) (figure 6.3).

In the pit the remains of at least four (possibly five) cattle skulls were found. The most complete of these skulls (find number 785) includes the maxilla with all the premolars and molars, and the mandibles with 8 of the 12 premolars and molars remaining (figure 6.4). The last and first premolars are in change which gives the age at death of the animal to between 29 and 34 months (Prummel *et al.* 2008: 243).

The remains of the second cattle skull (find number 787A) include a part of the right maxilla with the second molar and parts of both the mandibles with together five of the six molars. The tooth wear indicates that the animal was at least seven years old at the time of death.

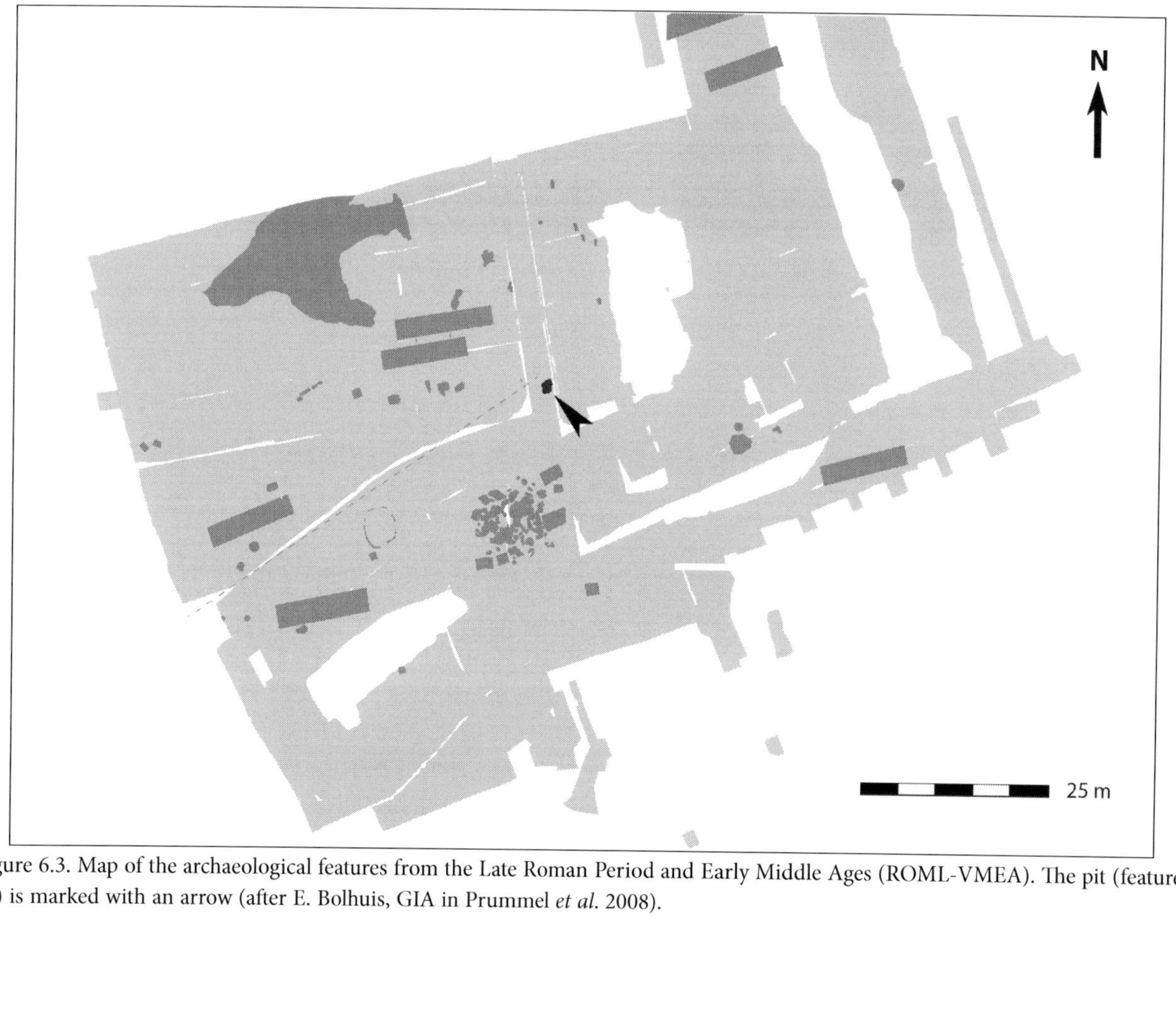

Figure 6.3. Map of the archaeological features from the Late Roman Period and Early Middle Ages (ROML-VMEA). The pit (feature 51) is marked with an arrow (after E. Bolhuis, GIA in Prummel *et al.* 2008).

Figure 6.4. The complete upper and lower jaws of the 29 to 34 months old cattle from feature 51 (find number 785) (photo by W. Prummel, GIA).

The third cattle skull (find number 787C) includes fragments from the maxilla with together five molars intact and fragments from both the mandibles, also with together five molars remaining. From the eruption and wear on the teeth the animal is determined to have been at least three years old at time of death.

Figure 6.5. The upper and lower jaws of the at least 7 year old cattle from feature 51 (find number 787D) (photo by W. Prummel, GIA).

The fourth cattle skull (find number 787D) includes parts of the right maxilla with three molars and both the mandibles with together four molars (see figure 6.5). From the tooth wear this animal is determined to have been at least seven years old when it died.

Two more cattle skull fragments were found in the same pit. One of these (find number 787B) is a part of a maxilla with the last premolar and the three molars still intact, belonging to an at least three year old animal. This

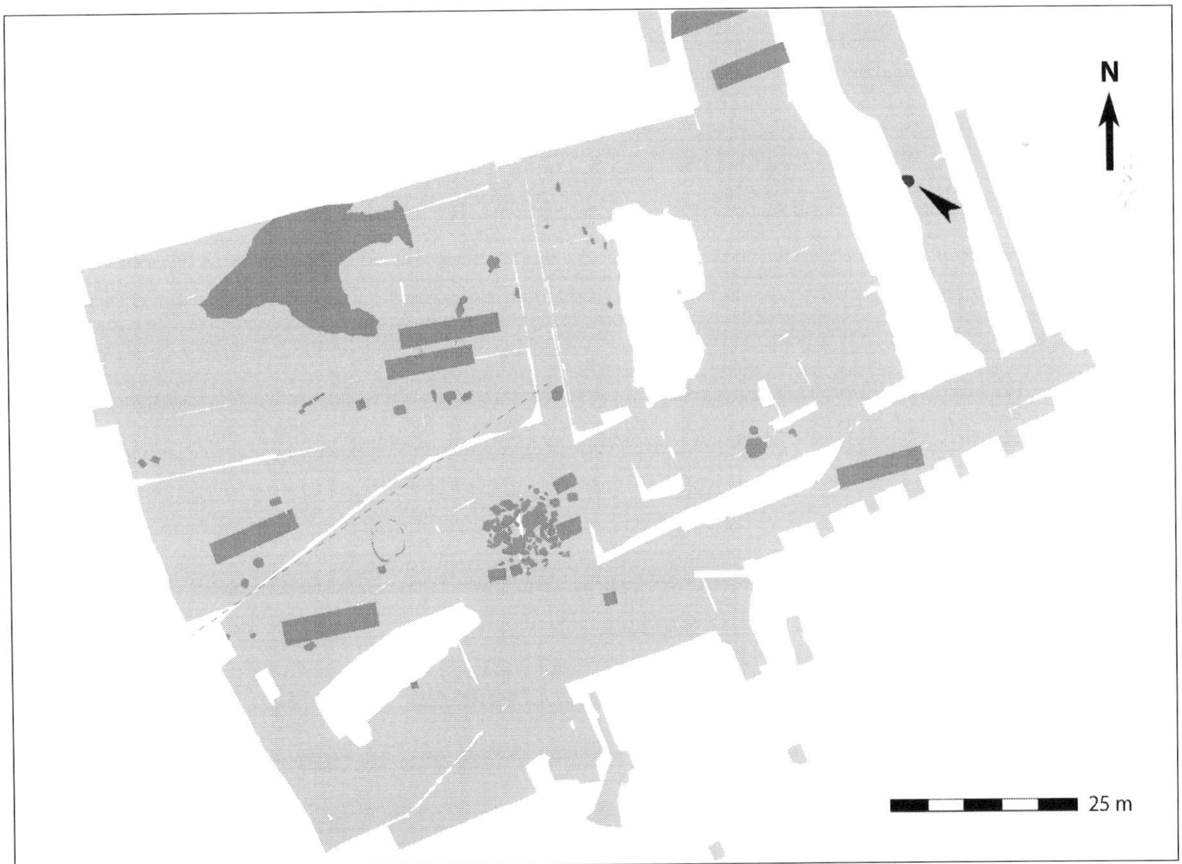

Figure 6.6. Map of the archaeological features from the Late Roman Period and Early Middle Ages (ROML-VMEA). The well (feature 3) is marked with an arrow (after E. Bolhuis, GIA in Prummel *et al.* 2008).

fragment could belong to one of the already described animals (either 787A or 787D) or constitute a separate individual. The other fragment was a single loose molar from a mandible (find number 787C) that could belong to any of the already described skulls.

It is very likely that the remains of at least four cattle skulls were deposited unbutchered (since they include both maxilla and mandible fragments) in the feature. All the animals were adult and at least two of them were old at the time of death. The age determinations are however not detailed enough to indicate any seasonality for the killings. Unfortunately the exact placement of these bone fragments in the pit was not registered during the excavation.

Despite the problems discussed above related to the bad preservation conditions, it is safe to categorise this deposit as special. The butchery of four or five cattle at the same time in such a small settlement as De Bloemert would in itself be a special occasion, and the placement of the unprocessed skulls in the same pit can hardly be a coincidence.

Feature 3 (cattle cranium)

This well, dated to the Late Roman Period or the Early Medieval Period (ROML-VMEA), contained five fillings and a shaft supported by an 80 cm wide, hollowed out tree trunk (figure 6.6).

Figure 6.7. The teeth from the 19-25 months old cattle from feature 3 (find number 1373) (photo by W. Prummel, GIA).

In the well several teeth from a cattle maxilla were found (see figure 6.7). The fragments originate from the top of the feature, situated in filling 1 which was composed of grey brown, fine grained sand. This filling was situated above the wooden well shaft and the deposit was thus laid down when the well was no longer in use. The animal is determined to have been 19-24 months old at the time of death and therefore belong to the category subadult.

In the actual well shaft (filling 6) one more bone was found, a tooth fragment from a large animal (roughly the size of a cow or horse).

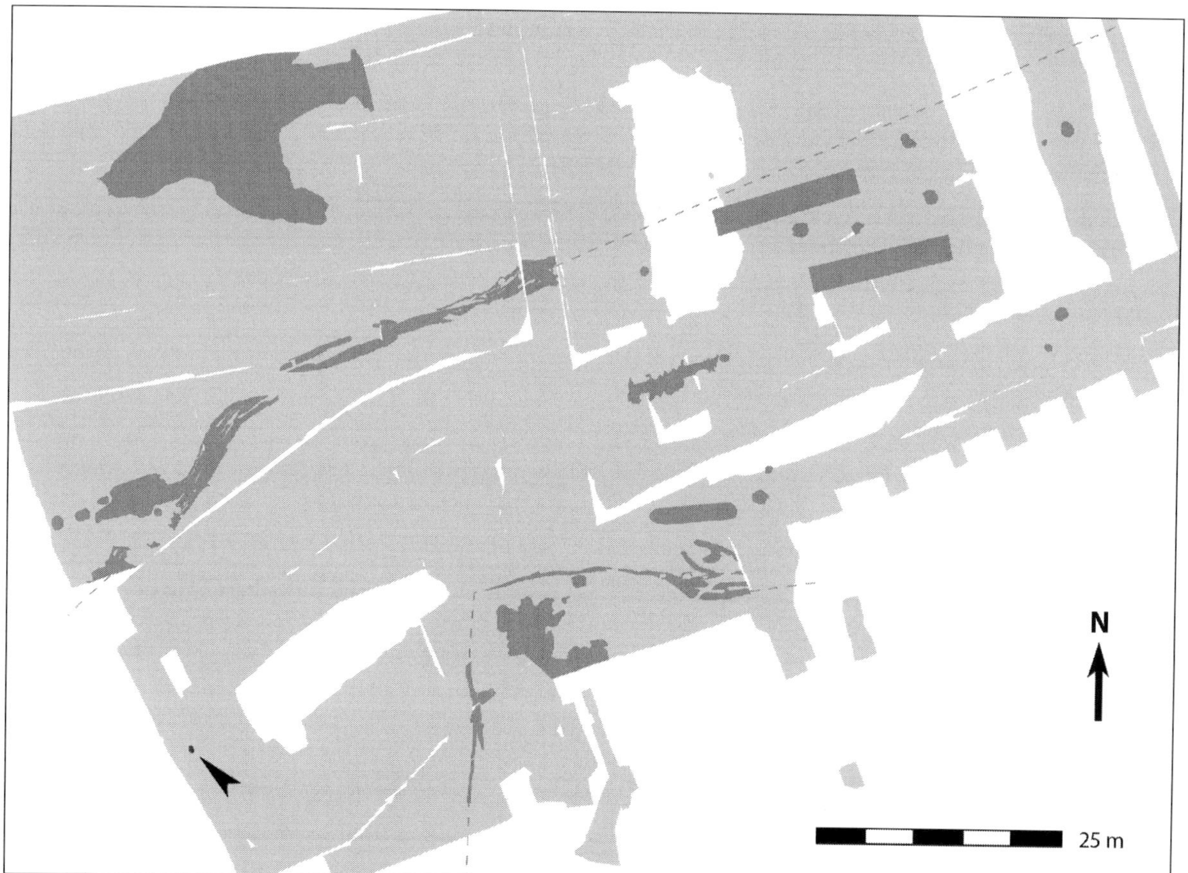

Figure 6.8. Map of the archaeological features from the Carolingian Period (VMEC). The cremation pit is marked with an arrow (after E. Bolhuis, GIA in Prummel *et al.* 2008).

This deposit resembles feature 51 but its speciality is less convincing since it only contains the remains of one animal and since the teeth only come from the maxilla. It is therefore possible that this deposit represents normal refuse in the form of a processed cattle cranium.

6.4.2 The cremation pit

The final deposit from Midlaren that will be discussed here is quite a unique feature in the form of a pit containing the cremated remains of a huge number of animal bones.

Feature 22 (cremation pit)

This oval shaped pit is 14C-dated to the Carolingian Period (VMEC). It was situated near the southwest corner of the excavation (figure 6.8) at a considerable distance from the contemporary farmhouses, something that indicates that the pit was not directly connected with these buildings.

The pit was 150 cm long, 90 cm wide and 20 cm deep. The filling of the feature was made up of spotted fine grained sand with a lot of animal disturbance. In the pit a round concentration, 65 cm in diameter and 10 cm deep, of calcinated bone fragments was found (figure 6.9).

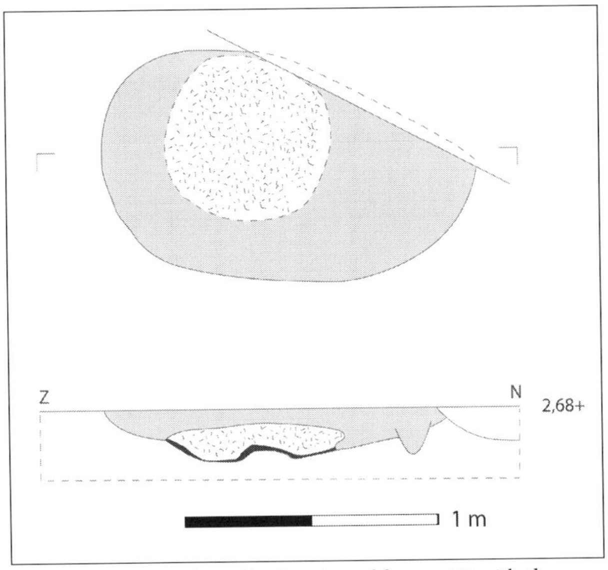

Figure 6.9. Plan and profile drawing of feature 22 with the bone concentration, scale 1:20 (after Nicolay 2008).

Below the feature a thin infiltration layer with charcoal coming from the bone concentration could be seen. The soil from the pit was sieved through 5 and 2 mm meshes resulting in an assemblage of almost 476,000 bone fragments. At first the feature was thought to be a human cremation but during the analysis of the assemblage not a single human bone was identified (Prummel *et al.* 2008: 248-250). Instead the remains of cattle, sheep/goat, pig and dog were found in percentages corresponding rather well with the rest of the material from the site, with the exception of the lack of horse fragments (see table 6.2)

A very large proportion of the material could not be determined to species. This should however be expected considering the extremely low medium fragment weight of 0.03 g for the feature (see table 6.2 and figure 6.10).

The low fragmentation weight is mainly the result of the high temperature at which the bones were burned. With the exception of 51 fragments that were black burned the complete material was white burned, thus heated to a temperature of more then 675 degrees (Prummel *et al.* 2008: 250).

The bone elements in the cremation pit come from all parts of the animals' skeletons. The age at death could be determined for three of the cattle and one of the sheep/goats. Of the cattle one was between 2.5 and 9 years old, one was more than 8 years old and one was a foetus or newborn calf. The sheep/goat that could be determined to age was more than 2.5 years old. With the exception of the foetus, which can still have been inside the cow

Figure 6.10. Unidentified fragments from the cremation pit (feature 22) (photo by J. Jans, GIA).

Species	NISP	Weight (g)	MNI
Cattle (*Bos taurus*)	312	627.6	4
Sheep/goat (*Ovis aries/Capra hircus*)	73	73.7	2
Pig (*Sus scrofa*)	48	26.8	2
Dog (*Canis familiaris*)	4	5.5	2
Unidentified	475440	11426.4	-
Total:	**475877**	**12160.0**	

Table 6.2. NISP, weight and MNI for the cremated material from feature 22.

during the cremation, all the animals identified to age were adult or old (Prummel *et al.* 2008: 252).

The manner in which the bones were placed in the pit is interesting in itself. The animals were almost certainly not cremated where they were found, but somewhere else and then collected and deposited in the pit. The shape and compactness of the bone concentration strongly indicate that the bones were placed wrapped in cloth or leather. The fact that 11 bone fragments had a green colouration connected with bronze corrosion indicates that at least one bronze object was deposited together with the bones, but this was unfortunately not preserved. Since one of the 11 coloured bone fragments was an unburned piece of red deer antler from the pit fill, it is reasonable to suggest that the bronze-object was not included in the actual cremation but added afterwards, possibly in the form of a fibula used to seal the cloth or leather sheet containing the cremated bones (Prummel *et al.* 2008: 252).

This deposit is considered special for a number of reasons. No less than nine adult domestic animals, three cattle, two sheep/goats, two pigs and two dogs were killed and cremated. Even if old animals were mainly chosen for the event, it would still be a substantial economic investment for a settlement the size of De Bloemert. These cremated remains were furthermore collected and neatly wrapped in a piece of cloth or leather. This package was finally placed in a pit together with a bronze object and covered with soil.

6.4.3 Final remarks

If the three special deposits in the Midlaren material are viewed in comparison with the complete Early Medieval material from the site it is obvious that we have a paradoxical situation not seen in the other materials. This situation is connected to the discussion in chapter 3 (section 3.1.3) about the problems involved in not looking thoroughly enough at the normal material, as well as the paradox of repetition. In Midlaren a majority of the assemblage is interpreted as not normal. The special deposits described in the sections above together contain close to half a million bone fragments weighing more than 12 kg while the remaining material contain only 112 fragments weighing a bit over 0.5 kg. These extreme numbers are of course mainly caused by the very large material from the cremation pit, but even if this feature is not taken into consideration the cattle elements from feature 51 and 3 would count for a large percentage of the total material (exactly how much is unfortunately impossible to calculate since the bones from feature 51 were excavated as block lifts).

Speciality is by its definition something that is out of the ordinary, but from the above exercise in numbers it is clear that even a majority of a material can be categorized as special. It is naturally possible that the categorization in itself is wrong, but the three features described in this chapter are actually very convincing. Two of them were observed already in the field and all three have been categorized as odd by several researchers (Holl 2005; Jans 2005; Nicolay & Den Hengst 2008: 594-595; Prummel *et al.* 2008).

If the initial definition of speciality (see chapter 1) is clear and the identification of these deposits are convincing, what is then the normality? The normal could of course have been defined by comparing with similar sites, but there are no sites with well preserved bone material from this specific area! My answer would be that the normality in a case such as this is some form of preconceived notion of how bone materials should be composed. This notion could be formed subconsciously by experience from numerous sites, but also consciously by comparison with materials that are situated spatially and temporally close to the site in question, in the case of Midlaren for example the *terp*-sites.

Since it is not at all certain that the *terp*-inhabitants had the same cultural habits as the people living in Midlaren it is wise to be cautious. As was stated in the discussion about the skulls and crania above it is possible that these deposits could represent normal behaviour in this area, even though this it is not likely from an economical point of view.

It is finally important to note that the cremation pit is not relevant for this discussion. Bones burned at such high temperatures are preserved well even in the sandy areas of Drenthe and if this type of deposit was common it is unlikely that it would not have been observed in other places.

7. Uppåkra

The fourth and last of the bone materials used as case studies in this theses is the only one that does not originate from the Netherlands. The animal bones come from Uppåkra in Southern Sweden and since the bone material retrieved from this site is much larger than from the other three used in this thesis, a different analysing approach had to be chosen. Instead of studying the complete material dated to the Early Middle Ages the bone assemblages from two smaller areas within the site were selected for analysis. The first assemblage was chosen for its direct connection to a large weapon deposit area while the other was selected for its peripheral placement within the site, in what has been described as an area with ordinary farms and few indications of unusual activity.

7.1 An overview of the site

Uppåkra is a central place located five kilometres south of Lund in the province of Skåne. It has been subject to a few small rescue excavations from the 1930s and onwards, and to yearly campaigns of archaeological research investigations since the mid 1990s. These excavations have been performed by the Department of Archaeology and Ancient History at Lund University.

Many years of intensive archaeological investigation at Uppåkra have given us a relatively good picture of the outline and structure of the site. The settlement was first established during the Late Iron Age on a low elevation 34 meters above sea level in the otherwise mainly flat landscape of South-western Skåne. Corings, detector-surveys and excavations have revealed a settlement that formed a roughly oval shape, 1100 meter from north to south and 600 meters from east to west. Within these over 40 hectares some areas seem to have been more intensively used than others, as is indicated by the variation in thickness of the occupation layers ranging from a few centimetres to about 2 meters (figure 7.1).

Numerous structures, including many houses, dating to the first millennium AD have been found during the excavations performed at the site, despite the fact that these investigations only covers a few percent of the total settlement area. The site seems to have functioned as a unit until the second half of the 10th century when the amount of finds is declining and the habitation was divided into two separate areas where the present day villages of Lilla and Stora Uppåkra are situated (Lenntorp & Lindell 2000: 4).

During several years of metal detecting campaigns over 20,000 metal finds have been recovered within the settlement area, despite the fact that iron finds in most cases were not recovered at Uppåkra (Hårdh 2003: 1). These massive amounts of objects, including many finds of gold and silver as well as other objects that indicate high status, suggest that the site was not an ordinary agrarian settlement. This conclusion is supported by the large size of the settlement, its long continuity, the existence of at least four burial mounds within the settlement, and the unearthing of several unusual structures. These structures include a high stave constructed building of very long duration and a clear ritual function, as well as weapon deposit areas.

Despite the clear outline of the settlement it is problematic to describe the internal structure of the site. It was definitely made up of a number of farms situated relatively close to each other, but it is not likely that the building arrangement resembled the tight structure of the earliest Scandinavian towns such as Birka, Hedeby or Ribe. The settlement structure could instead be described as proto-urban or even as "congested countryside", a term sometimes applied to the very first phase of some of the Early Medieval towns in Scandinavia (Andrén 1989: 588-591).

Maps from the 17th and 18th centuries as well as archaeological features and the local geography indicate that one of the main roads in the area before the High Middle Ages was running in a north-south direction straight through the settlement of Uppåkra, as is shown in figure 7.1 (Erikson 2001). If this hypothesis is correct the placement of the settlement is most likely connected to this road, but also to the favourable location on a slight elevation from the surrounding landscape. The settlement is however on a larger scale not very strategically placed in the landscape, being situated seven kilometres from the sea without easy access to major waterways leading to the coast (Larsson 1998: 111).

7.1.1 The lack of written sources

One peculiar circumstance regarding Uppåkra is the almost complete silence about it in the early written records. The site was situated in the eastern part of the kingdom of Denmark that was emerging during the 10th century, but the oldest text mentioning it under the name of Uppåkra is a letter from Knut the Holy dated to 1085 AD in which he donates several farms in the villages of

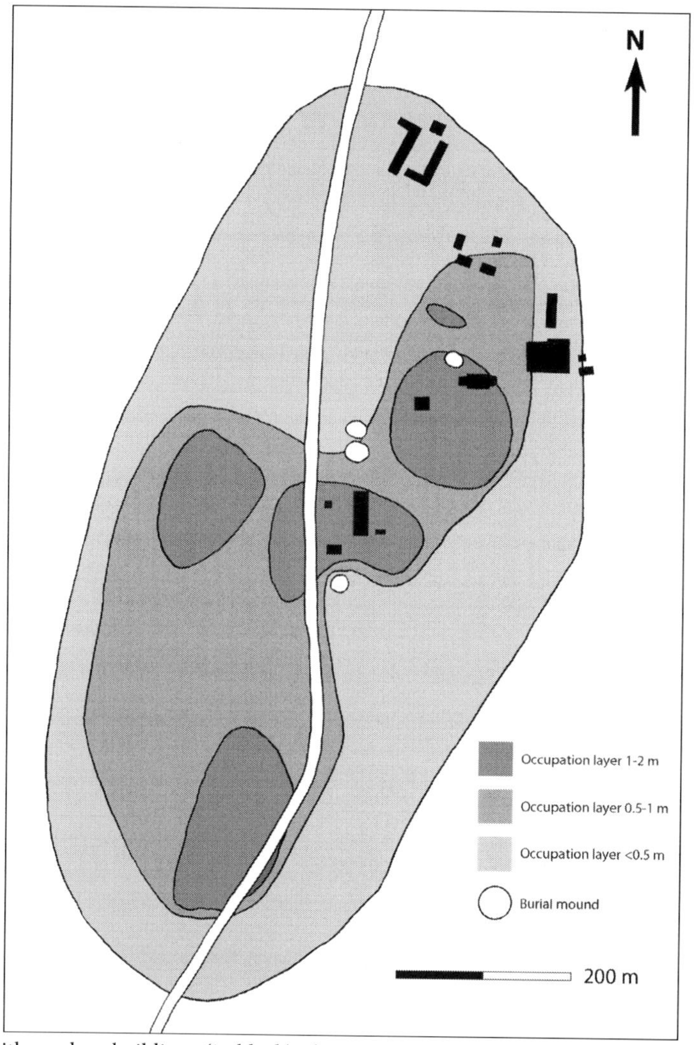

Figure 7.1. Map of Uppåkra with modern buildings (in black), the occupation layer thickness, burial mounds and possible prehistoric road marked (after Larsson 1998).

Lilla and Stora Uppåkra to the archbishopric in Lund (Riddersporre 1998: 173-175; Lenntorp & Lindell 2000: 4). At this time the settlement was thus already divided into two separate units and seems to have lost all political and symbolic importance, possibly after being on the loosing side in the political struggle surrounding the consolidation of the Danish kingdom. Despite some indications of missionary activity, Uppåkra seems to have been pagan until the very end of its existence as a single united settlement. Taking into account that the Christianisation of Denmark is often seen as an important tool in the unification of the different parts of the country, something that is indicated on Harald Bluetooth's famous rune stone in Jellinge, it seems logical that pagan central places would suffer consequences. The founding of Lund by the Danish king during the end of the 10th century, with its early churches within sight of the then still existing settlement in Uppåkra could also be taken as a sign of a power struggle (Hårdh & Larsson 2007: 85).

In addition to this letter two more medieval sources have been put forth as possibly dealing with Uppåkra, although both are dubious. In the first of these texts dated to around 1200 AD, the Danish historian Saxo Grammaticus writes about Ragnar Lodbroke, a semi-mythical Danish king. According to Saxo, Ragnar was waging a war against the provinces of Skåne, Halland and Jylland that tried to break free from his kingdom some time during the 9th century. In a battle somewhere in the flatland of Skåne one of the sons of Ragnar Lodbroke, Sigurd Snake-in-the-eye, was wounded. Sigurd was taken to a nearby town (*oppido*) where he was healed, supposedly by the direct intervention of the god Odin, and later became the next king of Denmark (Andrén 1998: 138). It has been suggested that the town mentioned by Saxo was Uppåkra, partly since few settlements large enough to be called towns existed in Skåne before the founding of Lund at the end of the 10th century, and partly because an Odin figurine and several surgical instruments have been found in Uppåkra, possibly indicating that the settlement was a centre for Odin worship, as well as known for advanced medical practices (Frölich 2007: 59-60; Hårdh & Larsson 2007: 99).

The second source is Egil Skallagrimson's saga that was also written around 1200 AD, but that deals with events supposed to have taken place during the 10th century. According to the saga, Egil landed with some ships on the western shore of Skåne some time around 960 AD and went "up on land" to Lund and plundered this "large trading town":

> *"But when they came to Eyrar-sound, then Aki said that up on land there was a large trading town named Lundr; there, he said, was hope of plunder, but 'twas likely that the townsmen would make resistance.*
>
> *The question was put before the men whether they should go up or not. Opinions were much divided, some liking, some letting it; then the matter was referred to the leaders. Thorolf was rather for going up. Then Egil was asked what counsel he thought good. He recited a stave:*
>
> *'Wolf-battening warrior,*
> *Wield we high gleaming swords.*
> *In snake-fostering summer*
> *Such deeds well beseem.*
> *Lead up to Lundr:*
> *Let laggards be none!*
> *Spear-music ungentle*
> *By sunset shall sound.'*
>
> *After that they made them ready to go up, and they came to the town. But when the townsmen were aware of the enemy's coming, they made against them. A wooden wall was round the town; they set men to guard this. A very fierce battle was there fought. Egil, with his following, charged fiercely on the gate nor spared himself. There was a great slaughter, the townsmen falling one upon another. It is said that Egil first entered the town, the others following. Then those of the town fled, and great was the slaughter. But Thorolf and his company plundered the town and took much wealth, and fired the buildings before they left."*
> (Egil Skallagrimson's Saga, chapter 47, translation by Green 1893).

It is important to note that the episode is described in both prose and in a skaldic poem in the saga. When something is written in a poem it is usually considered to originate from the time period described, rather than from when it was written down, since it is very difficult to change the poetry with its complex verse form (see section 3.4.5). Before Uppåkra was known historians used to dismiss this passage in the saga as a later addition or a confusion of names, since Lund did not yet exist at the time of Egil's attack. Now it has instead been suggested that it was actually Uppåkra that was described and not Lund (Andrén 1998: 138-139).

Beside these sources practically nothing about the settlement can be found in the texts. How is it possible that a very rich settlement and religious centre that most likely dominated large parts of Skåne politically is virtually not mentioned in the written sources? It is unlikely that all other sources dealing with Uppåkra have been lost, but there are other possible answers to this puzzling question. One option is that the settlement of Uppåkra was called Lund, and that this name was then transferred to the newly founded Christian town in the 10th century. This theory might explain the poem from Egil Skallagrimsson's Saga, and would also fit well with the idea of Lund as the successor of Uppåkra, in that case not only in function but also in name (Andrén 1998: 140-142).

Another reason for the silence of the sources could be the decreasing importance of the whole area of Skåne during the 12th and 13th century when most of the oldest preserved texts were written. Sjælland became more prominent as the political centre of Denmark and maybe the writing of the history of a former place of power was not in the interest of the then ruling class (Hårdh & Larsson 2007: 99).

A final reason might be found in the previously mentioned religious situation. If Uppåkra was a pagan stronghold until the end of the 10th century, early Christian scholars may have been reluctant to mention the place at all.

7.2 The archaeological research in Uppåkra

Uppåkra is today one of the most interesting sites from the first millennium AD in Southern Scandinavia, and considered to have been a supra-regional central place. It was already known as a rich Roman Period settlement in the mid 1930s when a new farm between the villages of Stora and Lilla Uppåkra was built and the construction workers unearthed a Late Roman Period grave. The resulting rescue excavation took place in 1934 and was supervised by Bror-Magnus Vifot, a researcher at the archaeological institute at Lund University.

The first thing that struck Vifot was the extremely thick culture layers within most of the 1200 m^2 large excavation area. These layers and the structures found within them produced large amounts of finds of various types. Ordinary settlement material was represented by more than 1700 pieces of bone and antler, some being the remains of bone and antler craft, 3400 pieces of pottery, as well as numerous grinding stones, loom weights, spindle whorls and several cubic meters of daub of which a majority was burned (Vifot 1936: 103, 105, 111-112, 126). These finds would be normal finds from any large Iron Age, Roman Period or Early Medieval settlement, apart from their massive amounts. Vifot however also uncovered things that are usually not found at agrarian settlements. These included several needles, combs, comb-sheets, three glass beads, five complete ceramic vessels, six fibulas, a bronze key, half an arm ring, an iron sickle, a sword hilt and a spear head, as well as numerous pieces of scrap metal and some slag (Vifot 1936: 105, 117-127).

The structures excavated include a few hearths and parts of a nine meters wide and at least nine meters long house. The existence of an ash layer as well as the abundance of burnt daub and fire damaged artefacts showed that the house had burned down. Three fire damaged fibulas found in the ash layer dated the house to the first half of the 5th century (Vifot 1936: 118). Vifot realised the great potential of the settlement of which he only had excavated a small part and entrusted the responsibility of the site to Berta Stjernquist, later professor in archaeology at the institute in Lund (Larsson & Hårdh 1998).

During the half century that followed little possibility for research appeared in Uppåkra, with the exception of a few smaller rescue excavations performed in connection with house building, laying of cables and road constructions. These sporadic and small scale excavations were carried out from the 1950s and onwards. They produced little information of value but confirmed the existence of thick and complex culture layers on the site (Larsson 1998: 97-98; Lindell 2001: 4).

It was not until the mid 1990s that Uppåkra again reached the top of the archaeological research agenda, with the initiation of the research project *Samhällsstrukturen i Sydsverige under järnålder* (The Social Structure of Southern Sweden during the Iron Age). This was a joint research project with the aim to study the development of society in the provinces of Skåne and Halland from 500 BC until 1000 AD. The project started in 1996 and included the Department of Archaeology and Ancient History at Lund University, Malmö Museum, Halland Museum and the National Heritage Board (Larsson 2003: 5).

Within the project Uppåkra functioned as a catalyst and eventually became a project of its own. At first the project mainly involved coring campaigns and yearly metal detector surveys performed by Danish amateur detectorists and people from the archaeological department in Lund. The detector campaigns produced large amounts of finds, many of a high-status character. It soon became clear that the site was much larger and had existed for a longer time period than Vifot's excavation had indicated (Hårdh 1998: 113; Branca *et al.* 1999: 59 Larsson 2003: 9-10).

At the end of the 1990s several research excavations were performed on the site (Lindell 2001: 4). Without secure funding these investigations were however small and therefore lacked the overview and structure required for a site with the size and complexity of Uppåkra. This problem was however solved when the company Tetra Pak sponsored a five year long large scale excavation campaign, beginning in the year 2000 (Larsson 2003: 10). The campaign was later extended with several more years through further donations from the Olle Tegstam Foundation.

With the finances secured the project could continue with stripping of the top soil along search trenches on a larger scale than before to get a better overview of the site. These test trenches were concentrated in two areas, one northern area directly east of Vifot's original trenches and one southern area a few hundred meters to the south (figure 7.2).

Both areas were chosen because of the high concentrations of metal finds discovered during the detector surveys and, for the northern area, the existence of thick occupation layers (Larsson & Lenntorp 2004: 3-4). Both areas contained numerous features, among them many houses. In the southern area finds and structures were mainly dated to the Late Roman Period and Early Middle Ages (ROML-VMEB), with a concentration of activity during the Migration Period. The structures from this part of the site seem to be of an agrarian character, with the exception of some indications of metal craft and a few high-status metal finds (Lenntorp & Lindell 2000: 132).

In the northern area structures from the Roman Period and the Early Middle Ages (ROMY-VMED) were found and the finds are in general more varied than in the southern area. Some of the most spectacular finds found so far at the site came from here, and therefore this area became the main focus of the excavations in the following years. From this time onwards all excavations were performed with the single-context method that is very useful when complex occupation layers such as those in Uppåkra are excavated and interpreted. A fully digital recording and documentation system was also introduced, something that was necessary considering the massive number of structures and finds that had to be recorded during the excavations (Larsson 2003: 11-12; Larsson & Lenntorp 2004: 5-6). Among the structures found in the northern trenches were the remains of a house with unusual features and a nearby weapon deposit area with a large animal bone layer. The animal bones from the weapon deposit area were chosen as the material for my case study and will be discussed in section 7.5. Given that this structure is situated only 30 meters north of the unusual house and since they existed at the same time a short presentation will be given of what has been called "the enigmatic house".

This house turned out to be not one, but a succession of houses that with very little variation had been built and rebuilt on the same spot for centuries. The basic outline was a 14 meters long and 6 meters wide building with slightly bent walls, four huge post-holes inside the building and three entrances, one to the north and two to the south (figure 7.3) (Larsson & Lenntorp 2004: 6).

The first house was most likely constructed in the 3rd century AD. The house was rebuilt at least seven times until it was finally dismantled in the 8th century. The house was constructed with stave technique and the impressive 2 meter deep post-holes that carried posts with a diameter of at least 0.7 meters indicate a very tall building, possibly as high as 5 or 6 meters. The stone constructions surrounding the doors as well as the door posts also indicate high portals, and thus a high building (Larsson & Lenntorp 2004: 7-12, 30).

The conservatism seen in the house in Uppåkra is

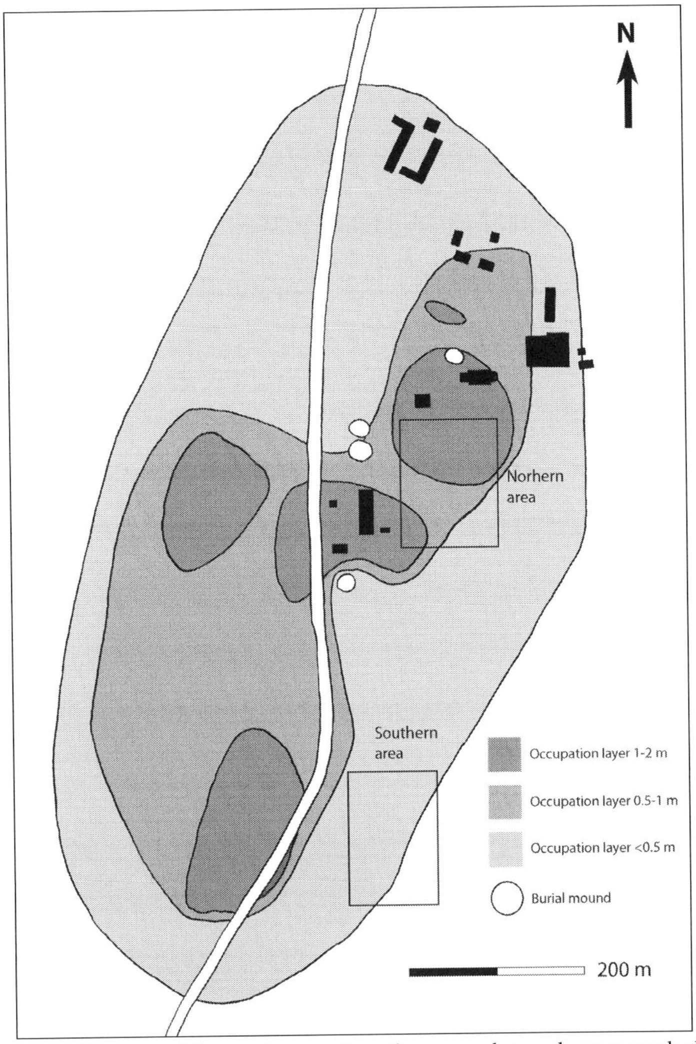

Figure 7.2. Map of Uppåkra with the location of the northern and southern search trench areas marked (after Larsson 1998).

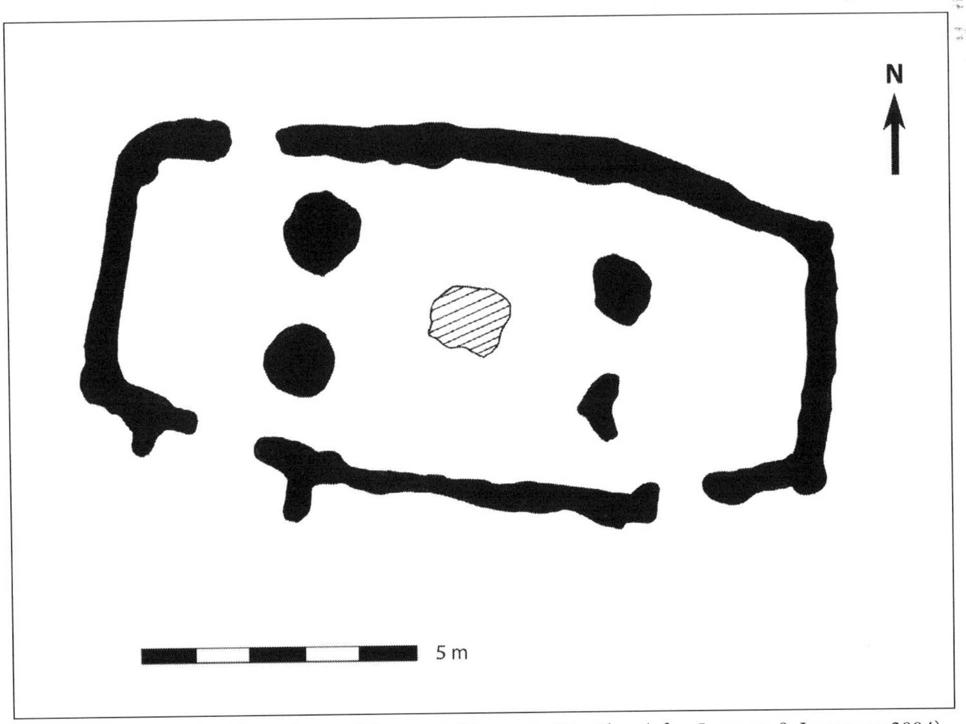

Figure 7.3. Outline of the youngest building phase of the special house in Uppåkra (after Larsson & Lenntorp 2004).

extraordinary. During the 600 years of its existence the only changes made to the construction were the position and number of hearths inside it, and the addition of a small antechamber in front of the south-western entrance. Considering the vulnerability of wooden buildings it is also notable that no episodes of fire were recorded, despite the fact that several burned down houses have been excavated only a few meters from of it (Larsson & Lenntorp 2004: 43; Larsson 2009: 14-15).

The unusual construction and long duration of the house would on its own be enough to set it apart as a building with special functions. The find material from the house and its immediate surroundings however further increases the significance of the building. In addition to several fibulas and beads, a number of very special finds were unearthed. As many as 111 gold-figure foils were found in the house. These artefacts were mainly dated to the 6th and 7th centuries and are closely associated with ritual activities (see section 3.3.1). The number of gold-figure foils found in the house is very high compared to those from other sites and makes it the second largest concentration in Scandinavia (Larsson 2007: 16). The figures were spread throughout the house and found in wall trenches, post-holes and floor layers. Two concentrations were nevertheless observed, one in the north-western post-hole and another one near the east gable of the house. It has been suggested that gold-figure foils were used as votive offerings by visitors at holy sites, possibly attaching them to trees, walls or high seats. The north-western post in the house in Uppåkra might have been a place where figures were attached in this manner (Larsson & Lenntorp 2004: 22; Watt 2004: 216).

A magnificent cache was found by metal detectoring underneath the floor of one of the later houses. The deposit contained a 16.5 centimetres high copper beaker with silver and gold decorations and a 9.5 centimetres high and 16.5 centimetres wide doubled layered glass bowl. The beaker and the glass bowl have been dated to around 500 AD and while the beaker might be of local production, the glass vessel was most likely imported from the Eastern Mediterranean or the Black Sea area (Hårdh 2004; Stjernquist 2004). The stratigraphy of the find indicates that the deposit was performed later, possibly several generations after the objects were made (Larsson & Lenntorp 2004: 43).

In addition to the glass bowl, glass shards from ten different vessels were found in the house (Larsson & Lenntorp 2004: 23-24). This is notable since finds of glass before the Late Middle Ages are rare in Scandinavia and are usually connected with the highest classes of society.

Two more unusual deposits were found in the building. The first is a ceramic vessel that was deposited intact below the floor of the second house. The vessel was found crushed and seems to have been empty when it was buried. The second deposit came from the north-western post-hole that also contained a gold-foil figure concentration, and was made up of large parts of a cow skull and a door ring handle made of iron (Larsson & Lenntorp 2004: 7-8, 24-29).

As a final note small amounts of slag and fragments of a crucible were found inside the building indicating that metal craft was performed here during some parts of its existence (Larsson & Lenntorp 2004: 189).

Directly outside the house a layer with large amounts of animal bones was found as well as some ceramics and a few metal artefacts such as fibulas and parts of a warrior's helmet. In the immediate surroundings several large houses of more normal character were found, as well as the already mentioned weapon deposit area that could be connected to the special house.

How should the "enigmatic house" be interpreted? It did not get this label without a reason; it is a difficult building to comprehend. It is a house with few parallels in Scandinavia and has been called a cult house, a special building, a ritual house or even a temple. In the terminology of the Early Medieval Scandinavian texts it could be called *hall*, *hov* or *hargh*. None of these designations have been entirely accepted however. The definitions for all these terms are vague and the function of the house may have changed during its 600 years of existence. The house most likely functioned as a meeting place and as a building where religious and ritual activity was performed during large parts of its history. It was at times a place where metal craft was carried out, but it was most of all a place so sacred and bound in tradition that it was kept safe and unaltered during centuries in an otherwise changing surrounding world. The exact terminology used to describe a unique house such as this might even be considered less important. This opinion is expressed by Lars Larsson in his article *The Iron Age ritual building at Uppåkra, southern Sweden.*

> *"Whether the building at Uppåkra should be regarded as a cult house, a hall or a temple might be a semantic detail. From a holistic perspective, it is rather a question of whether there are any real differences between these three hypothetical structures in terms of the role they would have played for the Iron Age societies at Uppåkra."*
> (Larsson 2007: 21)

7.3 The excavations

Instead of studying the complete Early Medieval bone material from Uppåkra that consists of hundreds of kilos of bone, the assemblage from the weapon deposit area was selected for this thesis. Because of the finds of weapons and other military equipment and the special character of these deposits, this specific area has already been interpreted as being the stage for ritual activity, something that is further emphasized by the proximity to the special house (figure 7.4).

The choice of this material for analysis should therefore be seen as a different strategy in how to approach ritual in that the interpretation of the context in this case is already

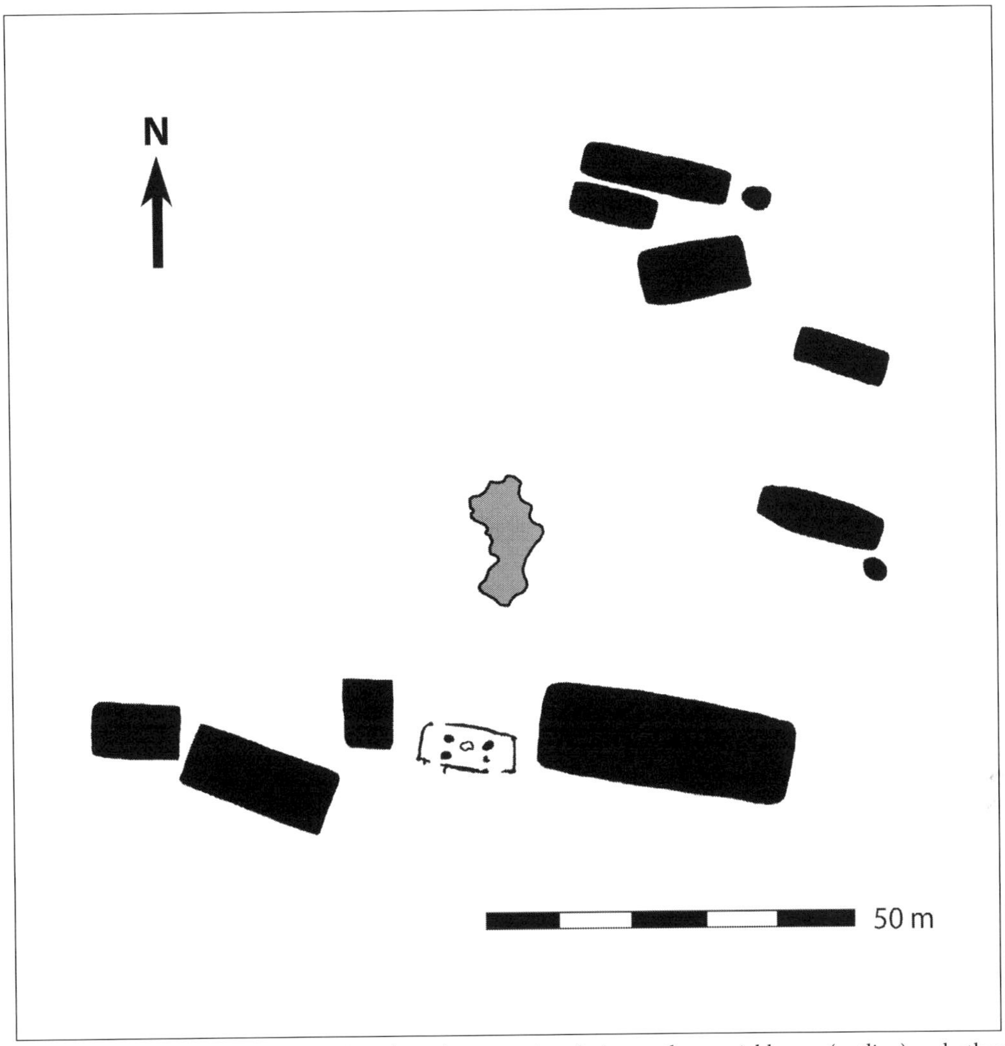

Figure 7.4. Map of the weapon deposit area (grey) and its placement in relation to the special house (outline) and other houses (black) in the vicinity (after Helgesson 2004; Hårdh & Larsson 2007; Larsson 2009).

given, but with the question remaining whether the bone material is of a similar character as the interpretations of the artefacts in the layer. The weapon deposit area was excavated 2001-2005 and the concentration of weapons, bones and other artefacts covered 86 m^2. The area was excavated using one meter squares and all the artefacts were digitally recorded with a total station. The bones were collected per square meter, but when concentrations of bones were observed these were either drawn or photographed.

As a complement to the weapon deposit area, an assemblage from the southern area was selected as a comparable material for this case study. This bone material mainly originates from refuse layers, pits and post holes in the vicinity of houses and is used as a contrasting material since the area has been interpreted as being of an agrarian character. This area was excavated from August to September 2000 and covered a total area of 1720 m^2. The excavation was performed in the form of a preliminary investigation with ten trenches that were extended when interesting features were discovered (figure 7.5) (Lenntorp & Lindell 2000: 8-10).

7.4 The archaeological results

7.4.1 The weapon deposit area

In the 86 m^2 large weapon deposit area situated on a small elevation ca 25 meters north of the special house about 300 objects were found, mainly weapons and parts of armour (figure 7.6). Lance- and spearheads dominated the material with 136 finds. In addition 10 shield bosses, 4 shield handles and a number of arrowheads and sling-stones were found, as well as various rivets and mountings (Helgesson 2004: 224-226). These metal finds were deposited together with massive amounts of animal bones and three human bones, a mandible, a tibia and a pelvis.

A majority of the bones are situated above the metal objects, but still within the same layer. The lance- and spearheads have been dated to the Roman Period until the Early Middle Ages, roughly 0-800 AD, but a majority of them could be dated to the later part of the Roman Period and the early part of the Early Middle Ages, roughly 250-550 AD (Helgesson 2004: 224-225).

In addition to the typological dating nine 14C-samples from cattle and human bones were taken and analysed at

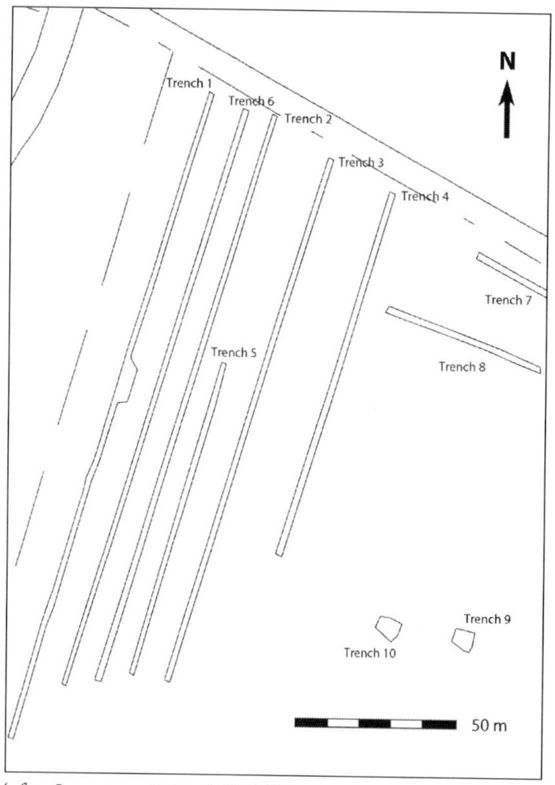

Figure 7.5. Map of the southern area (after Lenntorp & Lindell 2000).

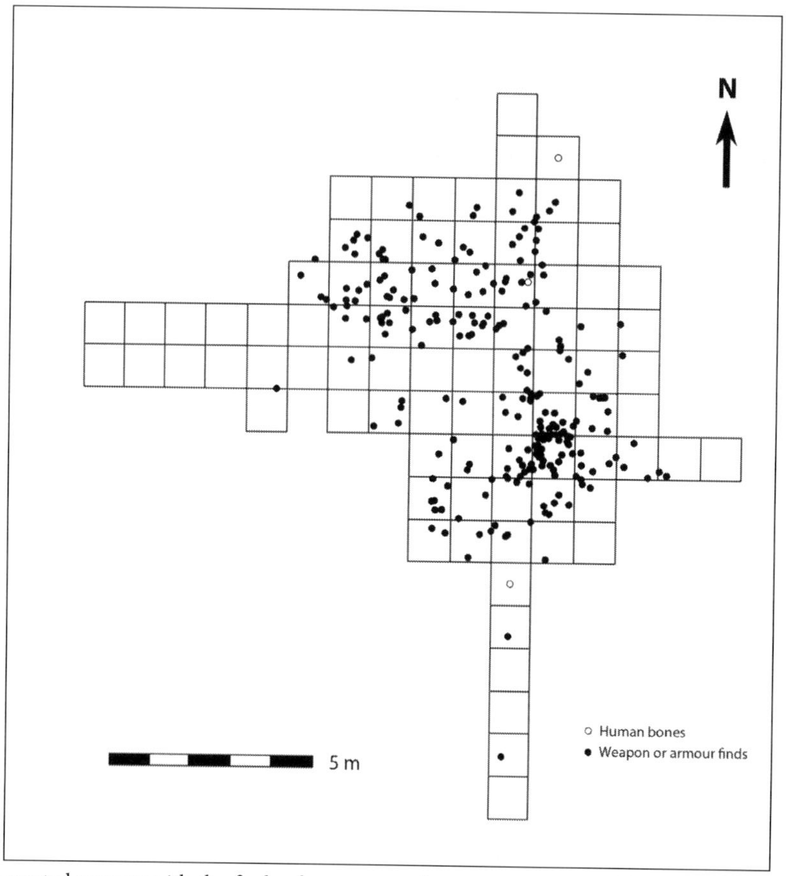

Figure 7.6. Map of the excavated squares with the finds of weapons and armour represented by dots and the three human bones marked with circles.

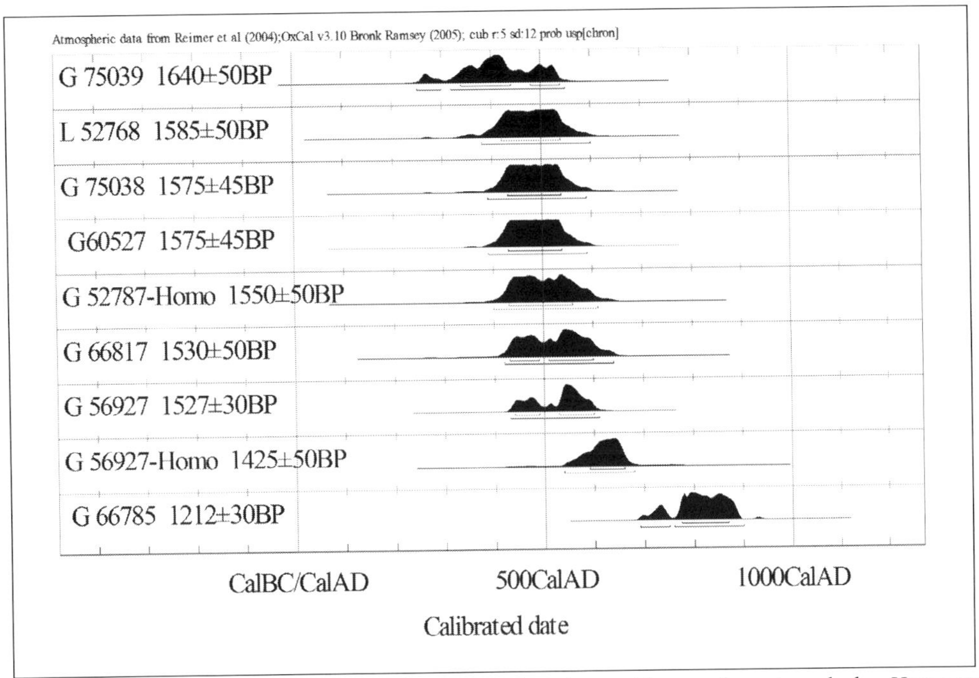

Figure 7.7. Calibration diagram for the 14C-dates from the weapon deposit area (the samples not marked as Homo are from cattle).

the Radiocarbon Dating Laboratory at the department of Quaternary Sciences in Lund (figure 7.7).

As can be seen in the diagram the datings mainly fall within the first half of the Early Middle Ages (VMEA-VMEB) with only one belonging to the second half of that period (VMEC-VMED). Since a number of the objects found in the layer are dated to the Roman Period the present interpretation is that the custom of depositing objects at the location begun in this period and continued well into the Early Middle Ages. In the later phase the weapons were deposited together with large amounts of animal bones and a few human bones. Occasional deposits continued to be made at least into the 8th century. The later spears and lances were deposited by digging small holes into the existing layer, explaining the apparent lack of a coherent stratigraphy for these objects within the layer (Magnell 2009, personal comment).

Since a lot of the equipment in the weapon deposit area were consciously destroyed, often in an elaborate way (figure 7.8), the find complex has been almost uniformly interpreted as a sacrificial place. The content in the form of weapons and other military equipment has resulted in the comparison with the large contemporary weapon sacrifices found in bogs in Northern Europe, probably best known from Danish sites such as Illerup, Vimose and Nydam (Helgesson 2004: 231-232; Larsson 2007: 19).

These weapon sacrifices have been interpreted as representing the loosing side's weapons and equipment that were sacrificed by the winners after a battle. There are however several major differences between the weapon

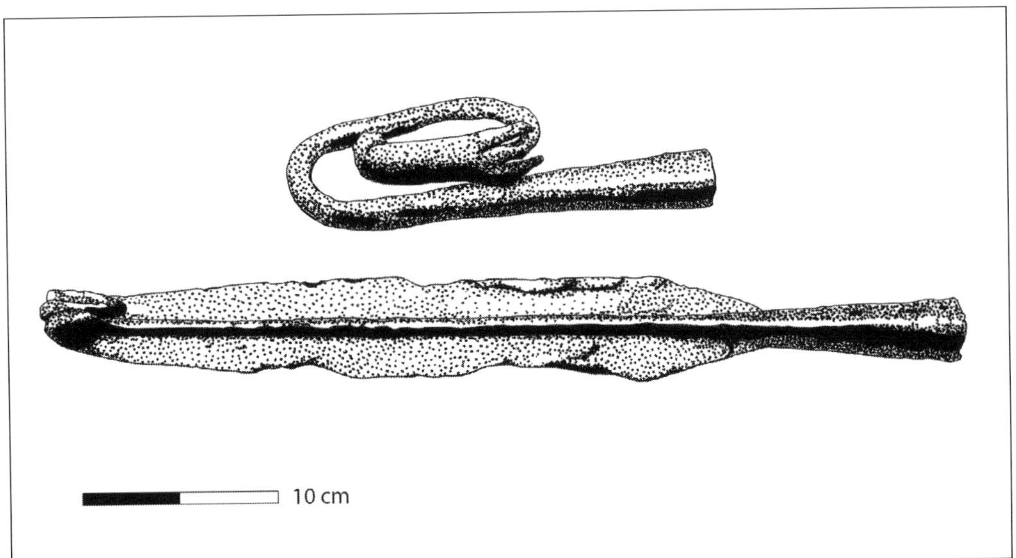

Figure 7.8. Two deliberately destroyed Early Medieval lance/spearheads from the weapon deposit area in Uppåkra (after B. Almgren, LUHM).

deposit area in Uppåkra and these sacrificial sites. First of all the number of objects deposited in Uppåkra was comparably small compared to most of the Danish bogs where often thousands of objects have been found. The range of objects is also smaller in Uppåkra than in the bog finds where often almost complete warriors' equipments have been found. The bog finds are also usually interpreted as being the result of a small number of large deposition-events while there are indications in Uppåkra for continuous burying of objects during several centuries. Furthermore the objects in Uppåkra are not deposited in a bog but on a slight elevation in a dry context in the middle of a settlement and close to a house with obvious special functions. Finally most of the Danish bog finds are from the Roman Period and it seems as if the practice of depositing weapons in bogs in Southern Scandinavia disappeared at the beginning of the Early Middle Ages, while weapons were still deposited in Uppåkra at least as late as the 8th century (Helgesson 2004: 235; Larsson 2007: 19).

An alternative explanation for the Uppåkra-finds is that the weapons should be seen not as the immediate result of a battle or raid, even though the objects still could originate from such events, but as sacrifices connected to the special house and possibly a warriors cult manifested in connection with this (Helgesson 2004: 237).

7.4.2 The southern area

The excavation trenches in the southern parts of the site showed that this area of the settlement was first inhabited during the Early Roman Period (ca 0-150 AD). From this period two houses were found and the area seems to have been mainly used for agrarian activities. At the transition to the Late Roman Period the buildings disappeared and were over-layered with soil. The area was then either left empty or used for agriculture activities.

Some time in the Late Roman Period the locality was again used as a settlement area with new houses being built, and continued to be used as such at least until the 7th century. 17 houses from this period could be identified as well as numerous other features such as refuse layers, pits, hearths, ovens, post-holes and a stone layer of unknown function. With the exception of a few indications of metal craft and a small number of high status metal finds (the majority found in the top soil with metal detectors), the locality seems to have been mainly used as a normal agricultural settlement area (Lenntorp & Lindell 2000: 131-132).

7.5 The archaeozoological material

7.5.1 The bone material from the weapon deposit area

Within the weapon deposit area the bone layer was so dense that almost 190 kg of animal bones were recovered during the excavation (see figure 7.9).

These bones were spread over most of the excavated area with two obvious concentrations (figure 7.10).

Because of the large volume of the material a sample had to be selected for my analysis. 26 squares and an irregular bone concentration near the south-western corner of the area were chosen for analysis (table 7.1).

The squares chosen for analysis were selected from different parts of the weapon deposit area. In an effort to get material not only from individual squares but also from larger areas that could be compared with each other, three portions of adjacent squares, a northern, a western and a southern, were chosen with four single squares in the area between them making up a central part (figure 7.11).

The northern portion was partly selected because of the human mandible and tibia found in the squares 52787 and 56927 respectively. The western portion was selected since this area contained virtually no weapons or other deposited objects and it would therefore be interesting to see if the bone material was different here. The southern portion finally, was selected because of the existence of what the archaeologists in the field interpreted as an irregular concentration of animal bones and because

Figure 7.9. Parts of the bone layer within the weapon deposit area (photos by O. Magnell, Lund University).

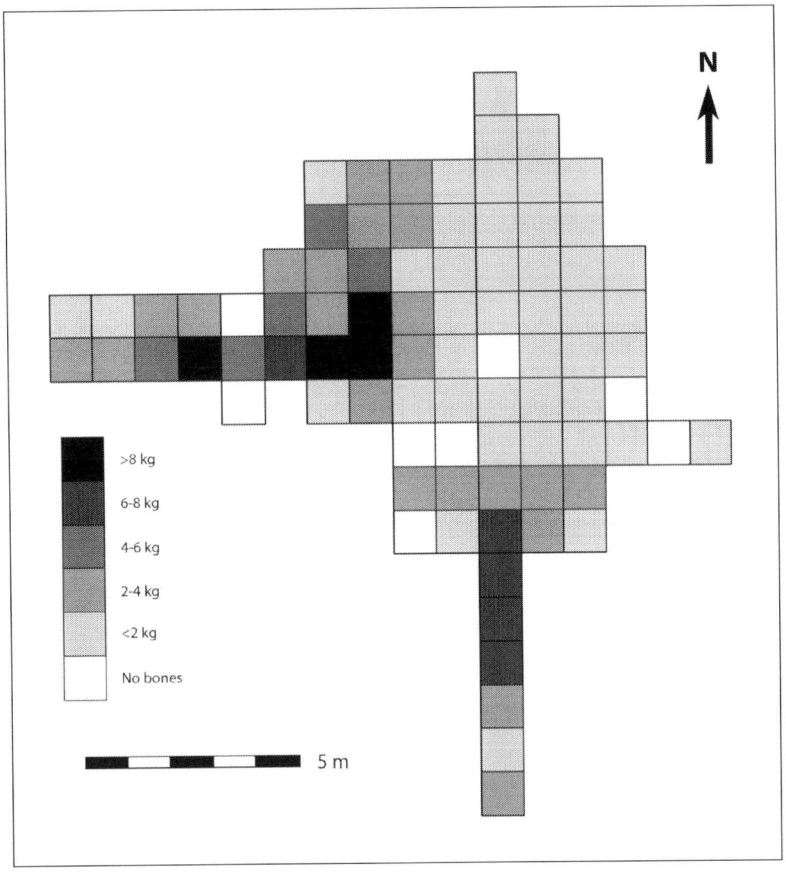

Figure 7.10. Map of the excavated squares and their bone density.

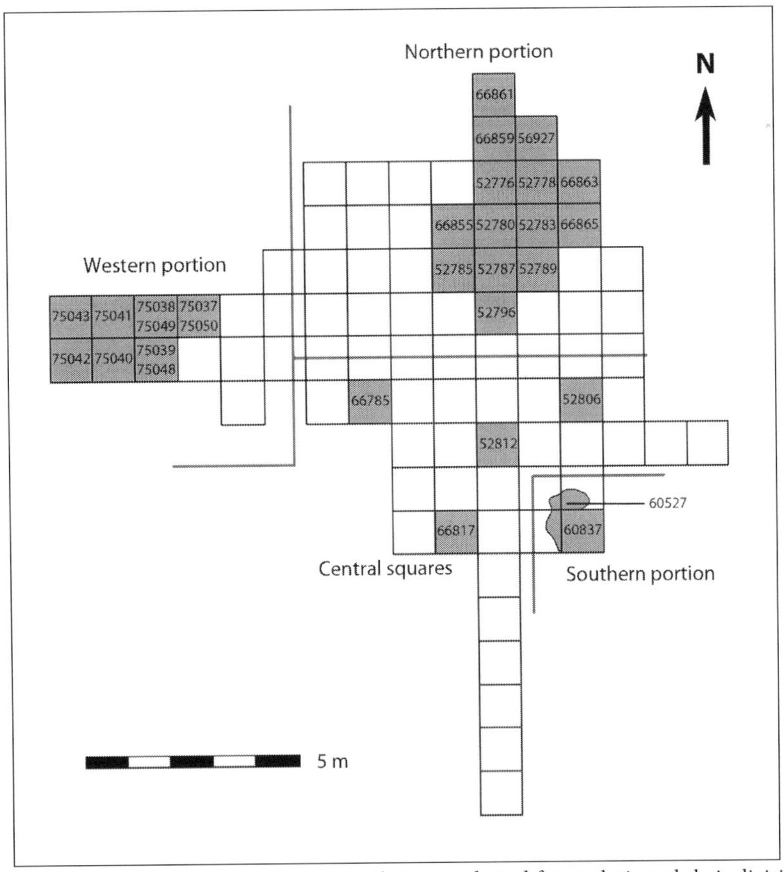

Figure 7.11. Map of the 26 squares and the bone concentration that was selected for analysis and their division into four portions.

Table 7.1. The squares selected for analysis and their bone content.

Square (context)	Weight (g)	NISP
52776	244.6	138
52778	419.3	129
52780	463.7	195
52783	298.1	122
52785	776.3	207
52787	354.3	147
52789	542.3	184
52796	648.4	130
52806	336.9	143
52812	121.6	347
56927	544.0	264
60527*	2941.4	213
60837	152.7	7
66785	2697.6	149
66817	837.9	101
66855	804.1	89
66859	325.4	42
66861	372.7	85
66863	133.6	30
66865	174.4	48
75037, 75050**	4669.5	1043
75038, 75049**	3733.8	2258
75039, 75048**	5893.9	2150
75040	3005.1	1329
75041	3612.5	938
75042	2647.9	499
75043	1982.7	892

* The irregular bone concentration in the south-eastern corner of the weapon deposit area.
** These contexts are situated above the squares in the western portion (see below). They were initially interpreted as a separate layer but were reinterpreted as belonging to the same layer that was disturbed by later agricultural activity.

it was situated close to the highest concentration of deposited metal objects within the area.

The selected areas contained a total of 11,879 bone fragments weighing 39,826 g. Of these, 1664 fragments weighing 27,425 g could be identified to species. The weight per fragment for the whole assemblage is only 3.4 g, indicating a highly fragmented bone material.

Only 49 bone fragments in the assemblage show signs of heating whereof 18 fragments were calcinated. 111 bone fragments were gnawed by animals and 29 of these fragments were heavily affected. Some 227 butchery marks were recorded on 172 bone fragments. Of these, 210 marks could be placed in one of the five animal processing categories presented in chapter 2. Cattle and sheep/goat were the only species that showed enough marks to give any significant indication regarding animal processing patterns, and as in the Dongjum and Leeuwarden assemblages disarticulation marks clearly dominated the assemblage (figure 7.12).

In figures 7.13, 7.14 and 7.15 the statistics for weathering, trampling and fracture freshness for the material are presented as number of identified fragments per stage or index number.

As can be seen in the diagrams the bone material from the weapon deposit area shows a lot of weathering, although mainly in the lighter stages. Since the preservation condition of bone material is generally good in Uppåkra this fragmentation most likely has other explanations beside diagenesis. One reason for the high degree of fragmentation could be that the bones were left in the open for some time and therefore subject to weathering. A second possible explanation is given by the fact that the later weapon deposits were dug into the bone layer, probably exposing and fragmenting bones in

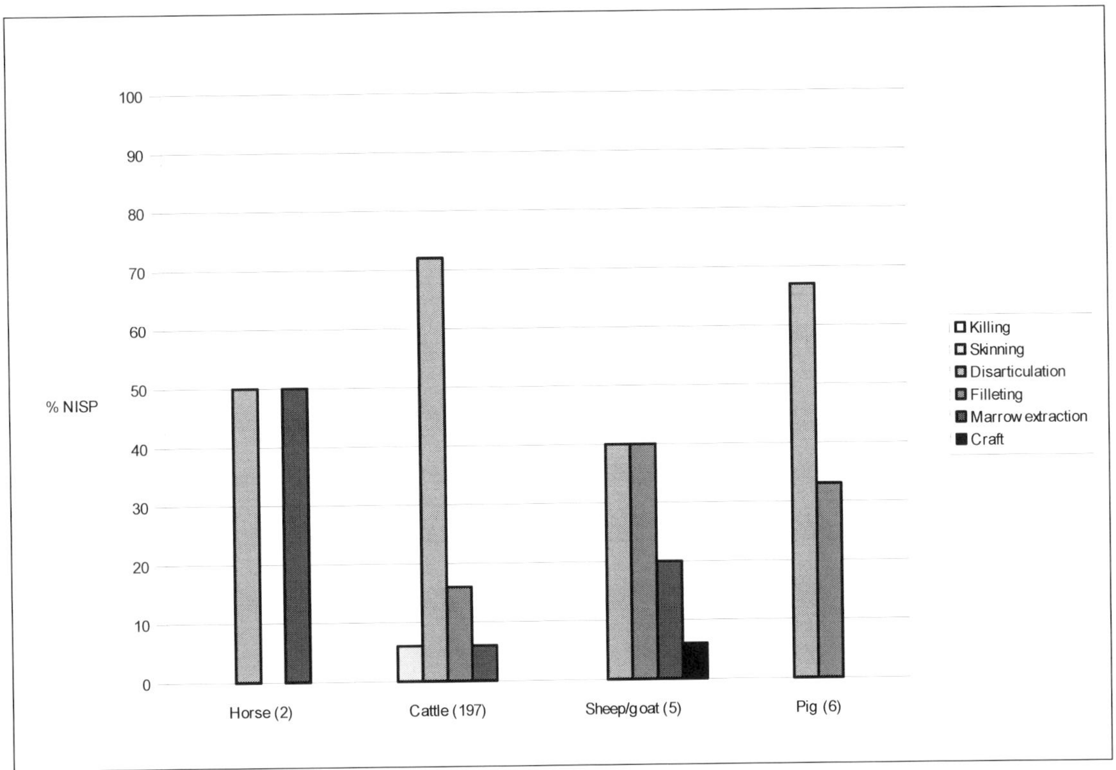

Figure 7.12. Percentage of fragments from the major species in the weapon deposit material from Uppåkra showing marks belonging to the different animal processing categories (the total number of fragments is presented within parenthesis).

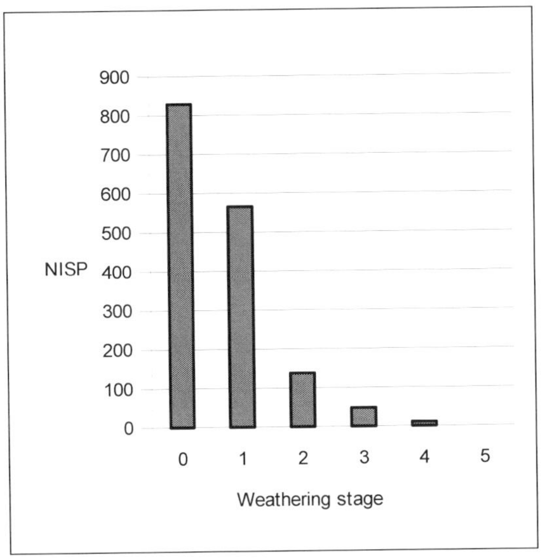

Figure 7.13. Weathering stage diagram for the weapon deposit material from Uppåkra (0: no cracking or flaking; 1: cracking parallel to fibre structure; 2: flaking of outer surface; 3: fibrous texture; 4: coarsely fibrous and rough surface, inner cavities visible; 5: bone falling apart).

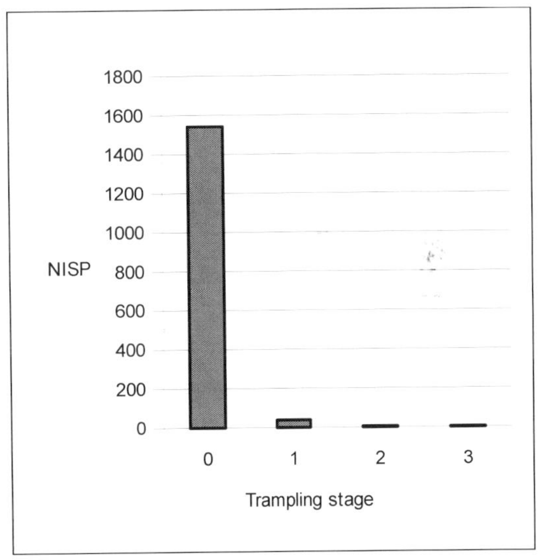

Figure 7.14. Trampling stage diagram for the weapon deposit material from Uppåkra (0: no marks; 1: few, but clear marks; 2: several and clear marks; 3: most of the surface is covered with marks).

the process. Only three squares were sieved during the excavation, with little result, so that cannot be the reason for the low medium fragment weight (Magnell 2009, personal comment).

The NISP-, weight- and MNI-numbers for the identified material from the weapon deposit area are presented in table 7.2. Beside the expected dominance of the major domestic species it is interesting to note the scarcity of both birds and fish in the material, something that however can be partly explained with the fact that only three squares in the material were sieved.

In figures 7.16 and 7.17 the proportions of the major species from the weapon deposit area are presented as NISP- and weight-percentage diagrams. The bone

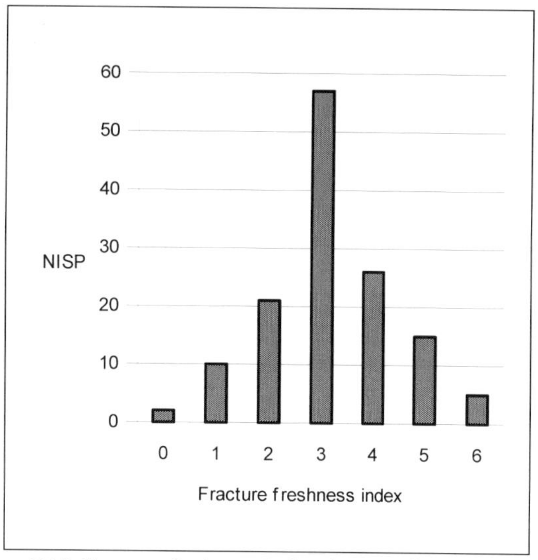

Figure 7.15. Fracture freshness index diagram for the weapon deposit material from Uppåkra (0-2: fresh fracture; 4-6: dry fracture).

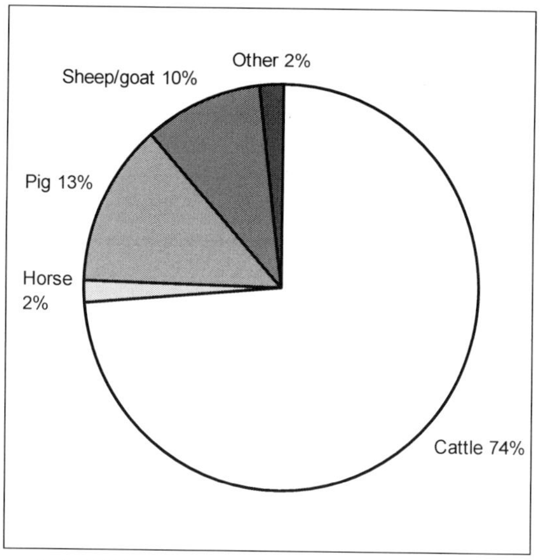

Figure 7.16. NISP-percentage diagram for the bone material from the weapon deposit area.

assemblage is totally dominated by cattle, even to a larger degree than the materials from Dongjum and Leeuwarden (figures 4.6, 4.7, 5.6 and 5.7). There is slightly more pig in the material than sheep/goat, also something that differs from Dongjum and Leeuwarden, but that is more expected since the *terp*-sites usually contain little pig. This can mainly be explained by that the salt marsh landscape is not as well suited for keeping pigs as the flatlands of South-western Skåne.

Fragments from the head and then especially loose teeth totally dominate the element distribution of the assemblage (figure 7.18). This is very different from the element distribution in Leeuwarden and Dongjum that gave a more even distribution and fewer loose teeth (figures 4.10 and 5.9). The many teeth might be explained by the high fragmentation degree of the weapon deposit material and also helps to explain the relatively small number of identified cranial fragments.

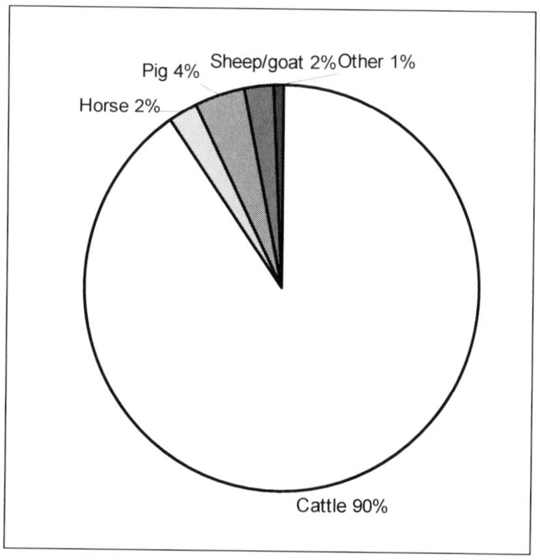

Figure 7.17. Weight-percentage diagram for the bone material from the weapon deposit area.

Table 7.2. NISP, weight and MNI for the identified material from the selected time period.

Species	NISP	Weight (g)	MNI
Cattle (*Bos taurus*)	1226	24,765.6	31
Sheep/goat (*Ovis aries/Capra hircus*)	161	656.9	8
Pig (*Sus scrofa domesticus*)	216	1126.4	7
Horse (*Equus caballus*)	29	656.2	2
Dog (*Canis familiaris*)	3	26.7	1
Elk (*Alces alces*)	1	139.5	1
Red deer (*Cervus elaphus*)	1	35.5	1
Roe deer (*Capreolus capreolus*)	1	14.0	1
Unspecified vole (Arvicolinae)	22	0.2	-
Greylag goose (*Anser anser*)	1	3.1	1
European perch (*Perca fluviatilis*)	2	0.2	1
Unspecified codfish (Gadidae)	1	0.5	-
Total:	**1664**	**27,424.8**	

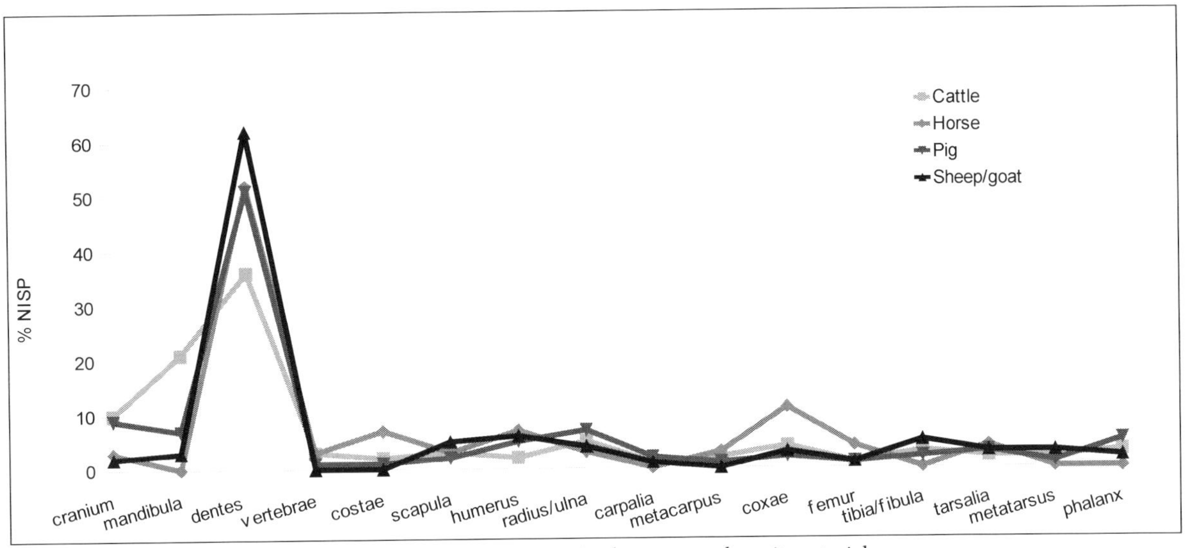

Figure 7.18. Element distribution for the different major species in the weapon deposit material.

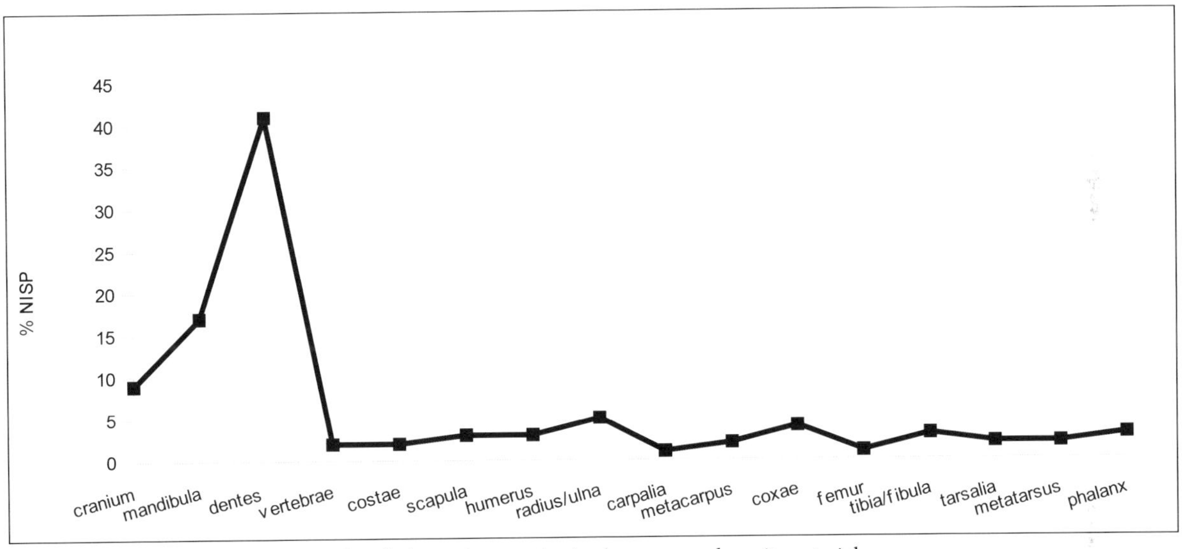

Figure 7.19. Total element distribution for all the major species in the weapon deposit material.

In a very fragmented material crania and mandibles tend to separate into small pieces of which the teeth can easily be determined to species while the other fragments are often unidentifiable.

The element distribution of the four major domestic species together (figure 7.19) explains the situation further. A comparison between the two diagrams shows that the dominance of crania and mandible fragments is mainly a phenomenon of cattle (and to a smaller degree pig), while loose teeth are a common feature for all species. The relatively low proportion of post-cranial elements is however the same for all species.

26 bone fragments of cattle and pig from the weapon deposit material could be identified to sex (table 7.3). These fragments show a perfectly even distribution of males and females for both species.

The age at death distribution in the weapon deposit material shows a majority of fully-grown animals (figure 7.20), as was also the case in both Dongjum and Leeuwarden (figures 4.9 and 5.7). The number of young pigs in the weapon deposit material, 30 percent being young adult or younger, indicates that the quality of the pork meat might have been more interesting than the maximum meat weight. Another interesting aspect of the age distribution is that very few young cattle were identified and that as many as almost 20 percent of the cattle where not killed until reaching old age (*senilis*). The corresponding figures for very old cattle from Dongjum

Table 7.3. Sex-distribution for the weapon deposit material.

Species	Male (fragments)	Female (fragments)
Cattle (*Bos taurus*)	9	9
Pig (*Sus scrofa domesticus*)	4	4

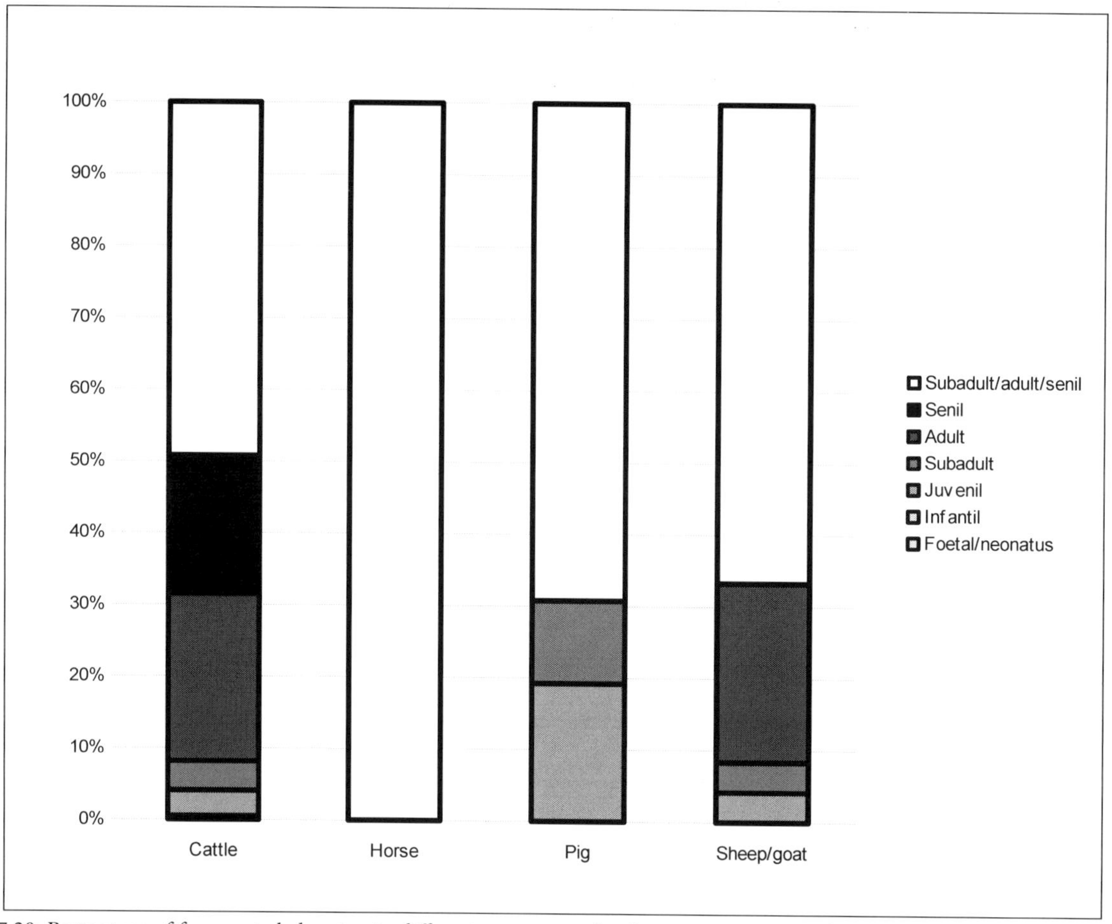

Figure 7.20. Percentage of fragments belonging to different age groups for the major species in the weapon deposit material.

and Leeuwarden are less than one percent. This could indicate that the cattle were to a larger degree used as milk producers and draught animals than in these sites. It is however important to remember that the weapon deposit material is suspected to be of an unusual character and that the age selection, as well as other characteristics of the assemblage, might be due to special circumstances and therefore not representative of the general situation on the settlement (see section 7.5.2 for a comparison).

7.5.2 The bone material from the southern area

The animal bone material from the southern area of the Uppåkra site (Lilla Uppåkra 4:1 & 8:1) used as a comparison originates from five different features (table 7.4). All of these features originated from the two westernmost search trenches of the excavation (figure 7.21).

The material from the five features consists of 6664 fragments weighing 11,528 g and was completely analysed. Of these 758 fragments weighing 5296 g could be identified to species.

The weight per fragment for the whole assemblage is only 1.7 g something that indicates a very fragmented bone material, even more so than for the assemblage from the weapon deposit area. This extremely low weight per fragment cannot be explained by taphonomic processes since the material showed no sign of being heavily affected by weathering, trampling or animal gnawing. One reason for the low fragment weight is however the fact that all the excavated features were dry sieved (3 mm), thus collecting many of the small bone fragments that would have been otherwise lost.

Six bone fragments in the assemblage show signs of heating but none were calcinated. 32 bone fragments were gnawed by animals and 15 of these were heavily affected. Some 37 butchery marks were recorded on 26 bone fragments. Of these, 30 marks could be placed in one of the five animal processing categories presented in chapter 2 (figure 7.22). Their distribution shows a similar picture as for the weapon deposit area (figure 7.12). No marks were identified on the few horse bones in the assemblage.

The diagrams in figures 7.23, 7.24 and 7.25 for the southern trench material show very little weathering and virtually no trampling. The fracture freshness index indicates slightly more dry fractures than in the weapon deposit area (figure 7.15).

The bone assemblage from the southern area differs from that of the weapon deposit area in that sheep/goat are represented by more fragments than cattle and that very few horse bones were found in the material. Cattle bones represent a majority of the bone weight, but are far

Table 7.4. The features with animal bones from the southern area.

Group	Context	Find number	Trench	Feature type	NISP	Weight (g)
35	4	1-14	1	Demolition layer	1205	3739.4
16	297	27	1	Pit (refuse)	196	520.3
14	52	40	1	Post-hole	270	335.1
139	118	43	6	Pit	208	458,4
15	28	88	1	Refuse layer	4785	6475.2

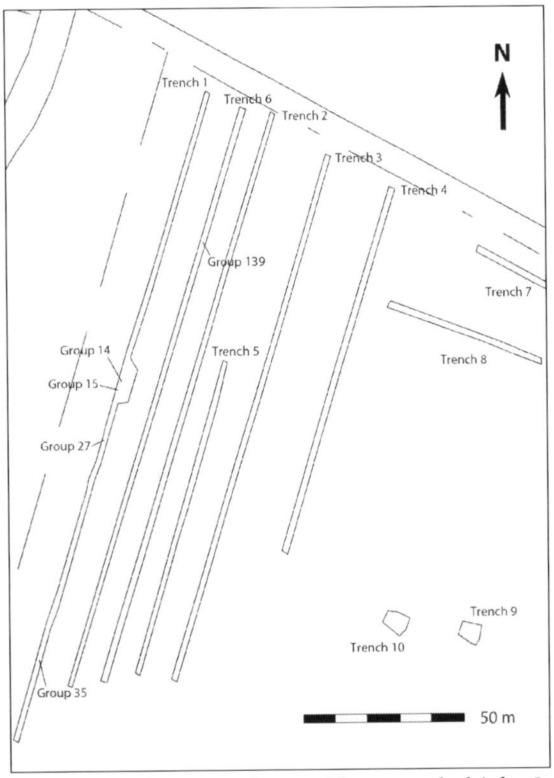

Figure 7.21. Map of the southern area with the five features with animal bones marked (after Lenntorp & Lindell 2000).

Table 7.5. NISP, weight and MNI for the identified material from the selected time period.

Species	NISP	Weight (g)	MNI
Cattle (*Bos taurus*)	218	3050.0	7
Sheep/goat (*Ovis aries/Capra hircus*)	331	1284.9	7
Pig (*Sus scrofa domesticus*)	153	714.6	5
Horse (*Equus caballus*)	5	209.0	1
Dog (*Canis familiaris*)	5	20.6	1
Cat (*Felis catus*)	1	1.4	1
Greylag goose (*Anser anser*)	1	2.1	1
Unspecified dabbling duck (Anas)	1	1.5	-
Cod (*Gadus morhua*)	30	10.3	1
European perch (*Perca fluviatilis*)	2	0.4	1
European plaice (*Pleuronectes platessa*)	1	0.1	1
Garfish (*Belone belone*)	3	0.3	1
Northern pike (*Esox lucius*)	3	0.1	1
Unspecified carp fish (Cyprinidae)	3	0.3	-
Unspecified codfish (Gadidae)	1	0.1	-
Total:	**758**	**5295.7**	

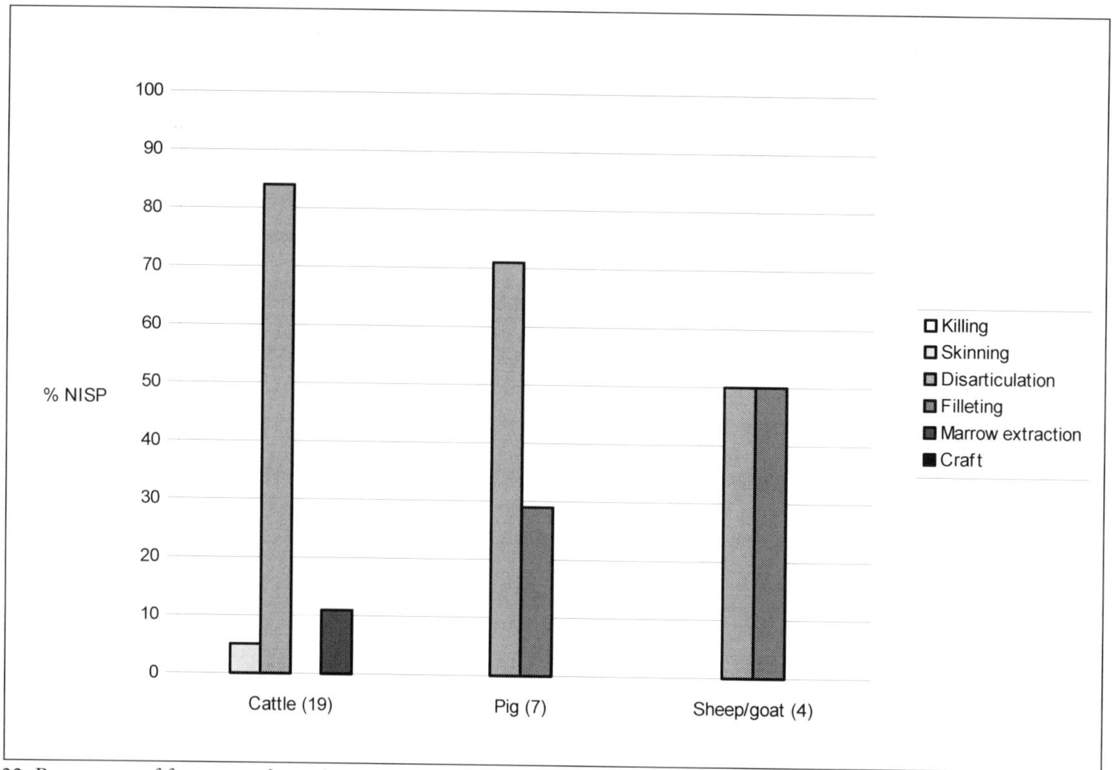

Figure 7.22. Percentage of fragments from the major species in the southern trench material showing marks belonging to the different animal processing categories.

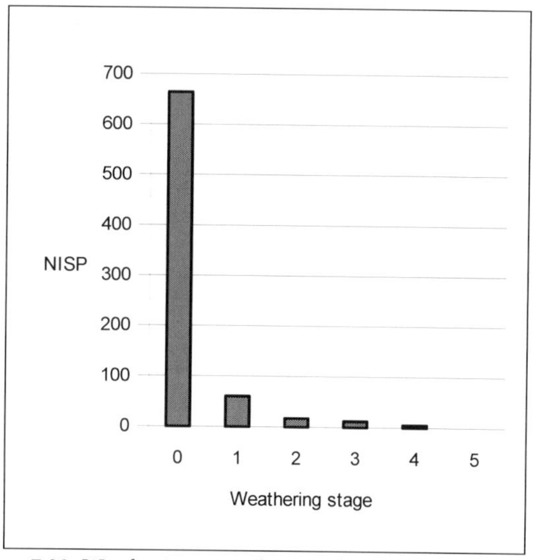

Figure 7.23. Weathering stage diagram for the southern trench material (0: no cracking or flaking; 1: cracking parallel to fibre structure; 2: flaking of outer surface; 3: fibrous texture; 4: coarsely fibrous and rough surface, inner cavities visible; 5: bone falling apart).

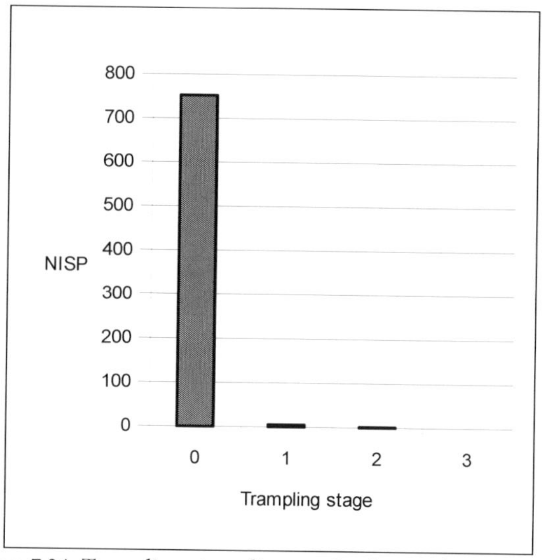

Figure 7.24. Trampling stage diagram for the southern trench material (0: no marks; 1: few, but clear marks; 2: several and clear marks; 3: most of the surface is covered with marks).

less dominating than in the weapon deposit area (table 7.5 and figures 7.26 and 7.27).

The element distribution shown in figure 7.28 and 7.29 is less extreme than that for the weapon deposit area. There are again many loose teeth in the assemblage, but less skull and especially mandible fragments. The high proportion of teeth could, as in the case of the weapon deposit area, be explained by the high fragmentation degree of the material were many crania and mandibles were broken up into unidentifiable fragments while the teeth remained intact.

Nine cattle and pig bone fragments from the southern trench material could be identified to sex (table 7.6). The number of fragments is however too low to say

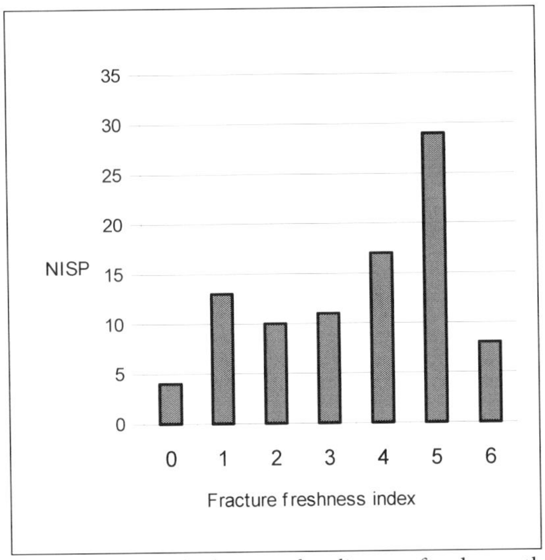

Figure 7.25. Fracture freshness index diagram for the southern trench material (0-2: fresh fracture; 4-6: dry fracture).

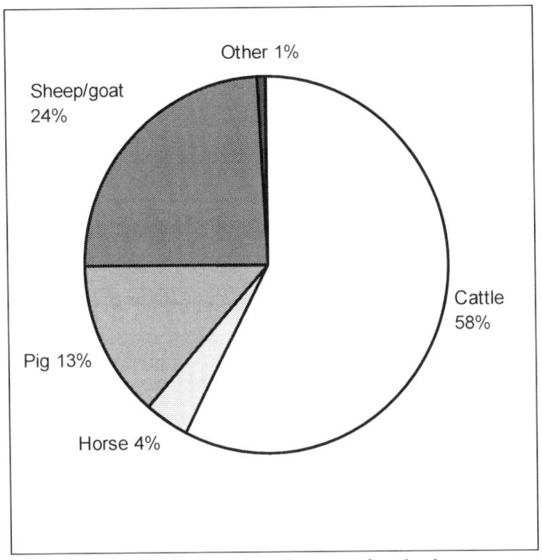

Figure 7.27. Weight-percentage diagram for the bone material from the southern area.

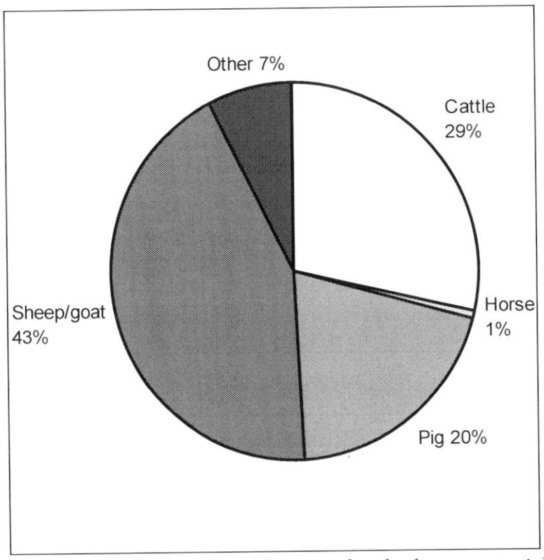

Figure 7.26. NISP-percentage diagram for the bone material from the southern area.

much more than that there seems to have been an even distribution between males and females for pig.

The age at death distribution in the southern assemblage shows a lot more younger individuals than is the case both for the weapon deposit area and the *terp*-materials (figure 7.30). Only 25 percent of the pigs with a determined age were fully grown and both sheep/goat and cattle showed a large number of very young individuals (*infantile*). No horse fragments could be determined to age at death. The age distribution in this assemblage indicates that a large number of animals, especially pigs, were killed at a young age for their meat, while some animals were kept to give wool, milk and as beasts of burden.

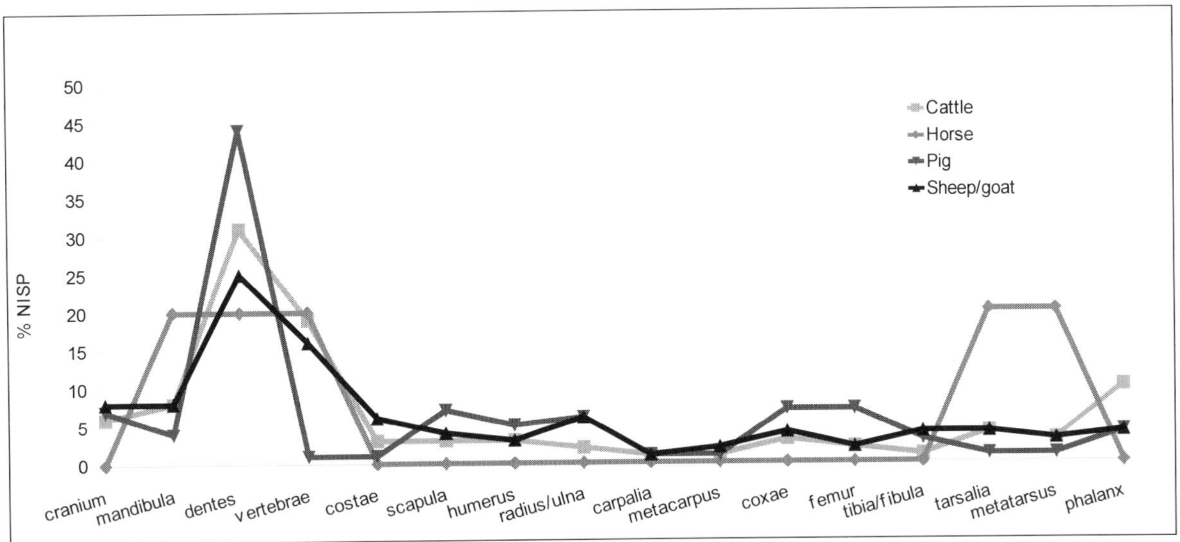

Figure 7.28. Element distribution for the different major species in the southern trench material.

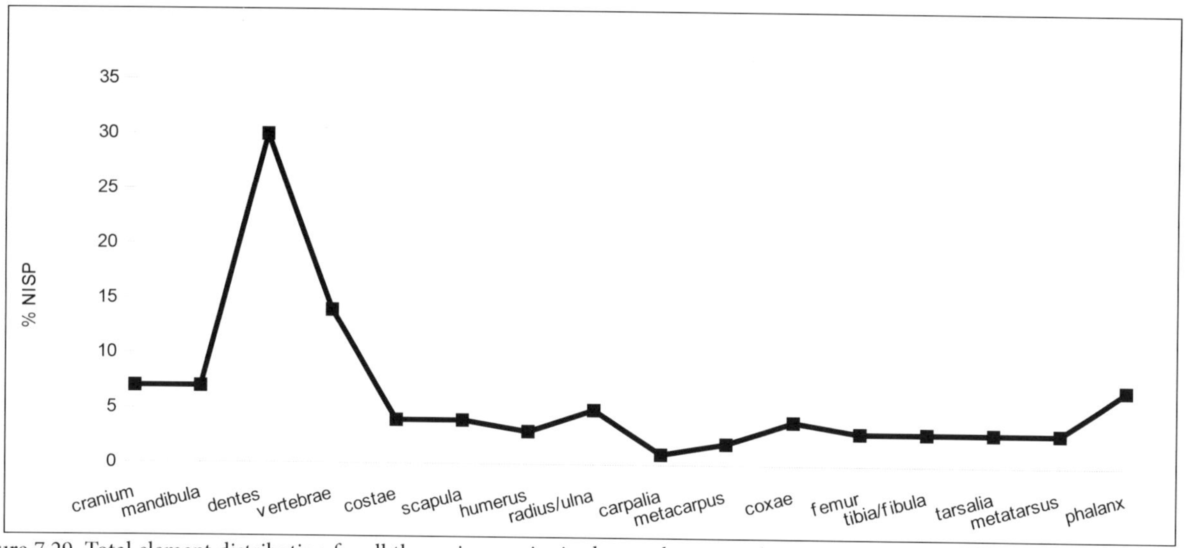

Figure 7.29. Total element distribution for all the major species in the southern trench material.

Table 7.6. Sex-distribution for the southern trench material.

Species	Male (fragments)	Female (fragments)
Cattle (*Bos taurus*)	0	2
Pig (*Sus scrofa domesticus*)	4	3

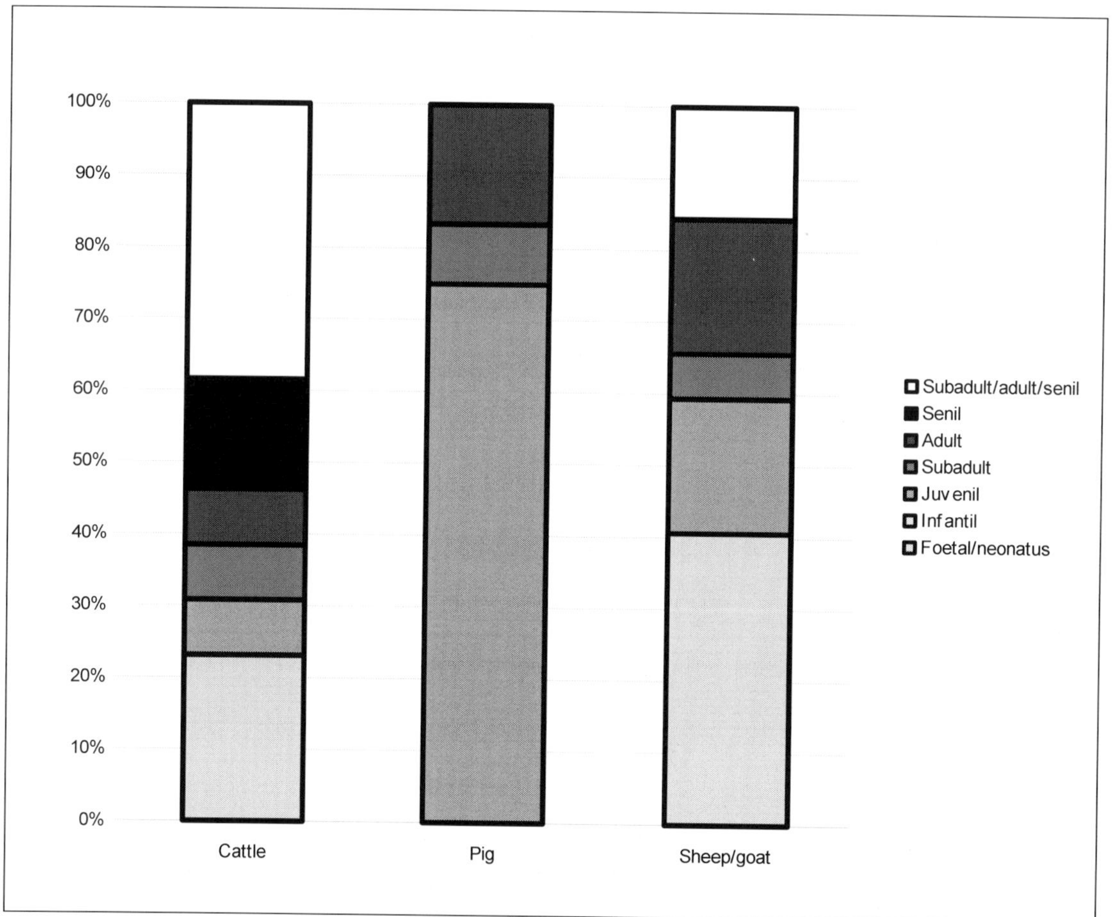

Figure 7.30. Percentage of fragments belonging to different age groups of the major species in the southern trench material (no age assessments could be made for any horse fragments).

7.6 The special assemblage

7.6.1 The archaeological context

The initial hypothesis about speciality of the weapon deposit area was formulated already during the excavation and was based on the archaeological artefacts found, the manner of their deposition in the ground, the spatial relationship with the special house and the finds of human bones. The artefacts, being weapons and parts of armour, are rare finds in Scandinavian Roman Period and Early Medieval settlements and did in themselves attract a lot of attention as well as the question why they would have been deposited in the ground. The manner in which many of these finds have been consciously destroyed before being buried is comparable with offerings described in Roman Period written sources dealing with Central and Northern Europe, and with sites interpreted as such by archaeologists (Ørsnes & Ilkjær 1993: 215-217; Grane 2003: 145-146; Ilkjær 2003).

The close proximity to a house interpreted as having ritual and religious functions during the same time as the weapons were deposited also strengthens the interpretation that this area was something out of the ordinary. The same is the case for the discovery of human bones among the weapons and animal bones, even though only three fragments have been identified and two of these were deposited outside the main concentrations of weapon finds. The relevant question is then whether the bone material in itself shows signs of speciality.

7.6.2 The complete bone assemblage from the weapon deposit

By comparing the analysed bone assemblage from the weapon deposit area with the material from the southern area it becomes clear that several aspects of the two assemblages are the same. From a taphonomic point of view both the processing patterns and the lack of trampling are identical, while the fracture freshness index is almost the same with only slightly more dry fractures for the southern material. The sex distribution is also identical even though the figures for the southern area are based on very few individuals.

The differences between the two assemblages are however larger than the similarities. There are more examples of weathering in the weapon deposit material than in the southern area, but the greatest differences are to be found in the age, species and element distributions. The weapon deposit assemblage mainly contains fully-grown and old animals while the southern trench material contains much more young individuals (see figures 7.20 and 7.30). Cattle totally dominate the weapon deposit material with 73 percent of the animal bone fragments (91 percent of the weight), compared to only 29 percent (58 percent of the weight) in the southern material (see figures 7.16, 7.17, 7.26 and 7.27).

The element distribution for the weapon deposit material shows a lot more cranial (almost twice as many) and mandible (almost four times as many) fragments from cattle than is the case for the southern area. A similar pattern is true for pig while sheep/goat actually is represented by less cranial and mandible fragments in the weapon deposit assemblage than in the southern material (see figures 7.18 and 7.28).

As a conclusion the complete assemblage from the weapon deposit area does on its own have several special characteristics. A large number of bones, mainly skulls, from adult or old cattle were deposited together with the weapons and armour parts in the Early Medieval Period. These deposits of animals were substantial with at least 31 individual cattle, in addition to 18 other domestic animals, identified in the analysed assemblage, which included only 20 percent of the total bone material.

7.6.3 The variation between the portions of adjacent squares

To investigate whether there was any variation within the weapon deposit area itself the three portions of adjacent squares chosen for analysis were compared with each other and with the four single squares situated in the middle of the excavated area.

In most respects the three analysed portions and the four central squares were similar (figure 7.31).

One of the differences observed is that the northern portion had relatively less cattle bones then the other portions. The cattle bones found in that portion also include less cranial and mandible fragments than those from the other portions (figure 7.32). The element distribution for cattle from the northern portion was in this respect more similar to that for the southern area (compare with figure 7.28).

This divergence is interesting but at the same time difficult to explain. One possibility is that the bone material from the northern portion, or at least a part of it represents different types of deposits than the rest of the area.

A final discrepancy between the different portions is that the southern one had a much higher medium fragment weight, 14.1 g compared to 2.8 g for the western portion, 3.4 g for the northern portion and 6.9 g for the central squares. This is also reflected in that the southern portion has a lot less loose teeth than the other parts. Exactly what the reasons are for the lower fragmentation degree for the southern portion of adjacent squares is unclear. The bone material from the southern portion shows less weathering than the western portion and the central squares, but almost exactly as much as the northern portion. The southern portion is situated partly outside the concentrations of weapon deposits (see figure 7.6) and this would support the hypothesis that the high fragmentation of the bone material in many

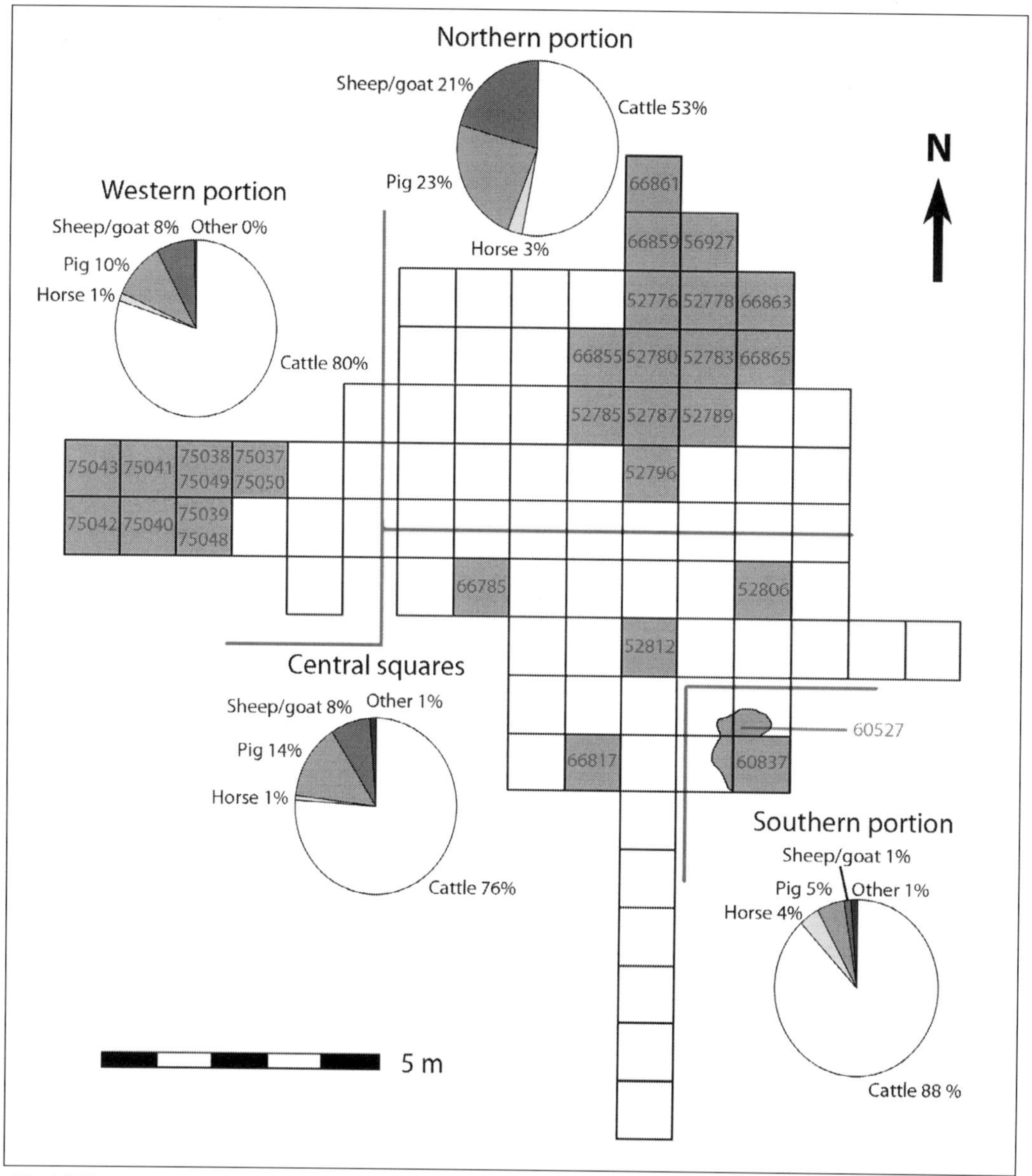

Figure 7.31. The NISP-percentages for the major domestic species in the three portions and central squares of the weapon deposit area in Uppåkra.

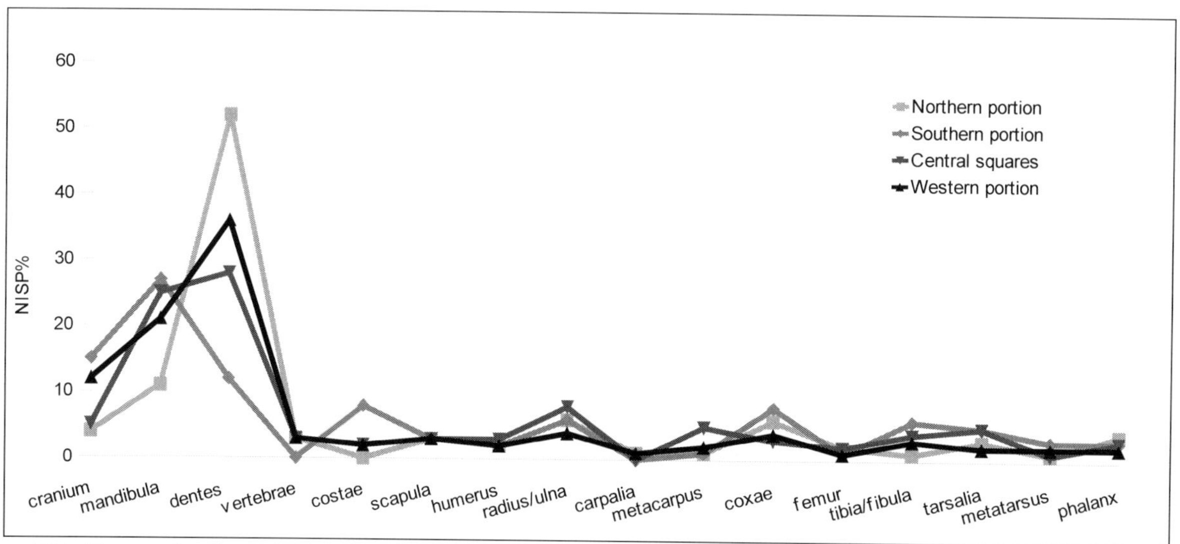

Figure 7.32. Element distribution for cattle from the four different portions of adjacent squares.

of the squares is the result of disturbance from later weapon deposits. The western portion however also has a high fragmentation degree and a high percentage of weathering, but is situated well outside of the weapon concentration. Here the high number of small fragments might instead be explained by the fact that some of the squares in this area were sieved, and that the upper part of the layer was affected by later agricultural activity that damaged the bone material.

7.6.3 The individual squares

A comparison between all the individual analysed squares was also attempted but gave no reliable results since most of the squares contain only a few hundred grams of bone material of which only a part could be identified to species. This resulted in crude diagrams and situations where a different identification of only a few bone fragments would dramatically change the observed patterns.

7.6.4 Conclusion

The final conclusion is that the bone material from the weapon deposit area is special in itself due to the unusually large amount of deposited bones from cattle, mainly in the form of cranial and mandible fragments from adult or old individuals. The bone material does therefore not contradict the interpretation of the weapon deposits as being the result of special activities (for example as has been suggested, offerings). The dating of the animal bones shows that these were deposited from around 400 AD and continued parallel with the weapon deposits, but that the majority of them probably were made during a limited time period (ca 400-600 AD).

8. Interpretation and explanation of activity

After the detailed presentations of the four case studies and the identification of special deposits within these materials, this chapter deals with the interpretations and explanations of the activities that lie behind these deposits. The methods described up to this point are meant to be universal in the sense that they are built on patterns of recognition basically independent from the contexts of the observed deposits. This is however only one step in a process that should not end with the identification of special deposits, but continue with the aim of interpreting and understanding the reasons behind these deviations from what we would call normal refuse (see section 3.1.5). Many researchers have pointed out the difficulties involved in making such interpretations, especially when dealing with rituals of past societies. As an example Christopher Hawkes placed religious institutions and spiritual life as the fourth and most difficult aspect of human behaviour to study as an archaeologist in his "ladder of reliability", after such things as technology, economy and even social structures and politics (Hawkes 1954: 161-162).

> *"Paleolithic art clearly has much to do with institutions of hunting-magic and, in the case of the so-called 'Venuses', with expressions of desire for human fertility. Grave goods, again, indicate a belief that the dead need material supplies or equipment, as though still alive. But how much further can one go than that?"*
> (Hawkes 1954: 162)

More than overestimating the difficulties in studying these things I think Hawkes underestimated the difficulties in understanding other aspects of past societies such as politics, economy and social structures. They may seem easier to grasp than spiritual life and religious institutions, but unfortunately only if we assume them to correspond more or less directly with our own society and hence by, often uncritically, using contemporary analogies. It should therefore be no more acceptable to refuse to interpret subjects of a spiritual nature by referring to the difficulty, than many other subjects regularly tackled by archaeologists, seemingly without much fear. Besides, if a deposit is only identified as ritual (or even worse special), without a more thorough interpretation, what use is this label really to us?

> *"After all, the question Is this ritual? can only be answered Yes or No. How much does it tell us about the actual nature and context of these deposits, or the societies who made them?"*
> (Hill 1996: 101)

For these reasons further steps of interpretation and understanding will be attempted for the special deposits identified in the four analysed bone materials. Since these steps are intimately connected with the context in which the deposits appear, the discussions in this chapter are not universal but mainly relevant for the circumstances presented by the geographical area, time period and society of the specific deposits. The manner in which these steps are performed could however be used as examples of how to approach the interpretation of deposits in general.

It is important to note that by definition a certain degree of speculation will be part of any exercise such as this. The interpretations and explanations presented in this chapter are all based on the archaeological material and on various analogies that I consider relevant for the specific situation. I regard these as the most likely interpretations, but this does not make them the only valid ones.

In table 8.1 all the identified special features from the four bone materials in this thesis are presented. In the following sections these will be discussed and, if possible, interpreted.

8.1 Articulated body parts

Articulated body parts from animals found in archaeological contexts are often associated with ritual activities, such as for example the partial offering of butchered animals. The reasoning behind such interpretations is mainly that a large unprocessed part of an animal would not have been wasted without a reason. For this line of thought to be logical the articulated body part however had to contain at least some meat or other form of resource (such as skin or sinews), and it also has to be shown that this specific part was not regularly discarded during normal butchery practice. Even if these circumstances are supported there is still a possibility that the articulated body part was discarded for some other reason, for example

Table 8.1. All the identified special deposits from the four bone materials.

Site	Feature	Feature type	Dating	Deposit type	Section
Dongjum	1022	Ditch	VMEA	Articulated body parts (horse)	8.1.1
Leeuwarden	55	Layer	VMEC	Articulated body parts (sheep)	8.1.2
Leeuwarden	146	Layer	VMEB	Articulated body parts (sheep/goat)	8.1.2
Leeuwarden	246	Layer	VMEC	Articulated body parts (sheep/goat)	8.1.2
Leeuwarden	127	Pit/ditch	VMEB	Selected elements (cattle and sheep astragali)	8.2.1
Leeuwarden	828	Layer	VMEB	Skulls/crania (horse)	8.2.2
Leeuwarden	1591	Layer	VMEB	Skulls/crania (sheep)	8.2.2
Midlaren	3	Well	ROML-VMEA	Skulls/crania (cattle)	8.2.2
Midlaren	51	Pit	VMEA	Skulls/crania (cattle)	8.2.2
Dongjum	1023	Ditch	VMEA	Selected elements (cattle and sheep cranial fragments)	8.2.3
Dongjum	1088	Well	VMEC	Age structure (infant and juvenile cattle and sheep/goat)	8.3.1
Leeuwarden	610	Sunken hut	VMEB	Selected species (geese and ducks)	8.4.1
Leeuwarden	611	Sunken hut	VMEB	Selected species (geese and ducks)	8.4.1
Leeuwarden	85	Pit	VMEC	Human bones (various)	8.5.1
Leeuwarden	90	Well	VMEC	Selected elements (metapodials)	8.5.1
Leeuwarden	182	Ditch	VMEC	Human bones (vertebra)	8.5.1
Leeuwarden	258	Pit	VMEC	Human bones (various)	8.5.1
Leeuwarden	555	Layer	VMEC	Human bones (two vertebra)	8.5.2
Midlaren	22	Pit	VMEC	Cremation pit (various animals)	8.6.1
Uppåkra	96	Layer	ROMY-VMEC	Bone layer (various animals but a majority of skulls/crania from fully grown or old cattle)	8.7.1

because it was spoiled or, as in the following example from Dongjum, left for the dogs to gnaw on. A thorough taphonomic analysis is therefore necessary to clarify that such factors were not involved.

8.1.1 Dongjum: Feature 1022 (left ankle of horse)

The horse ankle from a ditch system in Dongjum is probably not the remains of ritual activity. The ankle was heavily gnawed but not weathered, indicating that it was left for the dogs. Most likely the ankle, being a part with little meat, was given to the dogs after butchery and later deposited in the ditch together with various other refuse, among which the single pelvis fragment from a cattle foetus, in all probability a coincidence. It is also possible that the ankle was found by dogs some time after butchery and fractured further by their gnawing. This could be indicated by the dry fracture of the proximal part of the tibia. In that case the bones were deposited in the ditch later, for example when the area was cleaned.

Interpretation

I interpret this deposit as a piece of bone with little meat, either given to the dogs immediately after the slaughtering of the horse or as butchery refuse that was found or unearthed by the dogs later.

8.1.2 Leeuwarden: Features 55, 146 and 246 (front parts of sheep/goats)

These three deposits of articulated parts of sheep/goats are most likely examples of the same type of activity since there are obvious similarities between them. All three deposits included articulated front parts of the bodies from one, or in the case of feature 55, two sheep/goats and were all found in layers that could be dated to the Merovingian or Carolingian Periods. The pattern of deposited body parts is however not entirely similar. In all cases parts of the front of the bodies were deposited, but in two cases it was the spine and ribs together with the cranium, while in the other two it was the spine and ribs together with parts of the front legs.

The two single sheep/goats from feature 146 and 246 were both killed at roughly the same age, somewhere between 9 and 24 months, while the two sheep in feature 55 were subadult (6-24 months) and adult (older than 24 months) respectively.

The animal processing marks on the sheep/goat from feature 246 indicate that the animal was divided up into large parts with a cleaver or axe, but not processed any further. The other animals lacked butchery marks altogether, as well as signs of any other taphonomic process.

All these observations indicate a depositional pattern where several sheep/goats were coarsely divided up and parts of the front of the animals were deposited and quickly covered with soil.

The selected parts, the repetition and the coarse partitioning of the small bodies make it difficult to imagine that the deposits were made up of normal butchery refuse or were the result of carcass-removal.

Ethnographic descriptions

I have found no ethnographic texts that describe this type of phenomenon regarding sheep or goats. The practice of partial sacrifice where only a small part of an animal is given to the supernatural power and the remainder of the body is divided up and eaten, is however well described in various ethnographic sources and seems to have existed in many societies around the world (see for example Van Baal 1976: 161, 169-170; Lambert 1993: 294-295).

Archaeological parallels

There are remarkably few archaeological finds of sheep or goats on settlements in Northern Europe in the Early Medieval Period which have been interpreted as ritual deposits. This could be because these species for some reason were not considered suitable for this type of ritual use. It is however equally possible that the lack of written sources describing the use of these species in rituals have lulled the archaeologists' senses regarding deposits containing these animals. That sheep and goats really did play an important role in rituals in the society, although not necessarily in settlements, is indicated by the relatively high frequency of the species in offering bogs in Northern Europe (Carlie 2004: 120).

Among the settlement finds are the two deposits of sheep from a long house in the Early Medieval settlement of Bålsta in the Province of Uppland (Sweden). In two different post holes the front and rear parts of a young sheep, less than 2½ years old, were found. The front part was made up of the cranium, vertebrae and parts of the front legs while the rear part was made up of the femurs and other parts of the hind legs. The finds have been interpreted as some form of initiation offerings in connection with the building of the house (Carlie 2004: 120, 301).

An additional example comes from the Icelandic Early Medieval hall in Hofstaðir where the deposit of a complete female sheep has been interpreted as a closing sacrifice possibly connected with the ending of the pagan era and the introduction of Christianity (Lucas & McGovern 2008: 25)

Although only a few Early Medieval ritual deposits of sheep/goat in settlements have been identified by archaeologists in Northern Europe, a large number of deposits of other species interpreted as partial offerings have been found (see for example Klindt-Jensen 1957: 83-85; Müller-Wille 2002: 156; Carlie 2004: 127; Hamerow 2006: 4-7).

There are also a number of comparable deposits from earlier time periods, especially from the Roman Period. Examples include half a sheep found together with a human skull at the Churchill Hospital in Oxford (Great Britain) and the two sheep or goat skeletons, one complete and one partial, found in a Romano-British shrine dated to the late first or early second century AD in Haddenham, Cambridgeshire (Great Britain) (Merrifield 1987: 35, 46).

Historical sources

As was mentioned earlier very few written sources from Northern Europe mention sheep or goats in connection with rituals, and the species are equally seldom mentioned in the texts dealing with religious myths. This is somewhat surprising considering the obvious economic importance of these animals in the Early Medieval society.

In *Njal's Saga* a direct connection between goatskins and magic rituals is made on two occasions, first in chapter 12 when Swan adorns a goatskin to conjure a thick fog in order to discourage a search party, and then in Flosi's dream in chapter 132.

In chapter 14 of the *Saga of Håkon Góði* from *Heimskringla* it is mentioned that goats and sheep traditionally were slaughtered during offering feasts together with horses, but it is not stated whether the animals were actually offered or only eaten as a part of the feast.

One source that definitely describes the use of sheep in an offering ritual is Ibn Fadlān's *Kitāb's Rūsiyyah.* According to Ibn Fadlān it was a custom among the Rūs to tie the skulls of cows and sheep to poles as an offering of thanksgiving to the gods after a successful trade transaction. The rest of the meat from the animals was shared and eaten (see section 3.2.5). This is also a good example of the practice of partial offerings, which is highly relevant for these deposits. In partial sacrifices some of the offered objects are given to the gods and the remaining part is kept, eaten, or shared with other people. This type of offering is mentioned in several historical sources from Northern Europe, as well as in texts from other parts of the world and in some ethnographic sources (see for example Van Straten 1995: 3; Näsström 2001: 244-245; Steinsland 2005: 276). An obvious example of the practice, although of unspecified animals, comes from *Hrafnkel's Saga*, an Icelanders' saga probably written during the 13th century, but describing events that are supposed to have taken place in the 10th century:

"But when Hrafnkell had hallowed for himself the land of Aðalból, he held a great sacrificial feast, and a great temple, too, he reared up there. Hrafnkell loved no other god before Frey, and to him he made offerings of all the best things he had, going half-shares."
(Hrafnkel's Saga, chapter 2, translation by Coles 1882).

Interpretation

I interpret the three deposits as the remains of partial offerings in which a quantity of the front parts was given to some supernatural power, but the majority of the meat of the animals was eaten. It is possible that this offering ritual was repeated and performed at a certain time of the year, since two of the deposits include animals of the same age. The two animals in the third deposit were older and do therefore neither confirm nor contradict this. The age assessment is unfortunately not detailed enough to validate this possibility, or to give an indication of which season such a ritual could have been performed in.

8.2 Selected elements

The selection of certain bone elements from animals, and then especially skulls and crania, seems to be a recurring feature in various types of ritual activities around the world in different time periods. Several special deposits from the four case studies belong to this category and, with the exception of the astragali deposit from the Leeuwarden *terp*, they are all made up of crania and complete skulls.

8.2.1 Leeuwarden: Feature 127 (four left cattle and sheep astragali)

Four astragali from the left side of the body, three from cattle and one from sheep, made up the whole assemblage deposited in the bottom portion of a pit. Two of the cattle astragali show signs of light weathering while the third cattle bone displays cut marks possibly connected with the first stage of bone-craft. Several examples of worked bone were found during the excavation in Leeuwarden, including an Early Medieval die made from a piece of a sheep/goat metatarsus, and it is possible that the four astragali were discarded raw material from bone craft. The existence of only superficial modifications, and the unlikely coincidence that four complete astragali from the same side happen to end up in a single feature without any other material, however, make other interpretations than ordinary refuse likely, not the least since numerous sources show that astragali were used for activities such as games and divination.

Ethnographic descriptions

The use of bones for games and ritual purposes such as divination has been recorded by ethnographers at various places, for example in Southern Africa (Chidester *et al.* 1997: 21, 354, 366). The form of the astragali lends it well both as a gaming piece and as an object for producing random results and it is considered the direct precursor of the die (Holmgren 2004: 212). Astragali are also used in games of dexterity called knucklebones, jacks, fivebones or osselets (see figure 8.1). Such and similar styled games have been played in Europe at least since classical times and sets of five astragali can still be bought in game shops, although now made of plastic.

While knucklebones and comparable games are usually played with the smaller astragali from sheep, goats or calves, other types of games using both sheep and cattle astragali are played in various parts of Asia, for example in Turkey, Turkmenistan and Mongolia. In some of these hazard-like games the larger cattle astragali (sometimes fitted with a lead filling) are used to hit the smaller sheep bone, analogous with how the metal ball is used in some marble games. In the Turkish game the winner receives the astragali from the looser, also analogous with many marble game rules. In this way the astragali acquire the function of abstract money (Holmgren 2004: 214-215). Ethnographic observations of the Mongolian style game also, interestingly show that it has a strong ritual component and that the astragali besides being gaming pieces, also symbolises fertility. This link between games and ritual activity is not exclusive to Mongolia but is commonly seen between games of chance, divination and the casting of lots/throwing of bones (Pegg 2001: 233; Holmgren 2004: 213, 217).

Archaeological parallels

The archaeological record corresponds well with the ethnographic sources and also shows a long and widespread use of astragali in games and divination, at least in Eurasia. Modified astragali have been found at sites in Europe as early as the Neolithic. Finds of astragali, both modified and unaltered, in various non-refuse contexts became frequent around the Mediterranean during the Bronze Age and have often been interpreted as gaming pieces (Holmgren 2004: 213). The gaming interpretation is also supported through the finds of several figurines from the Classical Period portraying people that seem to be playing with astragali.

The astragali are found in a range of different features, including many with strong ritual connections such as graves and temples (Foster 1984: 79; Gilmour 1997; Bookidis & Stroud 1997: 159-160, 211). Astragali with inscriptions, for example the name of a god, have been interpreted as being used for divination, as charms or to give luck in games (Last 2000; Bar-Oz 2001).

Finds of astragali in various archaeological contexts exist around the world. This can to some extent be explained by the compactness of the bone element that make it less vulnerable to deteriorating taphonomic processes, but in many occasions the astragali are

Figure 8.1. A woman playing knucklebones (by J-B-S. Chardin 1734).

modified, decorated, worn by use, found in unusual contexts or in the form of isolated catches. This type of special deposit is however not very common in Northern Europe during the Early Middle Ages, something that seems strange considering the region's earlier connections with the Mediterranean cultural spheres, the distribution of knucklebone games during later historical periods and the rather frequent mentioning of the casting of lots in Early Medieval written sources.

One reason for the relative scarcity of such deposits could be that they are often not considered odd enough to be noted in excavation reports. Astragali often appear in great numbers at sites where the preservation condition for bone is decent. It would therefore require either heavily modified/worn astragali, or unusual find circumstances for an excavator to lift an eyebrow. I have personally observed deposits in pits at sites from the Roman and Early Medieval Periods in Southern Sweden exclusively made up of several sheep/goat and/or cattle astragali. These deposits were noted during the excavation, but were hardly mentioned in the final report, since no other unusual circumstances were present.

A number of Early Medieval deposits, especially from cremation graves have however been excavated in the terp-region of the Netherlands, Northern Germany, England and Sweden (Knol 1987; Göthberg et al. 2006: 24-25, 57).

In the Netherlands astragali from sheep have been found in five different urn cremations from the *terp* Hogebeintum dated to the 4th-7th centuries. The urns contained 12, 10, 9, 8 and 4 astragali respectively. In four of the urns the astragali were burned with the body and in several cases decoration and use-wear could be observed.

An urn dated to the 7th century with a single astragalus has also been found at the early medieval *terp* Driesumerterp.

In addition to the graves, two unstratified finds of astragali have been found in the area. Five knucklebones were unearthed at the *terp* Kalma near Stiens and a single one dated to the Middle Ages and showing signs of use-wear was found at Foudgum. As can be seen these finds solely contain sheep bones. Exceptions to this are the two cattle astragali with decoration that have been found at the *terps* Midlum and Holwerd.

If we look outside of the Netherlands an urn from the Migration period cemetery at Westerwanna in Niedersachsen, Northern Germany, contained five bone counters and some astragali fragments.

Another urn from an Early Medieval cemetery at Tating(-Esing) in Schleswig-Holstein, Germany contained the cremated remains of a child together with at least 15 knucklebones of which several were decorated or pierced.

An urn from Caistor-by-Norwich in Norfolk, England contained 33 bone counters and at least 35 knucklebones of which all but one were from sheep. The sole exception was a knucklebone from roe deer (*Capreolus capreolus*) with a runic inscription. Two other urns at the same site contained one astragalus each.

The deposits mentioned so far come almost exclusively from grave urns. An exception to this is a find from Gamla Uppsala and Vaksala in the province of Uppland, Sweden. Here three complete cattle astragali were found together as a deposit in a system of pits that had been in use during the Roman Period and Early Middle Ages (Göthberg *et al.* 2006: 24-25, 57).

Historical sources

The use of astragali as a form of dice or playing pieces is well documented in classical sources (Smith 1859: 1095-1096). Knucklebones, or a similar type of game, is mentioned as early as in *the Iliad* and *the Odyssey* (Grunfeld 1975: 162). In the Iliad (23.87-88) it is for example said that Patroclus as a youth killed another child *"in wrath over a game of knuckle-bones"* (Jebb 1898: 55-56).

In his *Histories*, Herodotus mentions the game of knucklebones and claims that the Lydians of Western Anatolia were its inventors (together with every other game except draughts):

> *"In the days of Atys, the son of Manes, there was great scarcity through the whole land of Lydia. For some time the Lydians bore the affliction patiently, but finding that it did not pass away, they set to work to devise remedies for the evil. Various expedients were discovered by various persons; dice, and knuckle-bones, and ball, and all such games were invented, except tables, the invention of which they do not claim as theirs. The plan adopted against the famine was to engage in games one day so entirely as not to feel any craving for food, and the next day to eat and abstain from games."*
> (Herodotus' The Histories, Book I, translation by Rawlingson 1858)

In the Hellenistic epic *Argonautica* by Apollonius Rhodius a game with "golden dice" is played between the gods Eros and Ganymedes where the winner receives the play pieces from the looser in a similar way as in some games of marbles:

> *"And those twain were sporting with golden dice, as youths alike in habits will. Now the one, even greedy Eros, held the palm of his left hand quite full already beneath his breast as he stood there upright; and a sweet blush was mantling, on the skin of his cheeks; but the other sat crouching near in moody silence, and he held two dice, casting one forth upon the other, where he sat, and he was angered at the loud laughter of Eros. Now when he had lost these at once as well as the first, away he went with empty hands, helpless, and he was not ware of Cypris as she drew nigh."*
> (The Argonautica, Book III, translation by Coleridge 1889)

Several different games using sheep or goat astragali are described in later Roman sources. One game called Tropa seems to have mainly been a test of dexterity since the astragali were to be thrown into a box, jar or a hole in the ground. This game is for example mentioned in several epigrams by the poet Marcus Valerius Martialis, mainly as an allegory.

A game of chance sometimes called Tali (from talus, latin for ankle, as well as an alternative name for astragalus) is also described in numerous sources. Suetonius even quotes a letter by Augustus in which he suggests an alternative system of scoring in the game that he obviously used to play. Cicero also mentions the game a couple of times as an example in his book about divination (*De Divinatione*). It is certainly no coincidence that Cicero uses Tali as an example in a text about divination, since the game of chance and foretelling using the casting of bones or lots conceptually lie close to each other.

While the mentioning of astragali used in games are common in the Classic Mediterranean sources, they are suspiciously absent from the sources dealing with Northern Europe, both from the Roman Period and from the Early Middle Ages. Games are mentioned occasionally, sometimes including the use of dice, but there are never any direct references to astragali. It is still likely that such games were practised in Northern Europe at this time, especially considering that they apparently were widely spread at the start of the Early Modern Period, as can be seen for example in 16th century paintings. The technique called "casting the bones" or "casting of lots" in which a number of objects such as bones or pieces of wood are thrown, seemingly at random, and where their positions are used to divine the future, or to make decisions, is however frequently mentioned in the written sources of the region (Spence 1945: 102; Näsström 2001: 242-244). One of many examples is *Vita Anskarii* where a man named Lie seeks out a soothsayer:

> *"Thereupon, following the local custom, he consulted a soothsayer and asked him to find out by the casting of lots which god Lie had offended and to explain how Lie might appease him. After performing all the customary ceremonies, the soothsayer said that all their gods were well disposed towards him, but that the God of the Christians was much incensed against him."*
> (Vita Anskarii, chapter XVIII, translation by Robinson 1921)

In none of the Northern European written sources the use of specifically astragali is mentioned in connection with the casting of lots. In the cases that the lots are at all described they are rather made of wood as in Tacitus'

Germania. The fact that divination rituals in neighbouring areas included the use of bones and dice does however still make this interpretation possible (Smith 1859: 1051-1052; Holmgren 2004: 213).

Interpretation

The various analogies present several possible interpretations for this interesting deposit. The relatively untouched state of the bones together with the obvious selection of left sided bones makes interpretations such as refuse from butchery, cooking, eating or even bone craft unlikely. Since an equal number of left and right astragali exists in all mammals it is only 12 percent chance that all the bones would come from only one side, if they where chosen randomly. With the tradition of astragali used in games and, most likely, divination and casting of lots in mind it is possible to imagine that the bones from feature 127 form a cache of game pieces or fortune telling bones that were discarded, hidden, or perhaps offered to supernatural beings.

In reality it is difficult both archaeologically but also conceptually to separate between games and ritual activities involving fortune telling, decision based on fate and a general fascination for chance. As several ethnographic and historical sources have shown, games often have a strong ritual component and the playing of games in important situations by gods, heroes and supernatural beings is a recurring theme in myths all over the world. The strong ritual ingredient is still evident in our modern society as well were games, gambling and sport are riddled with traditions, ritualistic behaviour and not the least much superstition.

If the four astragali were actually a set of game pieces, or part thereof, it was probably not meant for playing a game similar to knucklebones since the cattle bones should be too large for this dexterous type of game play. Other types of games such as the ethnographicly described boule/boccia or marble-like games from Asia would suit the composition of the set better. In these games the larger cattle astragali are thrown against a target in the form of a smaller sheep or goat astragalus. Despite the likeness of the bone composition there is of course a virtually endless number of possible games, or divination-like activities that the pieces could have been used for.

The scarcity of this type of deposit makes it extremely difficult to interpret it any further than this without venturing into outright speculation. It is however undoubtedly an interesting type of deposit and I am certain that many more similar ones lie hidden among already excavated osteological materials. The deposits' unassuming character makes them easy to overlook and a survey of old materials as well as further investigation of various ethnographic and historical sources would be an interesting future project.

8.2.2 Leeuwarden: Features 828 and 1591, Midlaren: Features 3 and 51 (skulls and crania)

The category of special deposits that are made up of more or less complete skulls and crania is represented in the case studies by four features from the sites of Leeuwarden and Midlaren. These four features include seven or eight different skulls and crania, five or six from cattle, one from horse and one from sheep. It was already suggested in the description of the special deposits from Leeuwarden (see section 5.4.4) that skulls from the domestic animals were regularly processed further after their initial disarticulation during the butchery process on that particular site. Since the horse cranium and sheep skull found at the settlement deviated from this practice they were interpreted as special deposits. In the Midlaren settlement the situation was more complicated due to the worse preservation conditions for bones on the site. In Midlaren there is a paradoxical situation with a majority of the bone material belonging to features that are identified as special. In such circumstances the speciality of the deposits is identified in comparison with the situation on other sites situated closely, both geographically and in time (see section 6.4.3 for further discussion).

The question that remains to be answered is what activities these four deposits represent, and if they can be seen as examples of the same type of deposit. There are several differences between the deposits, including the choice of species, the feature types and whether the mandibles were present or not. Common for the four deposits is that the animal skulls and crania were buried completely, without any signs of processing beyond disarticulation and without any other refuse material in the same feature (or in the case of the sheep skull from Leeuwarden, without any refuse material in the same filling). Moreover at least the horse cranium from Leeuwarden seems to have been carefully placed in the feature, indicating that the deposit had a symbolic meaning given that the cranium was placed in the same way as a living horse holds its head (see figure 5.16). At this point it should be noted that one more comparable deposit was found at the Leeuwarden site in the form of a horse cranium without mandibles placed at the bottom of a ditch, but unfortunately this feature could not be dated. Its existence does however show that more deposits of the same kind have existed within the settlement, even though we cannot know for certain that this one came from an Early Medieval feature.

In order to interpret the skull and crania deposits from the two sites we have to search for parallels and practices involving complete animal skulls or crania with no obvious functional use. Such deposits can be found both at other Early Medieval archaeological sites, in various historical records and in some ethnographic descriptions.

Ethnographic descriptions

The use of skulls and crania in rituals is both a very old and widespread phenomenon. The strong symbolism of the head of animals can be seen in cultures all over the world. Examples include the ritual hunt and subsequent use of animal skulls among various people in Southern Africa, the display of heads during the animal sacrifices in the *tivah*, that are a part of the reburial ceremony among the Ngaju Dayak people on Borneo, and the ritual bear killing ceremonies that were performed among several circumpolar societies such as the Sami and Ainu (Hallowell 1926; Pager 1982; Kimura 1999; Hertz 2004).

These specific rituals are however not very suitable as analogies for the skulls and crania from Leeuwarden and Midlaren, since they show many dissimilarities with the situation here. The ritual hunt and the bear killing ceremonies were performed by people that considered the wild animal in question to be holy and the ritual is firmly connected to the mindset of communities which hunting is, or was, an important part of the subsistence strategy. The animal sacrifice during the tivah is more interesting for us as an example both of the symbolic importance of displaying an offered creature's head, and of a partial offering ritual (Hertz 2004: 204-206). It is however also a part of a very specific reburial ceremony whilst the skulls and crania from Leeuwarden and Midlaren show no indication of being connected to human burials or funerary rites.

A couple of other ethnographic examples could however be of more interest. A custom of placing horse crania in the foundations of houses has been observed in modern times in Denmark and Southern Sweden. This tradition has been interpreted as being of a protective magical character (Carlie 2004: 124). This could be a valid explanation for many finds of crania in medieval Scandinavian houses, but does not work as well for the Dutch finds since these are primarily found in pits, wells, layers and ditches. It is however still possible that the deposits were meant to be of a protective nature, only not confined to a building but possibly instead for the whole farm or even settlement.

Another ethnographic example comes from the Zulu people in Southern Africa who in connection with the sacrifice of cattle to the shades (ancestor spirits) used to place the skull or horns of the animal on display in the village, either on the hut of the recipient of the ceremony, on the chief's hut or on the fence around the enclosure where the sacrifice was performed. Some of the remaining parts of the animal were offered to the spirits while the rest were eaten by the people of the community during a feast (Lambert 1993: 300, 306-308). This is again a good example of a partial offering that seems to be a common ritual phenomenon both in ancient and in more recent societies.

Archaeological parallels

When searching for archaeological parallels to the skull and crania deposits from Leeuwarden and Midlaren, we should look first at sites situated near the two settlements. A search among published excavations from Drenthe and the *terp*-area however reveal few comparable Early Medieval finds. This does however only really prove that such finds have not often been noted in excavation reports and articles, something that is not surprising since this type of deposit would attract little attention if it was not connected to some other special circumstance, or repeated several times at the same site as was the case in Midlaren.

One site where similar deposits have been found is the *terp*-settlement Feddersen Wierde in Niedersachsen (Germany). Two horse skulls and one cattle skull were found in two pits and a posthole, but only one of the horse deposits was dated to the latest phases of the settlement (around 350-450 AD) and hence the same period as the features from Leeuwarden and Midlaren (Haarnagel 1979: 223-227).

Another example comes from Elisenhof, an Early Medieval *terp*-settlement in Schleswig-Holstein (Germany), where a cattle skull was found near the southern wall of a house (Capelle 1987: 191).

If we widen our search and look at the rest of the Netherlands no Early Medieval sites displaying this type of deposits can be found. There are however quite a few deposits of animal skulls and crania from Roman Period settlements, for example from Heeten (province of Overijssel), Passewaaijse Hogeweg (province of Gelderland), Tiel-Bedrijvenpark (province of Gelderland), Midden-Delftland (province of Zuid-Holland) and Casticum-Oosterbuurt (province of Noord-Holland) (Lauwerier *et al.* 1999: 180; Groot 2008: 150-152). Of the 30 complete or nearly complete cattle, horse, dog, sheep or pig skulls deposited in pits, ditches and wells in Passewaaijse Hogeweg, Maaike Groot interprets 11, most of them horse skulls, as ritual deposits and suggests that they represent partial offerings made to ask the gods for favours, while the remaining parts of the animals were consumed by the inhabitants of the settlement (Groot 2008: 121; 157).

If we venture even further away, this type of deposit does appear during the Early Middle Ages in various places all over Northern Europe (see for example Müller-Wille 1971). In a survey of special deposits in Anglo Saxon settlements in Great Britain, Helena Hamerow concluded that almost a third of the deposits identified as special were made up of skulls or crania of various animals, mainly cattle, horses and dogs. 12 such deposits were identified at 8 different sites (Mucking, Pennyland, Sutton Courtenay, West Stow, Wharram Percy South Manor, Yarnton, Eye Kettleby and Yeavering). Seven of them were found in sunken huts, three in pits and two in houses (Hamerow 2006: 4-8). Comparable deposits dated to earlier time periods have been excavated in Great Britain, for example the numerous finds of skulls, mainly of cattle but also of dogs, horses and sheep, that have been found in wells in England dating to the later part of the Roman Period (Merrifield 1987: 47-48).

In Scandinavia a number of deposits including skulls or crania from especially horse and cattle, but also sheep and dogs have been identified. Many of these deposits appear in houses and have often been interpreted as the remains of protective magical rituals (see Ethnographic descriptions above) or in some cases as closing offerings when houses were abandoned (Carlie 2004: 110-129, 281). Another such house deposit that has already been mentioned in this thesis is the cattle cranium found in the north-western post-hole of the special house in Uppåkra (see section 7.2).

In other cases cattle skulls have been interpreted as the remains of offerings connected with episodes of large scale feasting, such as at Hofstaðir on Iceland where at least 23 cattle skulls, mainly from bulls, were first displayed and later deposited in a large 10th century hall building (Lucas & McGovern 2008: 7-12, 19-22).

Another example of a skull or cranial deposit, albeit of a slightly different character is the horse skull found in a water pond in the hill-fort Gamleborg on Bornholm (Denmark). The skull was deposited together with a sharpened stick with no other bones in the feature, despite the good preservation conditions, and was therefore interpreted as the remains of a magical cursing ritual (see Historical sources below) (Klindt-Jensen 1957: 154-155, 230-231).

If we leave Northern Europe it quickly becomes evident that the use of animal heads in rituals of various kinds is a world wide phenomenon that existed from at least the Early Neolithic Period and until modern times. A large number of finds have been made in Southern and Eastern Europe, as well as all over Africa and Asia (see for example Klindt-Jensen 1957: 85-86; Wilson 1999: 299-300).

The rituals that involved animal heads in different parts of the world were apparently diverse and included for example various ceremonies connected with holy places, such as the cattle heads at the Iron Age sanctuary of Gournay-sur-Aronde (France) (Green 1992: 119-121). The most common form of ritual use of heads was however in connection with burials, as can be seen for example in the Early Neolithic cemetery of Khvalynsk (Russia) (5000–4500 BC), where, among other animal offerings, seventeen sheep and nine cattle were buried as head and hoof deposits; or the similar head and hoof grave deposits of sheep and sometimes horse, that have been found in the Late Copper Age/Early Bronze Age Yamnaya-culture (Anthony 2004: 365; Lillie 2004a: 362-363). Other examples include the horse head found in a first century AD Kushan Period grave in Northern Afghanistan, the heads of dogs and pigs that often accompanied the dead in graves dated to the Lower Xiajiadian culture (ca 2200-1600 BC) in present day Northeast China, the various Bronze Age Kurgan burials in Central Asia with heads of cattle and sheep, and the somewhat later Early Medieval Magyar graves in Eastern Europe (Lillie 2004b: 370; Wells 2004: 98; Wigham 2004: 206, 350).

Historical sources

Only a few historical texts describe the ritual use of animal skulls in Northern Europe during the Early Middle Ages. Two such sources have already been mentioned in chapter 3. The first one is the description in Egil Skallagrimson's Saga of how Egil uses a horse skull attached to a stick in order to curse his enemies (see section 3.2.2). As was mentioned above, a deposit interpreted as the remains of such a cursing ritual was found inside an Early Medieval hill fort in Denmark. Without the sharpened stick that lay very close to the horse skull the interpretation of the deposit would be difficult to support, especially since the description of the ritual in the written sources indicates that it was an exceptional event only used when other means had failed. There are therefore no indications that the cranium and skull deposits in Leeuwarden are the result of this type of cursing ritual.

A more relevant historical source that describes the use of animal skulls is the already mentioned passage from Ibn Fadlān's *Kitāb's Rūsiyyah* about the custom among the Rūs along the Volga River to tie the skulls of cows and sheep to poles as a thanksgiving offering to the gods after a successful trade transaction (see section 3.2.5).

Interpretation

The examples above show that skulls and crania were used for a wide variety of rituals throughout the prehistory and early history of Northern Europe. With little more than a few skulls and crania buried with care and without first being fully processed it is unfortunately difficult to interpret the rituals behind them. It is however possible to use some of the analogies presented above to form a hypothesis.

Since the skulls and crania in Leeuwarden and Midlaren were not found in houses or seem to have been connected to burials, I consider the rituals in which animals were partially offered to be good analogies. In these rituals the heads, and sometimes other pieces of the animals, were given to the gods while most of the animals were eaten. Examples of this type of ritual are Ibn Fadlān's description of the Rūs trader's thanksgiving sacrifice and the Zulu offering ceremony.

Although skulls, as has been noted, do have an economic value it is a lot less costly for a family or a society to offer a skull compared to a complete animal. It has been observed both among present and historic societies that it is common to offer the resource poor parts of animals to supernatural beings while the rest is eaten. If a part of an animal is to be chosen as an offering, the head would also be a good symbolic representation of the whole body since it is easily recognisable.

In Ibn Fadlān's example the animals were sacrificed when trade was successful. The long range connections of the *terp*-region and the many descriptions of successful Frisian traders may suggest that similar rituals existed in the area. The specific motive for such an offering is however unknown to us and it could have been performed for a number of reasons.

8.2.3 Dongjum: Feature 1023 (cranial fragments of cattle and sheep/goat)

The bone assemblage from this ditch did in most respects resemble the rest of the material from the site, with the exception of a higher percentage of cranial fragments of cattle and sheep/goat. During the identification process five large mandible fragments stood out as unusual (see figure 4.17).

Interpretation

The bone assemblage from the ditch could be the remains of some form of skull/crania-deposit resembling those identified in the other materials (see section 8.2.2). This is however very difficult to determine since the cranial fragments only comprise a few pieces of the material which in all other aspects seems normal. The dominance of mandibles among the skull fragments is also clearly different from the Leeuwarden and Midlaren deposits, in which the mandibles were either lacking or deposited together with the rest of the skull. Because of this a likely explanation is that the bone assemblage from the ditch represents cooking and consumption waste.

8.3 Age structure

8.3.1 Dongjum: Feature 1088 (infant and juvenile cattle and sheep/goat)

This well contained a small animal bone assemblage deposited in the top and bottom layers of the feature with a thin empty layer in between. The only really unusual aspects of the feature is the high proportion of bones of infant and juvenile animals, and the presence of both a grinding stone and a sharpening stone in the well.

Ethnographic descriptions

The ethnographic literature gives few references to grinding and sharpening stones in ritual contexts, but they do appear in various cultures around the world. Grinding stones are sometimes used in rituals connected with marriage and with fertility in general, for example within Hinduism (see for example Abbott 1932: 247-248). In some places, as in rural Tanzania, grinding stones are strongly associated with female identity and specifically with female initiation rites (Fendin 2006: 160).

Sharpening stones are also sometimes used in rituals connected with fertility and agriculture in a similar manner as grinding stones (see for example Bonnefoy 1991: 174; Metcalf & Huntington 1991: 104). Ethnographers have however in other societies seen them as symbols of strength and power rather than of fertility (see for example Gilbert 1987: 318-319).

Archaeological parallels

That grinding and sharpening stones can have had ritual significance, often connected to transformation and fertility, has been suggested for various archaeological sites around the world dated to the Neolithic and onwards (Merrifield 1987: 33-34; Carlie 2004: 83-87; Fendin 2006).

In some cases grinding stones have been interpreted as ritual tools, such as the small stones found in houses in Oaxaca, Mexico, dated to the Early Formative Period (1500–1150 BC), and the slightly later finds from Oruro in Bolivia (Marcus & Flannery 2004: 18258; Bermann & Estévez Castillo 1995). In Central and Northern Europe grinding stones have been interpreted as offerings both at settlements and in bogs from the Neolithic and into the Middle Ages (Campbell 1991: 133; Carlie 2004: 83-95; Grant *et al.* 2008: 173-174; Nieweg 2009: 307-308).

Grinding stones of various sizes, from full scale millstones to small stones used to prepare cosmetic pigments or spices, have also been found in graves across Europe as well as in the Middle East, dating from the Iron Age Period and until the Early Middle Ages (see for example Naumann-Steckner 1997: 149-150, 161). Often these stones can be regarded as everyday objects deposited as grave gifts, but in some cases the depositional circumstances indicate a special ritual significance of the grinding stones beyond this (Selinsky 2005: 121).

Sharpening stones have also been found as grave gifts in burials from the Bronze Age until the Early Middle Ages in Europe and Asia. One example is an Early Skythian burial mound (7th century BC) in which a sharpening stone was found close to the body (Sulimirski & Taylor 2006: 563-564).

Other examples come from Northern Europe where sharpening stones are frequently found in predominantly rich graves mainly dated to the Early Middle Ages (see for example Tegnér 1992: 239; Vallet & Roesdahl 1992: 322; Appelgren 2007).

Some sharpening stones found in pits and houses have also been given ritual interpretations. In Iron Age Danebury a broken sharpening stone was for example placed against the jaw of a horse deposited in one of the numerous special pits at the site together with two young pigs and several other objects such as ceramics and sling stones (Grant 1989: 83).

The placement of sharpening stones in house foundations in Southern Scandinavia seems to be a recurring phenomenon during the Middle Ages (Falk 2008: 78-79). The practice even survived the transition into the Christian period and appear during the High Middle Ages and occasionally as late as the 17th century (Falk 2008: 96, 101, 115, 141).

Historical sources

Both sharpening and grinding stones appear, although infrequently, in symbolic contexts in religious and mythical literature. Sources dealing more specifically with these objects' use in rituals are however absent. The most common reference to the, often metaphorical, use of grinding stones in a ritual context is execution through drowning with a millstone as weight (see for example the New Testament, Luke 17:2, Mark 9:42 and Matthew 18:6).

The use of both grinding and sharpening stones as weapons is mentioned several times in Old Norse myths, for example in the story about the murder of Gilling or the part about Thor and Hrungner from *the Prose Edda* by Snorri Sturluson and in *Sverri's saga*, an Icelandic Royal Saga from the late 12th century.

Another ritual or magical use for a sharpening stone is shown in *Skáldskaparmál*. In this story Odin, in his role as a trickster god, uses his possibly magical whetstone to trick nine thralls into killing each other for no apparent reason:

> *"...Odin set out from home and came to a place where nine thralls were mowing hay. He asked them whether they would like to have him whet their scythes. To this they said yes. Then he took a whet-stone from his belt and whetted the scythes. They thought their scythes were much improved, and asked whether the whet-stone was for sale. He answered that he who would buy it must pay a fair price for it. All said they were willing to give the sum demanded, and each wanted Odin to sell it to him. But he threw the whet-stone up in the air, and when all wished to catch it they scrambled about it in such a manner that each brought his scythe onto the other's neck."*
> (Skáldskaparmál, chapter 5, translation by Anderson 1880)

In another part of *Skáldskaparmál* the story of Menja and Fenja tells about a grinding stone that also has magical properties. In this case the magic is benign, since the stone can grind fourth whatever the miller wishes.

Finally in *Gautrek's saga*, an Icelandic Fornaldarsaga from the late 13th century, a sharpening stone is used as a potent gift to a king. The cunning protagonist uses the stone to eventually gain great wealth through a number of gift exchanges. This part of the saga was most likely meant to illustrate the nature of gift exchange.

Interpretation

That two objects that bring such ritual connotations as grinding and sharpening stones should end up in the same well, one in each of the two find bearing layers (the sharpening stone at the very bottom of the feature) seems more than coincidental. Despite this a ritual interpretation is not very convincing, although possible. The bone assemblage in the feature deviates slightly from the rest of the material from the site, but the small number of bone fragments found in the feature makes this an inconclusive indication.

It is possible to imagine that the sharpening stone was deposited at the bottom of the well when the feature was no longer used as a water resource. The well was after this filled half way up with earth and refuse in several phases separated by episodes of water filling the feature. The former well remained in its half filled state for some time before it was completely filled, again with earth and refuse, but now most likely in one event. During this last episode the grinding stone was deposited. Again this could be the result of a ritual that marked the closing of the well, and was then possibly connected with fertility since both the deposited objects have such symbolic connotations. The grinding stone could on the other hand just as well be a mundane part of the refuse that filled the feature. If the exact placement of the grinding stone within the topmost layer had been determined and if more similar deposits in wells were known it is possible that the interpretation could be taken further, but now it remains inconclusive.

8.4 Selected species

8.4.1 Leeuwarden: Sunken hut 5 (feature 610 and 611, large number of bones from mainly wild geese and ducks)

In this sunken hut a large number of bones from wild birds were deposited at two separate occasions. In both cases the bird bones were rapidly covered with soil after being deposited. During the first instance the bones from most parts of the bodies of at least 27 individual birds, mainly greylag goose (*Anser anser*), white-fronted goose (*Anser albifrons*) and common teal (*Anas crecca*), were deposited. At the second occasion a smaller number of bones from at least nine individual birds were deposited. A few of the bones in the feature were black or white burnt (in one case only partially), something that could indicate roasting over open fire. No other taphonomic processes could be identified among the bird bones. The element distribution and the indication of roasting suggest that the bird bones are cooking or consumption refuse. It is important to note that especially the older deposit represents a very impressive amount of bird meat.

Ethnographic descriptions, archaeological parallels and historical sources

During the analysis it quickly became obvious that this was an unusual deposit. It is actually so unusual that I have been unable to find similar deposits or activities that could result in such a bone assemblage in archaeological reports, ethnographic texts or historical sources. The tradition of eating roasted bird does however exist, for example in the custom of eating turkey at thanksgiving in the US and Canada and during Christmas in a number of countries, and the tradition of eating goose or other birds on St. Martin's Day in many European countries.

The importance of feasting in general, although not necessarily including the eating of birds, is well documented in various ethnographic texts from around the world and in written sources dealing with Early Medieval Europe (see section 8.7.1).

Interpretation

The fact that the birds from the first deposition event at the very least could supply meat for forty people, in combination with the short duration and rapid covering of the deposit indicates that the feature represents the remains of a large feast, possibly of a ritual character and probably connected to bird hunting since only wild bird remains are present.

A connection of some sort with St. Martin and the feasting that surrounded the worship of this saint since the Early Middle Ages cannot be totally ruled out. The eating of many different bird species, all identified as wild and many rather small does in any case present a unique variant of bird feasting.

The bird species involved do not always migrate and cannot therefore give any clear indication of seasonality, a fact that of course does not rule out the possibility that the feast was a seasonal event. The two distinct deposits could theoretically be from the same feast, with the remains collected and deposited with some time in between. It is however equally likely that they represent two different occurrences of the same type of feast. The second deposit might either indicate a smaller scale feast, or the deposition of parts of the refuse in the sunken hut and the rest elsewhere.

A question that arises is if this specific type of feast was a tradition unique for this settlement, or if it is something that existed in a larger region? The abundance of wild birds in the salt marshes was of course not unique to this site, but since no similar deposits to my knowledge have been recognized at other *terp*-settlements it would be easy to state that this was a local custom. It is however important to note that the sheer scale of the bone material in the older deposit is the key factor in the identification of the event. If only the younger deposit would have been present, the feature would certainly not have been recognized as special. It is therefore possible that this type of feast existed at other places, but that the scale of the older of the two events that took place in Oldehoofsterkerkhof was unusual. The fragility of bird bones is also important in this discussion, since deposits of this kind might disappear through diagenesis, or be overlooked during excavation.

8.5 Human bones

8.5.1 Leeuwarden: Feature 85, 90, 182, 285 and 555 (human bones)

A number of human bones were found in five different features dated to the Early Medieval Period at the site of Oldehoofsterkerkhof. As was discussed in chapter 5 it is possible that some of the bones ended up in these features by coincidence and it is furthermore possible, although not likely, that feature 258 is the remains of a severely damaged grave. Even if this is taken into consideration, the number of human bones found in features not resembling disturbed graves, indicates that bones of humans sometimes were handled in such a way that they ended up in pits and ditches.

The deposit that gives the most convincing indication for the practice of using human bones in rituals is Well 39 (feature 90) in Leeuwarden. Three skulls and several other "large" human bones were found at the bottom of this feature with an upside down turned wooden trough placed over them. As has already been explained, the bones were unfortunately not recovered and no documentation of their exact composition was made, which makes the interpretation of this deposit difficult. The fact that a number of bones from several humans, but not the complete skeletons, were found deliberately placed together show that the bones did not end up at the bottom of the well by coincidences, and that the persons involved did not die in the well. It is of course possible that they represent remains of for example violence or human sacrifice that took place somewhere else. What we know of the bone composition suggests a secondary placement of human body parts, most likely in the form of clean bones, either from exhumed bodies or from persons that were never buried at all.

One of the most interesting aspects of this find is the very late date of the feature. The wood in the well is dated to 828 ± 6 AD which gives a *terminus post quem* for the deposit inside it. This date is well after Christianity was first established in the region and right before the *terp*-settlement was turned into a religious site with the first church probably built during the first half of the 10th century (Dijkstra *et al.* 2008: 320). It is intriguing that a deposit of seemingly pagan nature was made at such a late date at the very site where a church was to be built shortly afterwards.

Ethnographic descriptions

The most common forms of ritual handling of human bones are undoubtedly those connected with the varied funerary customs that can be seen around the world. Some of these rituals, such as the exhumation and reburial practices in for example parts of Greece, Southern China, Indonesia and Madagascar, include extensive moving and handling of the bones (see for example Grant *et al.* 2008: 158). The remains of the deceased do however in these cases always end up in some form of secondary burial, for example in urns or death houses.

Other types of ritual handling of human remains include some types of ancestor worship, cannibalism, trophy hunting, human sacrifice and various magical practices. Ancestor worship is a very common religious concept, but the actual use of human bones in the practice is rare. In some cultures ancestors were buried underneath houses, and the Merina of Madagascar kept the remains of their forefathers above ground as a part of their ancestor veneration. In Dobu in Melanesia people even used to wear the jawbone of certain ancestors around their necks as long as they were thought to bring good luck (Crump 1996: 433; Bloch 1996: 66-67).

Cannibalism has been recorded, albeit rarely, in ethnography. Usually the examples are old, such as the eating of enemies at feasts among the Huron Indians of the early 17th century, while one of the rare modern examples comes from the Bimin-Kuskusmin-people of New Guinea (Sanday 1986: 83-101; Tooker 1991: 73; Walker 1998: 283).

The Bimin-Kuskusmine also engaged in the conceptually related custom of trophy hunting, the practice of removing body parts from enemies. This custom of taking heads and other body parts of enemies to gain status and supernatural power has also been recorded during the early 20th century among various Philippine societies (Junker 2000: 345-346; Peralta & Scott 2001: 64-65). Occasional trophy hunting, usually connected to armed conflicts, seems to have been a relatively widespread phenomenon around the world, including our contemporary western world (Chacon & Dye 2007). The reasons behind such practices vary from ideas about magically stealing the power of the victim to more symbolic functions such as a show of strength and the intimidation of enemies. If trophy hunting that includes human bones is part of a society's culture, the bone fragments might eventually turn up in the archaeological record as stray human bones found in what we consider odd places.

Human sacrifice has been exceedingly rare in modern times, at least if we do not include the often highly ritualised forms of capital punishments used in some countries. One exception is the human sacrifices performed among the non-Islamic chiefdoms on the island of Mindanao in the Philippines at least until the early 20th century. A specific event of a human sacrifice was recorded there in 1907 during the installation of a regional chief (Junker 2000: 39, 137).

A related custom but with a somewhat different purpose is when humans are ritually killed in order to provide healing powers or ingredients for traditional medicine. This practice is nowadays mostly associated with the so called "Muti"-killings in Southern Africa. Earlier use of human bones in medicine, for example during the Chinese Ming Dynasty and the custom in sixteenth and seventeenth century Europe to steal or buy parts of the bodies of executed persons for medicinal or magical purposes, does however show that this is not a unique practice (Walker 1998:283; Magner 2005: 81).

Figure 8.2. Human cranium and dog maxilla from a Roman Period pit in Leeuwarden *in situ*, feature 1569 (photo by ADC Archeo-Projecten).

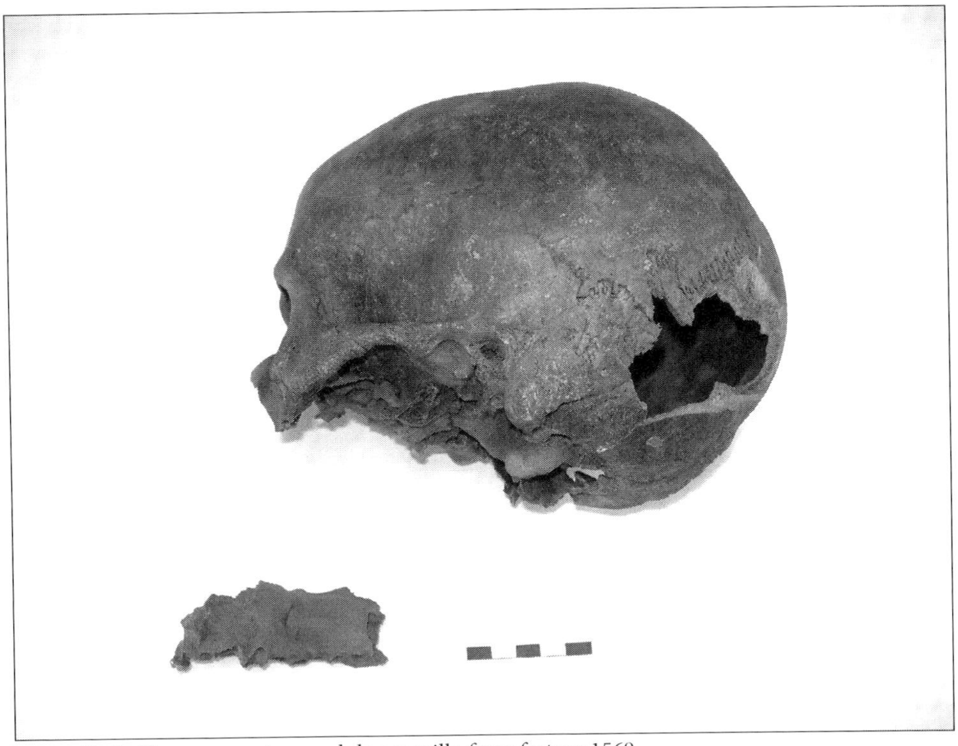

Figure 8.3. The Roman Period human cranium and dog maxilla from feature 1569.

Archaeological parallels

Human graves are a very common find category within archaeology, but human bones are often found outside of obvious burial contexts as well. These are usually interpreted as the remains of disturbed graves, but in a number of cases such explanations seem less likely.

One of the most well-known types of such finds in the *terp*-region are the infant burials found at several settlements. In the *terp* Wijnaldum the bones from five infants were found, one of these most likely in connection with the deposition of two horse foetuses. Similar finds have been found at other *terp*-sites, for example in Feddersen Wierde and Hessens in present day Northern Germany, and seem to be a recurring, if still rare, practice (Capelle 1987: 191-198; Haverkort *et al.* 1993; Cuijpers *et al.* 1999). The reason behind the infant burials placed in the middle of settlements that usually lack human graves is intriguing. Several possible explanations including separate burial customs for very small children, infanticide, sacrifice or magical rituals can be envisioned. The practice of finding infant bodies at settlements or even inside houses is however not confined to the *terp*-region or the Early Middle Ages, but appears in various places and time periods (see for example Scott 1991; Bailey 2000: 123; Carlie 2004: 139-141). In this case it is however sufficient to conclude that the essentially complete bodies of very small children have little in common with the finds of the seemingly randomly selected human bones from adults found in Leeuwarden.

Finds of complete human skeletons are occasionally made on *terp*-settlements, and are then usually interpreted as graves (Haverkort *et al.* 1993). In some cases, such as in that of the skeleton of an adult human found near one of the entrances of House 20 in Feddersen Wierde, the circumstances are unusual enough for a ritual explanation (Capelle 1987: 191).

These finds of the remains of whole adult humans are however still not comparable to the human bones found in Leeuwarden. Finds of stray human bones clearly outside of grave contexts are not very frequent in the *terp*-region during the Early Middle Ages. During earlier periods some interesting finds have however been found. One of the most obvious examples of human bones used in rituals on a *terp*-settlement is the Iron Age deposit from Englum containing eight human skulls in a circle with pottery, ashes, manure and animal bones inside it that was briefly discussed earlier in this thesis (see section 3.3.5).

Another example, dated to the Middle Roman Period, was actually found at Oldehofsterkerkhof in the form of a human cranium deposited in a small pit together with the upper jaw of a dog (figures 8.2 and 8.3).

In the rest of the Netherlands and Northern Europe there is a number of finds of human bones that have not been interpreted as normal burials. In general it seems as if small numbers of loose human bones are normal finds in settlements from the Iron Age, Roman Period and Early Middle Ages, but they seldom receive much attention (see for example Groot 2008: 120, 156)

Most of the more well-known examples of human bones outside of grave contexts, such as the bog bodies, building sacrifices and the finds of human bones in special pit deposits are dated to the Iron Age (Merrifield 1987: 23-24, 50-51; Grant *et al.* 2008: 172). Some finds however originate from the Roman Period, for example the numerous human heads deposited in British rivers, ditches, wells and buildings, often together with other objects or animal bones. Sometimes other parts of humans were also deposited, such as the human femur and a horse skull found under a part of the city wall of London which was dated to the late Roman Period, or the human skull and a femur that were found in a ritual pit in the Romano-Celtic temple at Colchester (Merrifield 1987: 27, 37-38, 45-46, 54; Scott 1991: 116).

Such finds where certainly not unique for Roman Britain. In Bavay, Nord (France) close to the Belgian border three humans, a woman, a man and a child with a dog, had been deposited one above the other in a well dated to the early first century AD. Several similar deposits have been found in Gallo-Roman shafts in France, for example in two of the ritual shafts excavated in the Luxembourg Gardens in Paris (Merrifield 1987: 43).

Finds of loose human bones from the Early Middle Ages are however comparably few in Central and Northern Europe, most likely because of the spread of Christianity which both standardized burial practices and suppressed or transformed earlier pagan and magical rituals. It is therefore not surprising that the unusual finds of human bones in this time period are mainly found in Scandinavia and the Slavic region where the new religion was not yet dominating.

A prominent Slavic example are the human bones from the temple fortress of Arkona on Rügen (Germany) that have been interpreted as the remains of human sacrifices, an interpretation that is supported by written sources (Müller 1974: 291-293).

There are also several archaeological finds of human sacrifices in connection with burials in Early Medieval Scandinavia as well as finds of stray human bones in settlements. One obvious ritual example is the human skull fragment from Ribe on Jutland (Denmark) that has the name of Odin inscribed on it (Steinsland 1992: 146; Oestigaard 2000: 42).

Another example of human bones deposited in settlements is Uppåkra where a number of bone fragments have been found outside of graves. Three of these bones were found in the special context of the weapon deposit area that is analysed in this thesis and presented in chapter 7.

If we leave Northern Europe, single or small numbers of human bones deposited in various non-grave ritual contexts have been found all over Europe from the Mesolithic and onwards (see for example Jameson & Shaw 1999; Andersen 2004; Bonsall 2004; Grant *et al.* 2008: 150-158).

Human bones deliberately deposited in Mesolithic settlements have for example been excavated in present day Romania and Serbia (Bonsall 2004). Other examples of human bones found in ritual contexts come from the various Neolithic causewayed enclosure sites across Europe (see for example Jameson & Shaw 1999; Andersen 2004). From a later date, unbroken human bones have for example been found mixed with settlement materials at several Scythian sites in the area north of the Black Sea dated to the 6th-4th centuries BC (Sulimirski & Taylor 2006: 586-587).

If we finally look at the whole world it is apparent that human bones found within settlements is certainly not a European phenomenon. The occurrences of such finds are abundant, appear at many different places and span virtually all archaeological periods. Examples include human bones found in Mayan wells and other features that have been interpreted as the remains of human sacrifices, and the bones found in various contexts, often mixed with refuse, in other pre-Columbian Central American sites such as San Lorenzo, Yucuñudahui, Yucuita and Teotihuacan interpreted as the remains of both human sacrifices and other ritual activities, including cannibalism (Kyle 1998: 138; Sugiyama 2005: 224-229; Headrick 2007: 84-85).

Another example is the humans buried beneath or inside buildings in the Middle Preclassic Period (1000-400 BC) of the Maya culture as a part of ancestor worship (Kelly & Thomas 2006: 335-336). A final example is from the Roviana Period (ca 1600-1900 AD) of New Georgia (Solomon Islands). Here the shrines of the ancestors were placed in the middle of settlements with the bones of enemies as well as other ritual objects scattered on the ground around the house platforms (Walter & Sheppard 2006: 152-154).

Historical sources

Historical sources covering Northern Europe do occasionally talk about human sacrifice or other forms of ritual handling of human remains. In most of these sources such practices are only mentioned very briefly, and are used as warning examples or to show the cruelty of pagan people. It is important to remember that almost all the sources dealing with this subject are written from an outsider's perspective and by a "civilised" observer of what they conceive to be "barbaric" societies. This in combination with the fact that much of the information the authors used was exaggerated, came from secondary sources or was no more than hearsay, forces us to be sceptical.

Some Roman sources dealing with the Germanic tribes, such as Tacitus' *Germania*, mention human sacrifice, but do not say anything about how these were performed (see section 3.2.2). In contrast the East Roman bureaucrat Jordanes who in the 6th century AD wrote *The Origin and Deeds of the Goths* describes in more detail how this East Germanic tribe sacrificed war captives, cut of their arms and hanged these in trees:

"Now Mars has always been worshipped by the Goths with cruel rites, and captives were slain as his victims. They thought that he who is the lord of war ought to be appeased

by the shedding of human blood. To him they devoted the first share of the spoil, and in his honour arms stripped from the foe were suspended from trees."
(Jordanes, The Origin and Deeds of the Goths 41, translation by Mierow 1915)

A few Roman sources such as Lucan describe human offerings performed by the people of Gaul. Later commentators to Lucan's texts describe the practice of burning humans sacrificed to the god Tarais, drowning those meant for Teutates and hanging those meant for Esus. The works of the Greek historian Strabo and his well-known contemporary Julius Caesar also describe various forms of human sacrifice among the Gauls, including striking with sword, the killing with bow and arrow as well as burning in an elaborate human shaped wickerman-construction:

"The Romans put a stop both to these customs and to the ones connected with sacrifice and divination, as they were in conflict with our own ways: for example, they would strike a man who had been consecrated for sacrifice in the back with a sword, and make prophecies based on his death-spasms; and they would not sacrifice without the presence of the Druids. Other kinds of human sacrifices have been reported as well: some men they would shoot dead with arrows and impale in the temples; or they would construct a huge figure of straw and wood, and having thrown cattle and all manner of wild animals and humans into it, they would make a burnt offering of the whole thing."
(Strabo, Geography IV, translation by Koch & Carey 1995)

"All the people of Gaul are completely devoted to religion, and for this reason those who are greatly affected by diseases and in the dangers of battle either sacrifice human victims or vow to do so using the Druids as administrators to these sacrifices, since it is judged that unless for a man's life a man's life is given back, the will of the immortal gods cannot be placated. In public affairs they have instituted the same kind of sacrifice. Others have effigies of great size interwoven with twigs, the limbs of which are filled up with living people which are set on fire from below, and the people are deprived of life surrounded by flames. It is judged that the punishment of those who participated in theft or brigandage or other crimes are more pleasing to the immortal gods; but when the supplies of this kind fail, they even go so low as to inflict punishment on the innocent."
(Julius Caesar, De Bello Gallico VI.16, translation by Koch & Carey 1995)

Classical authors also attributed the people of Gaul and Britain with the tradition of trophy hunting. The 1st century AD Sicilian historian Diodorus Siculus described this practice in his *The Library of History*:

"When at any time they cut off their enemies' heads, they hang them about their horses' necks. They deliver their spoils to their servants, all besmeared with blood, to be carried before them in triumph, they themselves in the meantime singing the triumphant paean. And as the chief of their spoils, they fasten those they have killed, over the doors of their houses, as if they were so many wild beasts taken in hunting. The heads of their enemies that were the chiefest persons of quality, they carefully deposit in chests, embalming them with the oil of cedars, and shewing them to strangers, glory and boast how that some of their ancestors, their fathers, or themselves, (though great sums of money have been offered for them), yet have refused to accept them. Some glory so much on this account, that they refuse to take for one of these heads its weight in gold; in this manner exposing their barbarous magnanimity. For it is brave and generous indeed not to sell the ensigns of true valour; but to fight with the dead bodies of those that were men like ourselves, resembles the cruelty of wild beasts."
(Diodorus Siculus, The Library of History V, translation by Booth 1814)

If the classical sources rarely talk about human sacrifices among the North European cultures and describe them as barbaric when they do, the Christian Early Medieval sources deal with the subject even more seldom and contribute even fewer details. The authors are often content with only mentioning the existence of such pagan practices, too horrible to describe further. One of the more prominent exceptions is Adam of Bremen's brief description of the great sacrificial feast in Old Uppsala where the human bodies were hanged in the trees together with animal carcases (see section 3.2.2). This practice corresponds to some degree with Jordanes earlier description, but since Adam of Bremen never saw the actual event the report is not entirely reliable.

Another Northern European example is from the *Vatnsdœla saga*, one of the Sagas about the Icelanders, that was written down in the mid 13th century. It is said in this saga that people thought that Torolf sacrificed both humans and animals in pits (see section 3.3.2). Exactly how this sacrifice was performed is unknown, but the bones would in any case end up buried in the ground. The statement is said to be only a rumour and there is therefore cause for scepticism, even if it might be Torolf's personal guilt rather than the tradition in itself that was unclear in this case.

Finally two Arabic scholars, usually considered as quite reliable sources, describe human sacrifice. The first one, Ibn Rusta writes about the hanging of sacrificed humans by the Rūs of the town of Novgorod, but nothing about what happens with the body afterwards (see section 3.2.2). The second more famous one, Ibn Fadlān, describes the sacrifice of a female slave as a part of a chieftain's burial among the Rūs that took place somewhere along the Volga River in the early 10th century. In this case the sacrifice, as well as the deceased and all other offerings, were however cremated onboard a ship on the river and does thus not correspond with the finds of human bones at settlements (Montgomery 2000).

Interpretation

Since the only two characteristics shared by the five deposits are that they are bones from humans and that they were, seemingly, not placed in normal graves it is quite possible that they might be the result of different processes. Because of this several possible interpretations will be discussed.

The most common interpretation regarding finds of single or small quantities of human bones at settlements, is that they are remains of disturbed graves from earlier habitation phases. In the case of the finds from Leeuwarden this interpretation is not suitable for most of the deposits. The composition and placement of the human bones in the well is certainly too elaborate to be the result of disturbance, and regarding the remaining four deposits the stratigraphical situations do in most cases not indicate that the features are the remains of graves, even though this cannot be completely ruled out. A further circumstance pointing against the disturbed grave theory is that no graves from the Early Middle Ages dated before the 10th century have been identified at the site.

An alternative explanation could be that the deposits, or some of them, are the remains of secondary burials. The tradition of secondary burial appears in various cultures around the world but as was seen above for the ethnographic examples, secondary burials usually include a careful handling of the bones of the deceased. The Olderhoofsterkerkhof deposits however only contain few bones from each individual, which often seem to have been placed in random features without much care.

Another possible interpretation is human offerings. As the various finds, analogies and written sources show, human offerings have been a rare, but existing practice in pre-Christian Northern Europe. It is not possible to rule out this theory for the Leeuwarden finds, but a few human bones are in themselves not strong evidence for such an interpretation. A human offering will result in a complete body to be disposed of. As the excerpt from Jordanes indicates it is possible that parts of an offered body were treated in dissimilar ways. That such rituals should involve the deposition of various, small parts of the human body in different types of features with no trace of the majority of the bones, seems to be far fetched, with the exception of the bones in the well.

Some ethnographic observations as well as the passage about Britain and Gaul from Diodorus Siculus' *The Library of History*, describe the custom of keeping trophies in the form of body parts of enemies killed in battle. The human bones found in Leeuwarden could be the result of such an activity. The well deposit is then again the most convincing of the finds since it included human skulls, the most common bone element used as a war trophy according to both ethnographic and historical sources. This interpretation still does not explain why the bones were discarded in the way they were.

An alternative explanation to trophy hunting is that the human bones were relics used in ancestor worship. In that case the bone fragments would be from the ancestors of the people living in the settlement. The bones would then symbolise, or even be thought to possess, the spirits of the ancestors. Again the interpretation does however not necessarily explain why the bones were discarded even though it could be that the bones should be buried within the settlement in order to stay near their descendants, or that the tradition of keeping the bones of the ancestors changed.

Finally the human bones could have been used as tools for religious or magical purposes. This would explain why small and varied bone elements were used and why they could be disposed of in a fairly undignified manner.

To conclude I consider it unlikely that all the human bone deposits are coincidental. At least some of the bones are most likely the result of trophy hunting, ancestor worship or are disposed religious or magical tools.

The well deposit is especially intriguing since it is dated right before the official Christianisation of the site. An interesting parallel to this find is the already mentioned complete sheep deposited at Hofstaðir on Iceland at the very end of the pagan period before the site was abandoned. This has been interpreted as a very conscious performance of a pre-Christian ritual before the final transfer to Christendom (Lucas & McGovern 2008: 25). Possibly the human bones in the well at Oldehofsterkerkhof are the result of a similar final pagan ritual before the site was taken over by the church. The ritual could in that case be interpreted either as a last defiant resistance, as a way to get rid of incriminating evidence of pagan rituals, or as a respectful act towards the forefathers or the old gods, soon to be forgotten.

Alternatively, the deposits could be connected to the belief that existed in some societies that objects used for religious or magical purposes are inherently dangerous and have to be disposed of in a certain manner not to bring harm (Merrifield 1987: 44). This practice can be seen for example in *Egil's Saga* in which the main character removes a rune-inscribed whale bone/fish gill/baleen (the translators disagree on the correct meaning of the Old Icelandic word *tálkn* in the Saga) used in a failed magical healing spell in an intricate way. Egil then continues by composing a short poem warning about the dangers of ignorant use of sorcery:

> *"Next he searched the bed in which she had lain, and there he found a piece of whalebone whereon were runes. Egil read them, then cut the runes and scraped them off into the fire. He burned the whole piece of whalebone, and had the bed-clothes that she had used hung out to air. Then Egil sang:*

'Runes none should grave ever
Who knows not to read them;
Of dark spell full many
The meaning may miss.
Ten spell-words writ wrongly
On whale-bone were graven:
Whence to leek-tending maiden,
Long sorrow and pain.'"
(Egil Skallagrimson's Saga, chapter 75, translation by Green 1893).

In the saga the disposal of the magical object was done by fire and air, but it could be envisioned that the human bones in Leeuwarden, and then especially the human skulls and the bones from the well, were instead safely deposited and covered deep in the earth when they were no longer of use, or when they could no longer be utilized because of the spread of the new religion.

8.6 Cremation pit

8.6.1 Midlaren: Feature 22 (cremation pit)

The arguably most intriguing deposit discussed in this thesis is the Carolingian Period cremation pit from Midlaren, a find that has few parallels and therefore has been placed in a separate find category.

The cremation pit was estimated to contain nearly half a million bone fragments, weighing more than 12 kg that had been thoroughly burned at high temperatures and later collected, probably in a large piece of cloth or hide and deposited in a pit together with one or more bronze objects. Bones from at least four cattle (one of these a foetus), three sheep/goats, two pigs and two dogs were identified in the material, but since more than 99 percent of the bone fragments could not be identified these should certainly be considered minimum figures (Prummel *et al.* 2008: 248-252).

Since the bone fragments come from all parts of the animals' bodies, there is reason to believe that complete or large parts of the animals have been burned. Compared with the amounts of burnt bones collected after experimental cremations of animals, and with the experience from studies of human cremations, the 12 kg found in the deposit corresponds well with the expected bone weight for the number of completely cremated animals identified among the fragments (see for example McKinley 1993; Sigvallius 1994: 15-32; Jonuks & Konsa 2007: 104-105).

Ethnographic descriptions

Cremation is today a common way of getting rid of dead animals, both when pets are buried and when animal carcasses are destroyed for health reasons. The cremation of a loved pet or a treasured domestic animal can be an important event with ritual content. I have however been unable to find any ethnographic texts that describe a ritual where several animals were cremated for other reasons than as a part of a burial of a pet or the removal of a carcass.

Archaeological parallels

Archaeological finds comparable to the cremation pit in Midlaren are few and far between, both geographically and time wise. To my knowledge no similar deposits dated to the Early Middle Ages have been identified in settlement contexts in Northern Europe. There are however some Early Medieval deposits that share aspects with the Midlaren pit. The inclusion of burned animal bones in human cremation graves is for example a common feature on cemeteries from the Bronze Age and onwards in Northern Europe (see for example Jonsson 1972; Sten & Vretemark 1988; Iregren 1997). In most cases the amounts of burnt animal bones are small in such cremation graves, usually only a few grams. An exception is the rich burials interpreted as chieftains' graves found in Central Sweden where the animal bone material sometimes can be counted in tens of litres and contain the remains of over 40 individual animals (Sten & Vretemark 1988).

Occasionally pits have been found at Early Medieval cemeteries containing cremated animal bones and totally lacking human bones. Ten such cremation pits were observed in the grave fields at Norra Spånga in Central Sweden containing together 187 bone fragments from at least 13 different animals; 5 horses, 4 dogs, 2 cattle, 1 sheep and 1 lynx (Sigvallius 1994: 7, 197-212). One cremation pit was excavated at the cemetery of the *terp* Oosterbeintum in the Northern Netherlands that contained 293 burned animal bone fragments from at least one lamb/kid and one teal (Knol *et al.* 1996: 379-380).

All these deposits however contain a comparably small amount of bones, and come from cemeteries. Comparative finds from settlement contexts are a lot scarcer. A few deposits interpreted as building offerings have been observed in Southern Scandinavia, but these usually contained even smaller amounts of burned bone than the cemetery finds, never more than a few hundred grams (Carlie 2004: 110-111).

If earlier periods are taken into consideration a few interesting deposits however exist. A site that contains features with some similarities to the Midlaren cremation pit is the excavation of an Iron Age (1-2nd century AD) wheelhouse at Sollas on North Uist in the Outer Hebrides (Great Britain). During the excavation as many as 150 pits belonging to the site's early phase were discovered and 50 of these contained both burnt and unburned animal remains from cattle, sheep, pig and deer. The burned bones, sometimes mixed with unburned bones, came from 20 pits. Some of the animal deposits contained complete cremated animals while others only included parts of bodies. In most cases only one animal could be identified per pit, but sometimes parts from as many as

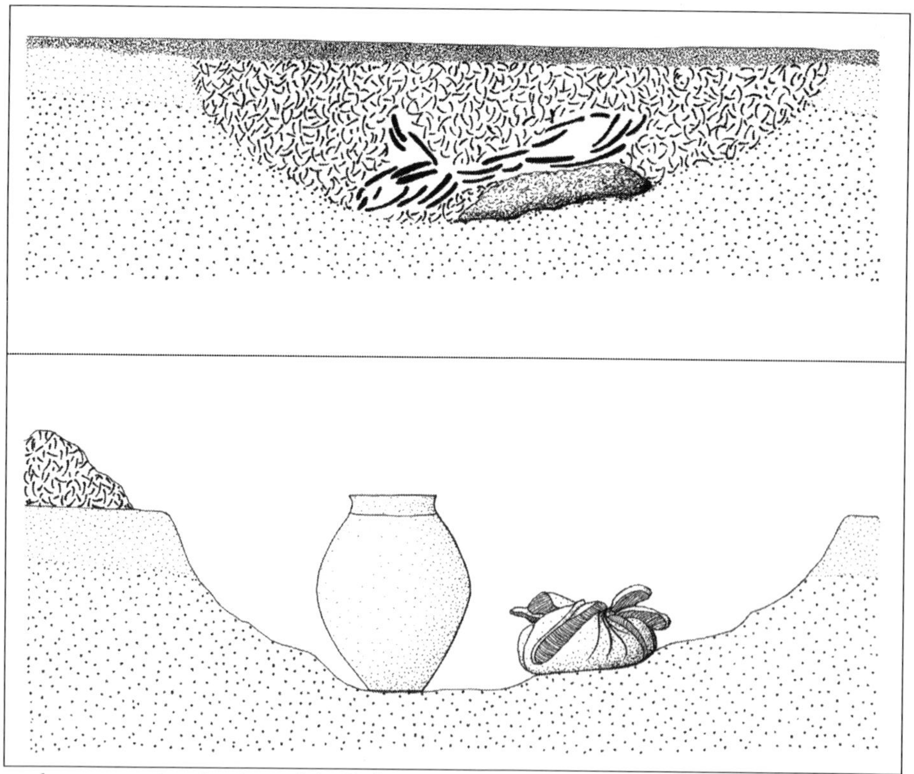

Figure 8.3. Section and reconstruction drawings of the Early Iron Age ceramic and animal bone deposit from Oldenburg (after Gabriel & Heinrich 1976: 124).

four animals had been deposited in the same feature. Calculations show that the remains from at least 100 animals were deposited in the pits and were interpreted as the results of ritual activity, probably in the form of votive foundation deposits (Campbell 1991: 132, 141-147).

Another parallel may come from a recently unearthed pit dated to the Bronze Age or Early Iron Age in Helden-Schrames in the province of Limburg (the Netherlands). From 2004-2008 BAAC BV excavated a large settlement area with traces from the Late Neolithic until the Modern Period. A deposit of 4.5 kg burned bones was found in a 32 cm deep, 160 cm wide and 170 cm long pit found near several Early and Middle Bronze Age houses. In addition to some charcoal and two pieces of burned clay, 33 fragments of Late Bronze - Early Iron Age ceramics were found, providing a broad dating. The bones are mainly calcinated and the size of the bone fragments varies, but is mostly small. Since no actual traces of burning were found in or near the pit the excavators assume that the bones were burned somewhere else and were then collected and deposited in the feature. It has been confirmed that no human bones are included in the deposit, but the bone material has however not yet been further analysed and it is therefore still not known which animals are represented and if they come from several individuals or not (de Winter 2010, personal comment).

The find that is most comparable with the Midlaren pit comes from the Slavic fortress in Oldenburg, Schleswig-Holstein (Germany). At the end of the excavation campaign of 1974 an unusual pit dated to the Early Iron Age was found. The oval shaped feature was 2 meter long and 1.1 meter wide and was filled with dark brown clay mixed with pieces of charcoal. In the pit potsherds from a large (62.5 cm high and 53.5 cm wide) vessel of early Slavic type was found. This pot has been deposited whole but later collapsed on top of a concentration of burned bones. The pile of bones was so dense that it was interpreted as being deposited wrapped in a piece of cloth or leather (figure 8.3) (Gabriel & Heinrich 1976: 123-124).

The bone assemblage constituted a great number of burned bones weighing 3.1 kg of which only 71 fragments could be identified. 62 fragments came from at least two cattle of which one was older than 4-5 years and thus fully grown (adult or senile). The remaining 9 bone fragments were identified as coming from at least one subadult (15-24 months old) pig. An interesting observation regarding the cattle bones was that no cranial fragments were identified while most parts of the postcranial skeleton were represented. The characteristics of the find reminded the archaeologists of a cremation grave and it was interpreted as an offering pit constructed as part of a death or burial ritual of some sort (Gabriel & Heinrich 1976: 126-128).

Despite the difference in dating both the cremation pits from Helden-Schrames and Oldenburg are good comparative finds to the Midlaren deposit. The features share several characteristics regarding their content and how the bones were handled until the time of deposit.

This shows that this ritual way of treating and depositing animal bones within a settlement context is not unique to Midlaren, or to the Middle Ages.

Historical sources

The burning of animals or portions of animals as part of an offering is known from various cultures in different time periods. A large number of Roman and Greek sources show that it was regularly done around the Mediterranean in classical times and the custom also appears for example in the Old Testament of the Bible (Smith 1859: 998-1000; Lambert 1993: 294-297, 300). It is often, albeit slightly simplistically assumed that the smoke from a burnt offering is supposed to rise towards the sky and the heavenly gods, in contrast to the burial or drowning of offerings that were meant for supernatural entities of the underworld or the sea (compare with for example Merrifield 1987: 41; Lambert 1993: 301).

Interpretations

When large amounts of cremated animal bones are found without any human bones the interpretation of the deposit as the result of carcass removal is close at hand. In the case of this specific feature the fact that at least ten animals from four different domesticated species are present strongly argues against this interpretation, as does the great care taken in collecting and depositing the burned bones.

A ritual interpretation of some kind seems more convincing in this case. The shortage of both comparative deposits and analogies however makes it difficult to interpret the Midlaren feature further than this. The situation is in addition complicated by the fact that the feature was found some distance from the houses dated to the Carolingian Period and close to the edge of the excavation. This makes it difficult to put the feature into a context since it is not known whether it was placed a short distance away from the settlement, or close to some other structure situated outside the excavated area, possibly another farm, an activity area or even a cemetery. In such situations the only option is to use the available information, but to keep in mind that the unknowns might radically change the interpretation. Because of this I will suggest two possible explanations for the cremation pit.

The element distribution and the inclusion of dog bones in combination with the lack of evidence for consumption of dogs in the region indicate that the bone material does not represent food refuse (although it cannot be ruled out that dog meat was eaten during special ceremonies). The fact that the animals were most likely burned as complete bodies, or at least in large parts, leads to the interpretation of the deposit as an offering to supernatural beings. If this interpretation is correct the sheer amount of animals involved would make it a very large offering and a tremendous symbolic act for people both in and outside the community.

The other possible interpretation is based on the fact that most of the comparable finds, especially those from the Early Middle Ages, are found in cemeteries and that the cremation pit in Midlaren therefore also could be viewed as part of a burial ritual or at least a ritual that somehow dealt with the dead. It could be a burial where the person died elsewhere and a body was lacking, or a ritual connected to a burial outside the excavated area.

It is finally possible that both suggested interpretations are correct and that the cremation represents a grand offering to the dead, either in a very specific sense or as a more general part of ancestor worship.

Regardless of the reasons behind the ritual the cremation was certainly a spectacular event. Experiments have shown that a pyre with at least ten cubic meters of firewood is required to fully cremate this amount of body mass and that such a process takes at least six hours before the fire dies out and all the soft tissue of the animals is burned away (Sigvallius 1994: 15-32; Jonuks & Konsa 2007: 104-105).

8.7 Bone layer

8.7.1 Uppåkra: Feature (group) 96 (bone layer with deliberately destroyed weapons)

The interpretative circumstances for the studied bone assemblage from Uppåkra are, as has previously been discussed, slightly different from that of the other three materials. The deposition area in question had already been identified as a special feature and even been given a ritual interpretation as the place for repeated weapon sacrifices. The details of the ritual interpretation, such as whether the weapons came from vanquished enemies or not and whether they were first stored in the ceremonial house before being destroyed and buried is still debated, but the general ritual character of the area is rarely questioned. The bone material is, despite making up the vast majority of the find material in the feature, however mostly mentioned in passing and not interpreted in any detail.

Since the bones were deposited at the same place and during the same time period as the weapons, even though weapons were buried before the period when bones were deposited, it has simply been assumed by most researchers that there is a direct connection between the two find categories. In chapter 7 the bone material from the feature was compared with a bone assemblage from another part of the settlement that lacked obvious special deposits, signs of ritual behaviour or high status activities. This comparison showed that the bone material from the weapon deposit layer differed significantly from the other material and could be labelled special in its own right. The question what precisely the function of the bones in the elaborate rituals that took place in the area during the Migration Period was, does however need to be addressed.

The destroyed warriors' equipment from feature 96 has put a lot of focus on weapon sacrifice sites when analogies have been discussed in connection with this find. The large animal bone layer that is the main interest here, does however require a slightly broader approach, including comparisons with other offering sites as well as various feasting practices.

Ethnographic descriptions

I have not been able to find any modern ethnographic description of offerings that can be directly compared with the deposited bones, weapons and other objects from the layer in Uppåkra. There is however a large ethnographic record that deals with two of the main aspects usually discussed in connection with the weapon deposit area, sacrifice and feasting. The first aspect is indicated by the consciously destroyed weapons and equipment while the second is connected both with the large bone material and with the close proximity to the ceremonial building and several other large hall-like houses that are, among other things, associated with feasting in the Scandinavian Roman Period and Early Middle Ages (see for example Fabech 1998: 152; Jørgensen 2002: 218; Larsson & Lentorp 2004: 43).

The very term sacrifice implies an offering of something that is ritually killed, either in practice or at a symbolic level. This type of offering is known from countless societies around the world and while a sacrifice might often be associated with the killing of an animal, or more rarely a human, it is also frequently linked to the destruction of non-living things (Van Baaren 1964: 7-8).

The custom of feasting, the gathering together of many people that consume large quantities of food and drink for a specific reason, is also a common feature in ethnographic descriptions, and is something that still exists in contemporary societies for example in the form of celebrations in connection with weddings, birthdays, Christmas, Easter, Midsummer or Thanksgiving. Feasting can also be performed for various reasons beyond the official, and often symbolic, motivation. Such reasons include gaining prestige (for the sponsor of the feast), strengthening social bonds, keeping traditions and upholding social hierarchies.

Ethnographic examples include the ritual feasts among the non-Islamic chiefdoms on the island of Mindanao in the Philippines that have been described as the cornerstones of social, political, economic and religious life in these societies. The feasts were traditionally held at occasions such as child birth, death, illness, marriage, chief succession, *rites de passage*, raids, the forming of alliances and at critical points in the agricultural annual cycle. Several key features of these feasts where that they were sponsored by elite individuals to achieve social status, that the performance included animal sacrifices (usually pigs, chickens and water buffaloes), the exchange of gifts and the sharing of meat in accordance to kinship ties and social rank (Junker 2000: 314-318; Peralta & Scott 2001: 55-56, 61).

Among the Huron Indians of the early 17th century feasts gave both prestige and gifts to the sponsor. The official reasons for the feasts would be farewells, thanksgiving, deaths or life changing events such as the changing of ones name (Tooker 1991: 72-76). Ritual feasting with the main aim to gain prestige has also been observed among households of various Naga tribes in Northeast India (Kirsch 1997: 18).

Finally feasting is also often connected with mortuary practices as can be seen for example in the mortuary feasting observed on the island of New Ireland in Papua New Guinea or in the great feasts held by the Ngad'a of Flores in East Indonesia when a chief is deceased. At such occasions numerous buffaloes are slaughtered and the meat distributed among the participants (Van Baal 1976: 169-170; Bolyanatz 2000).

Archaeological parallels

As was discussed in chapter 7 the artefacts from the weapon deposit area show similarities with the large Roman and Migration Period weapon offering sites from Southern Scandinavia. The artefact material from Uppåkra does however differ from these sites in some respects, mainly in the relatively small number of artefacts, the location on a small elevation within a settlement instead of in a bog, the composition of many small deposits rather than a few large events, and the continuation of the weapon deposits well into the 8th century, albeit more seldom than during the earlier periods.

Another difference between the weapon deposit area in Uppåkra and the large Roman and Migration Period weapon offering sites is the character of the bone material. Animal bones have indeed been retrieved from several of the large weapon sacrifices, but in these cases the composition is quite different from that of the bone layer in Uppåkra. In sites such as Hjortspring, Vimose and Illerup Ådal bones from several animals were found, but these came from complete or almost complete animals that in many cases had been killed with a blow to the forehead and left lying where they fell. The animal species found in the large Roman and Migration Period weapon sites are also different from Uppåkra in that they are totally dominated by horses and dogs with only occasional examples of other animals. The composition of the animal bones from these deposits has resulted in interpretations along the same lines as those for the destroyed artefacts, that these were mainly animals that belonged to the loosing side in a battle (Ilkjær 2002: 203; Jensen 2003: 232; Kaul 2003: 215-216).

Considering the differences between the Uppåkra weapon deposit area and the large scale battle sacrifices, a comparison with the more "mundane" sacrificial sites used in Southern Scandinavia during the same time period are interesting. Some of these sites, such as Kärringsjön in Halland (Sweden) lack a substantial bone material, but at sites such as the waterhole at Eketorp on the island of Öland (Sweden), Röekillorna in Southern Skåne (Sweden) and Valmose on Sjælland (Denmark) large amounts of bones were found together with everyday objects such as grinding stones and ceramics, as well as occasional more valuable artefacts such as fibulas.

In these sites complete or large articulated sections of animals are part of the assemblages, but a majority of the material shows more resemblance with food refuse. The materials however often have a high proportion of skull fragments and contain occasional human bones just as the Uppåkra bone layer. The species distribution of the materials is on the other hand, like that of the large weapon offerings, dominated by bones from horses and dogs, or at least have a much higher frequency of these species than contemporary settlement finds (Ferdinand & Ferdinand 1962: 49-71; Møhl 1962; Backe *et al.* 1993; Stjernquist 1997: 78-79; Gräslund 2004: 167).

Historical sources

The practice among the people of Northern Europe of sacrificing the weapons of defeated enemies after a battle has been described by several classic authors. The earliest is probably Julius Caesar in *De Bello Gallico* written around 50 BC where he describes how the Gauls vow the possessions of the enemies to a god seen as equivalent to the Roman god Mars, and how they afterwards sacrifice all surviving animals and objects to this god:

> *"To him [Mars] when they have determined to engage in battle, they commonly vow those things they shall take in war. When they have conquered, they sacrifice whatever captured animals may have survived the conflict, and collect the other things into one place. In many states you may see piles of these things heaped up in their consecrated spots; nor does it often happen that any one, disregarding the sanctity of the case, dares either to secrete in his house things captured, or take away those deposited; and the most severe punishment, with torture, has been established for such a deed."*
> (Caius Julius Caesar, De Bello Gallico 6, XVII, translation by Macdevitt 1915)

Julius Caesar can be considered a first hand witness to this Gallic custom, and other authors have described similar events among Germanic tribes. Tacitus' *Annales*, written in the beginning of the 2nd century AD, does for example tell of how the spoils of a battle, including the captured enemies, were offered to the gods that Tacitus meant was equivalent to the Roman gods Mars and Mercury:

> *"The war was a success for the Hermunduri, and the more disastrous to the Chatti because they had devoted, in the event of victory, the enemy's army to Mars and Mercury, a vow which consigns horses, men, everything indeed on the vanquished side to destruction."*
> (Tacitus, Annales, XIII, 57.2, translation by Church and Brodribb 1942)

The same custom is described by Osirius in his *Historiae Adversum Paganos* written in the beginning of the 5th century AD. Osirius describes the Battle of Arausio that took place in 105 BC between a Germanic force mainly constituted by Cimbri, and a Roman army. The battle ended in a catastrophic Roman defeat and the victorious army killed and destroyed everyone and everything in the two conquered Roman war camps:

> *"The enemy, after gaining possession of both camps and great booty, by a certain strange and unusual bitterness completely destroyed all that they had captured; clothing was cut to pieces and thrown about, gold and silver were thrown into the river, corselets of men were cut up, trappings of horses were destroyed, and the horses themselves were drowned in whirlpools, and men with fetters tied around their necks were hung from trees so that the victor laid claim to no booty, and the conquered to no mercy."*
> (Osirius, Historiae Adversum Paganos, V,16, translation by Deferrari 1964)

Regarding feasting, the importance of this custom as a social and political event has long been an established idea in the study of Iron Age, Roman Period and Early Medieval Scandinavia. The idea has its roots in the many references made in early Scandinavian historical documents, especially in the Sagas, and in comparisons with other societies (see for example Graham-Campbell & Kidd 1980: 82-84; Bagge 1998: 37; Solberg 2002: 224). To be a generous giver in general and especially to arrange feasts was a very positive trait often attributed to nobility and royalty as well as to heroes and normal honest people in the Sagas. This trait is also repeatedly used to praise people on rune stones (Backe *et al.* 1993: 339-341).

According to the Sagas, feasting often had a religious content, but was also a custom that held societies together and manifested power structures and loyalties. In *Njal's Saga* Gunhild, the mother of the Norwegian king Harald Grey-Cloak offered a feast for Hrut when he was about to become a retainer in the king's retinue. Later in the same saga a feast was held by a man called Mord, officially to commemorate his dead father, but in reality to gain the confidence of the sons of Njal. In this episode the giving of precious gifts is also an important theme:

> *"On one occasion when Mord was visiting Bergthorsknoll, he said to Njalssons, 'I have arranged a feast at Hof to honour my father's memory. I want to invite your brothers and Kari to attend, and I can promise that you will not leave without gifts.'*
> *They promised to go, and Mord rode home to prepare the feast. He invited many farmers, and there was a large crowd at the feast. The Njalssons and Kari where also*

there. Mord gave Skarp-Hedin a large gold brooch, Kari a silver belt and Grim and Helgi fine gifts."
(Njal's Saga, chapter 108, translation by Magnusson & Pálsson 1960)

The connection between feasting and offering is for example shown in *Egil's Saga* which describes a feast held in connection with the disablot:

"The same evening, King Eirik and Gunnhild came to Atley Isle, and Bard had prepared a feast for the King, as sacrifice were to be held there to the Fates. Nothing but the best was provided, with plenty to drink in the main hall."*
* Diser
(Egil's Saga, chapter 44, translation by Pálsson & Edwards 1976)

The same theme is present in *the History of the Gotlanders* (*Gutasaga*), in this case also with a direct link between feasting/offering and the Thing, the governing assemblies in Scandinavia during the Early Middle Ages:

"In those days, and for long afterwards, men believed in holt and howe (grove and grave-mound), sanctuaries and sacred enclosures, and in the heathen gods. They made offerings of their sons and daughters and cattle, with feasting and drinking. They did that in their error. The chief sacrifice among the people was the one for the whole land, but each Third had its own sacrifice, and the smaller assemblies had lesser sacrifices with cattle, food and ale. They were called suth-nautar, that is 'Brethren of the Boiling', because they cooked [the sacrificial feast] together."
(The History of the Gotlanders (Gutasaga), translation by Tunstall 2004)

Interpretations

If the archaeological parallels presented above are used for comparison, the bones from the mundane offering sites such as Eketorp, Röekillorna and Valmose show many similarities with those from Uppåkra, while the large weapon offering sites such as Hjortspring, Vimose and Illerup Ådal instead contain a comparable artefact material. In both cases there are however also aspects that differ: for the artefacts the focus on spear and lance heads in Uppåkra compared with most of the weapon offering sites, and in the case of the bones the high proportion of cattle in Uppåkra compared to most of the mundane offering sites.

These observations could indicate that the weapon deposits in Uppåkra are combinations of elements from these two different types of offering sites. It is however also possible that we are looking at the remains of a unique form of activity, and it is therefore necessary to return to the actual material from Uppåkra and review it once more.

Large amounts of parts of animals were, during a few hundred years starting in the Migration Period, deposited at the same location as ritually destroyed weapons and warrior equipment. These objects had already been regularly buried here for several hundred years. The hundreds of animals deposited were mainly cattle, but all domesticated species common in the area are present. The bone material and especially the cattle fragments, display a high degree of cranial elements and of adult individuals. The taphonomic history of the assemblage indicates that the bones were covered soon after being deposited. The light weathering that is nevertheless visible on some of the bones is most likely the result of disturbance of the bone layer by later weapon deposits and, in some parts of the area, by later agricultural activity.

The species, element composition and taphonomy of the bones indicate that a majority of the material represent disarticulated parts of animals (mainly cattle skulls or mandibles), but that a substantial part also could be considered as food refuse.

Taking the different aspects of the material and the relevant analogies presented earlier in the chapter into consideration, I interpret the bone layer as being mainly the result of repeated feasting in combination with partial offerings of animals, not unlike the scenario Charlotte Fabech presents in her article *Kult og Samfund i yngre jernalder – Ravlunda som eksempel*:

"Foregik offerhandlinger i flere tempi, hvor dyrene først blev ofret på/i kultpladser/bygninger, hvorefter kødet indgik i et rituellt måltid i hallen, efter hvilket knoglerne blev samlet sammen og deponert på særlige pladser på eller ved bopladsen?"

"Did the offering actions take place in several steps where the animals were first killed at the cult site/building and the meat thereafter being a part of a ritual meal in the hall, after which the bones were collected and deposited on special locations at or right outside the settlement?"
(Fabech 1998: 153, translation by the author)

During such an event one or several animals (usually cattle) were butchered. Some parts of the animal, especially skulls or mandibles, were offered to the supernatural entities and deposited in the same place where weapons were buried. This event might also have included weapon offerings. The feast may well have taken place in the ceremonial house or in the larger houses just south of the weapon deposit area. If that was the case the remaining parts of the animals were probably eaten there and some of the refuse from this episode might have been dumped at the weapon deposit site, but the majority were disposed of elsewhere. Some of the bones found in the weapon deposit area could be the remains of separate food offerings, but such a practice would be difficult to distinguish from the rest of the material.

Since the studied bone assemblage, that consists of only one fifth of the total excavated material, includes at least 50 animals the total number of deposited animals might have been at least 250, a large investment even if the long time period is taken into consideration.

The interpretation of the bone assemblage from the weapon deposit area given here does however pose two important questions that have to be addressed. First if feasting and accompanying partial animal offerings were the main source of the animal bone deposits, why was this habit not performed during the whole time period that the weapon sacrifices were carried out?

There are several possible answers to this question. Either the feasting and offerings were also performed outside of the time frame we can observe in the layer, but the bones were in that case deposited elsewhere. It is however also possible that we are observing a real change in the ritual practice at the settlement. Maybe this specific event was not performed before ca 400 AD. If this is the case the question of why it started should be posed. Again there are many possible explanations for why a ritual practice is initiated, changed or given up. The point in time when the practice started in Uppåkra is however interesting and might give us a clue to at least one possible explanation. The social, economic and political effects on Europe, and especially on the Western Roman Empire, by the great migrations of people beginning in the fourth century have in recent years been toned down and treated as the complex thing it is. Despite this it is still accepted that some of the large migrations, as for example the Hunnic invasion at the end of the fourth century AD, created great changes in the societies of Western Europe (see for example Todd 1994: 481-482).

It is unclear how strong the impact of these events was on societies in Southern Scandinavia. Some researchers suggest that the region was actually invaded by the Huns, but even if this was not the case the turmoil was certainly not far away (Hedeager 2007). It is striking that many of the large weapon sacrifices in Southern Scandinavia were deposited during this very period. The number and size of these sacrifices indicate a far from peaceful time with major battles being fought in the region at least every few years (Jensen 1995: 262; Jørgensen 2003: 16).

In such a situation it is easy to imagine that the ruling families of an important settlement like Uppåkra would initiate new or strengthen existing religious and political rituals in order to maintain their position. The same strategy could of course be applied by a new ruling family to secure their recently established rank. This could have happened in the form of the introduction or "upgrading" of a, possibly yearly, feast by adding animal offerings to an existing tradition of "killing" and burying weapons near the special house, possibly the very seat of power of the rulers. This connection between animal offerings, weapons and warrior symbolism fits well with the state of affairs in the surrounding world.

A large scale change in the religious customs beginning around 400 AD in Southern Scandinavia has already been suggested by Charlotte Fabech. She has proposed that the main religious practice during this period moved from bogs and lakes outside the settlement to the hall or ceremonial house in the settlement as the religion became more institutionalised and fell under the direct control of the ruling class (Fabech 1998: 149-153). This model is supported by the decline of depositions in bogs in Southern Scandinavia roughly at the same time that ritual activity becomes more visible at settlements. It does however not fit entirely in the case of Uppåkra. Here the ceremonial house and the weapon sacrifice areas had already been in use for several hundred years at the heart of the settlement, despite the fact that a sacrificial bog, Gullåkra Mosse, is situated only a couple of kilometres away. The fact that this change of practice can be observed in many other places in Southern Scandinavia at this time nevertheless indicates that it was a period of religious transformation, even if this was manifested differently in a central place such as Uppåkra than in other settlements.

The second problem regarding the interpretation is the focus on cattle in the bone assemblage. In most offering sites found in Scandinavia, regardless of whether they are weapon sacrifice bogs such as Illerup and Vimose, or more mundane sites such as Röekillorna and Valmose, horse is either the most frequent animal in the material or at least appears a lot more often than in nearby settlement materials. The same pattern can be seen among animals deposited in Early Medieval cemeteries, and the custom to sacrifice horses and to eat horse meat at religious feasts is documented in several written sources such as Håkon den Godes Saga, in which the king, being a Christian, initially refuses to eat the horse meat that was given to him during the autumn sacrifice feast (Näsström 2001: 246).

There could be several reasons why the people who produced the bone layer in Uppåkra mainly selected cattle instead of horses for their rituals. One reason could be that we are viewing partial offerings and remains from feasts, and not animals taken from enemies and sacrificed as complete animals to the gods, as was the case in many of the weapon sacrifice sites. It is also important to remember that both the archaeological record and the written sources are often more complex than is suggested in a summary on offering practices. In many of the sacrificial sites where horse is dominating, cattle is present as well and often appears in large numbers. There are also several sites where cattle was the dominating species in ritual activity, for example at the already mentioned settlement of Hofstaðir on Iceland. Other examples include the sites Bukkerup and Turup in Denmark where articulated body parts of 15 and 7 cattle respectively were deposited during the Roman Period, as well as Nørrekær and Grøntoft where horns of several cattle dated to the Iron Age were found (Møhl 1972; Hatting 1993; Lucas & McGovern 2008)

In the written sources the already mentioned phrase about the bull as a sacrificial animal in *the Ynglinga Saga* (see section 3.2.4) as well as Ibn Fadlān's descriptions of both the merchants offering ritual and the chieftain burial (see section 3.2.5) show that cattle according to the texts were used as offerings.

With these questions addressed I consider the bone layer in Uppåkra as the result of repeated feasting with accompanying partial offerings of mainly cattle to be a viable interpretation.

9. Conclusions

9.1 Conclusions regarding methods, analogies and interpretations

9.1.1 Introduction

A number of methods were applied on the four bone materials and the main conclusion is that no single, or even a set of methods stood out as a flawless tool for identifying ritual deposits. Even if similarities did exist between different human societies in the past, either because of direct connections between them, or as a result of the influences and constraints of human biology and the surrounding physical world, the variation in human behaviour is simply too great for any method to be generally applicable.

When a material from an area, period or culture of which no previous studies of unusual deposits have been performed is to be analysed, a suitable set of methods of identification can only be put together through experimentation and evaluation. In the same way relevant analogies to be used for interpretation have to be established for each type of deposit within such a society. Because of this the methods used in this thesis should mainly be seen as an array of what could be included in the methodological framework for a certain material or culture area.

This however does not mean that all methods and analogies are equally suitable for identifying and interpreting special deposits. Of the methods that were applied on the four bone materials some showed more potential than others.

9.1.2 Archaeozoological methods

The archaeozoological methods turned out to have a high potential for identifying special deposits and as many as 19 of the 20 identified deposits showed anomalies within species selection, age structure, seasonal analysis, sexual selection or element distribution, or a combination of these. The methods that was most frequently used were element distribution and, to a lesser degree, age structure and species selection. Seasonal analysis and sexual selection did not come into play in the identification of any of the special deposits in the four case studies, but it has been shown for other animal bone materials that these two methods can be viable tools for identifying such deposits (see for example Lindeblad & Nielsen 1997; Legge *et al.* 2000).

9.1.3 Archaeological methods

The archaeological methods of artefact context, feature type, stratigraphical position, spatial context and general find context showed some potential. With the exception of the bone layer from Uppåkra these methods did not provide the primary signs of special deposits, something that however is natural considering the archaeozoological focus of the research, but they nevertheless gave supporting indications for most of the identified deposits. Especially the artefact context, stratigraphical position and general find contexts supported the identification of special deposits, while the feature type and spatial context were of less use.

The feature type could have been more useful if larger materials had been studied. The 20 special deposits however originated from layers, pits, wells, ditches and a sunken hut. This basically includes all the common feature types from Northern European Early Medieval settlements, with the exception of post holes.

The spatial context on the other hand was problematic to use for the three materials from the Netherlands. None of these three settlements were completely excavated and in the case of the two *terp*-sites the identification and dating of houses were difficult. Dongjum was especially complicated in this respect since the excavation of the *terp*-section was made up of very long and narrow trenches that uncovered few large structures and gave virtually no spatial overview of the site.

9.1.4 Taphonomic history

The observation and recording of the various taphonomy related traces on the bones were time consuming and acted as the primary indicator for only one special deposit, the cremation pit from Leeuwarden. Despite this, taphonomic history produced supporting indications for all the identified special deposits. In addition, by striving to formulate a taphonomic history for the bones in a deposit a deeper understanding of the human and natural agencies involved in its formation processes is gained.

A good example of the important supporting role of taphonomy for identifying special deposits, and indeed to interpret these further, are the bird bones from the sunken hut in Leeuwarden. The massive amounts of wild bird bones from all parts of the bodies deposited in two separate filling of the feature were the prime indicators of the special deposit. The taphonomic history of the bone material did however also show that the bones most likely had been rapidly covered after being deposited, since they lacked any traces of weathering and gnawing, hence indicating two single events. It also indicated that at least some of the birds had been roasted, in view of the fact that some bones where partially black burned.

9.1.5 Quantification methods

In a very early stage of the thesis I chose not to use any advanced quantification methods since initial tests had shown that these could not be applied with any reliability on the small bone assemblages that the individual deposits contained. Therefore only NISP, fragment weight and MNI were recorded for the materials from the case studies. NISP and fragment weight are basic recording methods within archaeozoology and are essential for understanding and comparing the size and composition of the studied materials, as well as the individual deposits. The MNI-method was of some use in giving a rough estimate of the number of animals in a deposit. It for example showed that a large number of animals had been involved in the events that produced the bone layer in Uppåkra, as well as the cremation pit from Midlaren and the wild bird deposit from Leeuwarden.

9.1.6 Analogies

As was concluded in chapter 3 the interpretation of the meaning of human behaviour in the past requires either access to direct historical sources, or the use of analogies. Since direct historical sources dealing with the exact places and time periods in question were in effect absent, analogies had to be used in the process of interpretation. In chapter 3 I defined three types of analogies: historical, ethnographical and contemporary of which all were used in the interpretations of the special deposits.

Historical analogies were used, sometimes together with ethnographical analogies, to interpret several of the identified special deposits, despite the fact that few historical sources from the time period and geographical area suitable for analogical reasoning were available. I consider an assessment of the available written sources an essential part of this kind of study, partly to find those useful sources that do exist and partly to give general insights into the studied societies, that in turn can be used for evaluating the potential of other analogies.

Ethnographical analogies were used to interpret several deposits and could in those cases provide some information. Relevant detailed ethnographical examples of ritual behaviour including animal bones were however less common than expected. In many cases the same situation as for the written sources was true, that rituals were mentioned in the ethnographical report, but without the detailed descriptions necessary for using them as analogies. This could partly be the result of that ethnographers, having an existing society to study and interact with, are less interested in the minuscule details of the handling of objects that archaeologist needs in order to interpret the fragmented remains of a material culture. It could however also be explained with the fact that a thorough knowledge of ethnographical literature is needed to find good analogies and that archaeologists without an anthropological background often lack this. One remedy for this is for archaeologists to always clearly present the ethnographic analogies they are using in interpretations, so that these later can be used as inspiration for researchers that deal with other materials and problems.

Contemporary analogies were most certainly used in the interpretation of the deposits since these are integrated in the general human thought process. Much of what we take for granted about meanings behind human behaviour in past societies is actually based on similarities with our own world. I did however try to limit unconscious and generic inference from modern society as far as this was possible. This was mainly done by always trying to identify the actual sources used in interpretations, and and by constantly questioning ideas, both my own and other researchers', that are based on "common sense".

9.1.7 The interpretations

Even though the methods leading to the interpretation of deposits are the main focus of this thesis, it is interesting to analyse how the special deposits were interpreted and explained. Of the 20 special deposits identified in the case study materials, 3 came from Dongjum, 13 from Leeuwarden, 3 from Midlaren and 1 from Uppåkra. Of these 3 were considered normal refuse and 17 were interpreted as ritual.

The most common ritual interpretation was partial offering. This was the case for the three articulated sheep/goat deposits from Leeuwarden, the four skull/crania deposits from Midlaren and Leeuwarden, and the bone layer from Uppåkra, even though this interpretation also included a connection to feasting. The five features with human bones found in Leeuwarden were interpreted as consciously deposited and could be the remains of trophy hunting, ancestor worship or the disposal of magical tools. The large amount of wild bird bones found in a sunken hut in Leeuwarden was interpreted as the remains of feasting at two different occasions. The four astragali from Leeuwarden were interpreted as a set of pieces used for gaming and/or divination that may have been deposited as an offering. Finally the large cremation pit from Midlaren was interpreted as a burnt offering, possibly connected to a burial, or some other ritual of the dead.

The fact that partial offering is by far the most common ritual interpretation is really not surprising considering what we know from historical and ethnographic sources. It seems to be common practice not to spend complete animals on offerings to supernatural powers, but to consume most of the meat. Only if the ritual also functioned as an important indicator of wealth and status, complete animals were given to the supernatural powers.

The five features with human bones found in Leeuwarden, and especially the lost deposit from the well show that such finds at settlements should not be habitually dismissed as the remains of disturbed graves, but seen as potential ritual deposits.

The interpretations in general show that we have to be very observant of the composition of deposits. If not we risk missing the less obvious ones, such as the four astragali and the large amount of bird bones in Leeuwarden. It is important to always take a step back and think about what was actually found and why it ended up in that place.

9.2 A holistic approach

The discussion in the previous section leads to my suggested methodological framework for understanding individual deposits. This approach is based on a holistic perspective where all information should be regarded as potentially valuable and where every possible scrap of data and every idea have to be assessed. This does not imply that the researcher should produce incredibly detailed studies and get bogged down in a never ending search for information. Rather it suggests openness in the approach, to take various methods and ideas into consideration and to avoid formalisation, simplification and stereotyping. It is just impossible to approach such a complex problem as this from only one perspective, or with the use of only one source material.

Rituals and their place in society is a very good example of why a holistic approach is needed for the interpretation of individual deposits. As with most human activity, rituals are not something that can be viewed and understood in separation. They were, and still are an integrated part of life. The very formalised and detached role of religion in the western world might lead us to think otherwise, but modern religious historical, anthropological and archaeological research indicates a close integration of most rituals in everyday life (see for example Carlie 2004: 9; Insoll 2004: 154-159; Steinsland 2005: 261-273).

If we lift our gaze from religious activities and also view the countless everyday rituals performed for various reasons today, from birthday customs to academic ceremonies, conventions for greeting someone or eating procedures, it is quite apparent that most of them are integrated parts of our lives. They are often deemed important but their meaning and origin are frequently obscure, or at least difficult to logically motivate, even for the performer self.

9.3 The responsibilities of the excavator and the expert

9.3.1 Introduction

For a holistic approach to work regarding the interpretation of individual deposits some responsibilities of the researchers involved in the archaeological process have to be fulfilled. As always the real world presents many practical problems and puts constraints on the researchers. These constraints usually involve a shortage of time and money, something that is unavoidable and has to be dealt with on a case to case basis. The point here is therefore not to propose that the archaeologists involved must perform miracles, but rather that they should keep this aspect of archaeological research in mind when performing excavations and material analysis, and use the available resources accordingly.

9.3.2 The responsibilities of the excavator

The excavating archaeologist is the most important person for making the approach suggested in this chapter possible. The information and material that is not excavated and recorded in the field will be irretrievably lost regardless of how thorough and methodical later analyses are. The experience from the four materials analysed in this thesis is that there is a direct correlation between the amount of information recorded about a feature during the excavation and the extent and reliability of later interpretations. Seemingly simple things such as the recording of a feature through good descriptions, field drawings and photographs are often what separates a wild hypothesis from a convincing explanation in the analysing and interpreting phases. The following points are the most important responsibilities of the excavator in order to make detailed intra-site analyses of deposits possible:

- A thorough archaeological methodology and knowledge of the theoretical possibilities and limitations of the discipline.
- An appreciation of the importance of individual features for understanding human behaviour.
- An attentiveness for find circumstances, also when mass materials are excavated.
- A good field documentation that is made available to the expert as early as possible.
- An openness towards, and integration of, the expert early in a project, preferably already in the planning stages, and a willingness to make use of his or her skills both in the field and in the archaeological interpretation phase.

9.4.2 The responsibilities of the expert

In addition to the duties of the field archaeologist, the expert analysing the material also have corresponding responsibilities if the suggested approach should be made possible. In theory the expert's effort could be recreated by another researcher since his or her work, in contrast to an excavation, is non-destructive. In practice a limited amount of time and money makes repeated analyses of the same material an utopian situation for all but the most high profile sites. Furthermore if the expert is to play a direct part during the excavation, as is envisaged above, the role of the expert as a field advisor suddenly forces him or her to make decisions as irreversible as any other during an excavation. With this in mind I consider the following points the most important duties of the expert, both as a potential participant in the archaeological field work, as well as during the later stages of analysis and interpretation:

- A willingness to learn the theory, methodology and limitations of archaeology as a discipline, and not only the specifics of the expert's own sub-discipline.

- A readiness to study the area, time period and society from where the material originates, also outside of the expert's sub-discipline.

- A willingness to participate both in the planning phase and in the field part of an excavation in order to support the field archaeologists.

- The delivery of a report of the material analysis focusing on the archaeological problems, including interpretations based on the expert's knowledge, and not only a technical report.

Appendix I – Additional butchery marks

CRANIUM

13 14	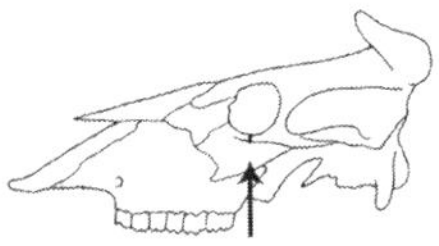	lateral	13 14	chop cut	on zygomaticum
15	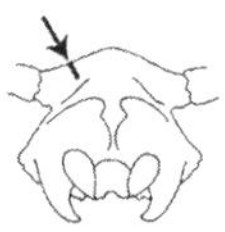	aboral	15	chop	on the rear part of frontale
16	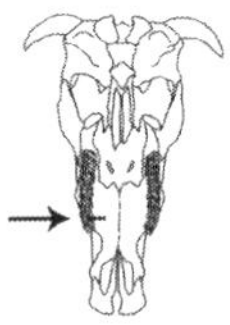	basal	16	chop	inside the dental row
17	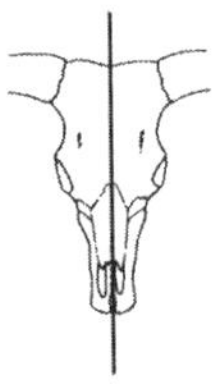	dorsal	17	chop	chopped through
18	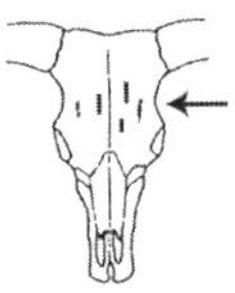	dorsal	18	chop	on frontale
19 20	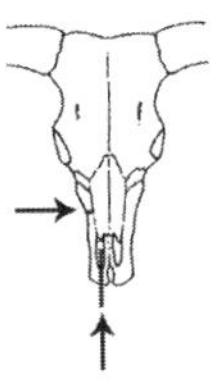	dorsal	19 20	chop cut	incisivum choped through on incisivum
21	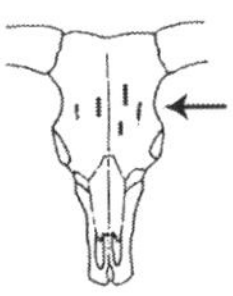	dorsal	21	cut	as 18, but cut

HORN-CORE

9	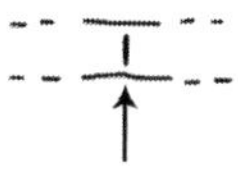		9	chop	as 7/8, but not cut through

MANDIBULA

34	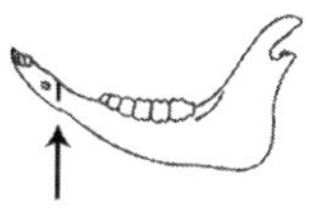	lateral	34	chop	on lateral side of the diastema
35	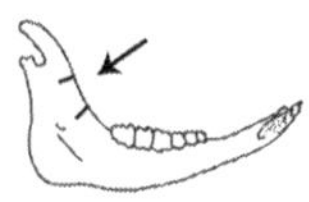	medial	35	cut	on medial side of the ramus
36 37	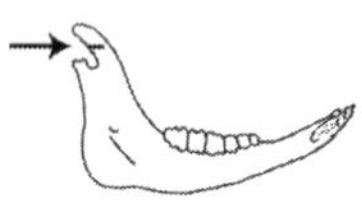	medial	36 37	chop cut	on medial side of processus coronoides as 36, but cut
38	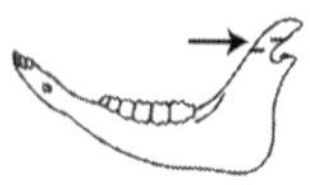	lateral	38	chop	on lateral side of processus coronoideus
39 40	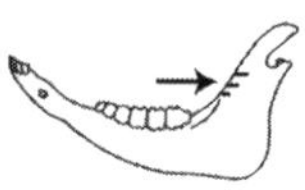	lateral	39 40	cut chop	on lateral side of the ramus as 39, but chop
41	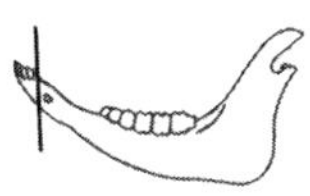	lateral	41	chop	mandible chopped through near the incisors

ATLAS

19	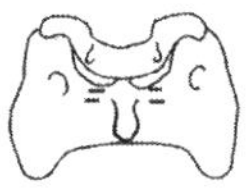	ventral	19	cut	on cranial part
20	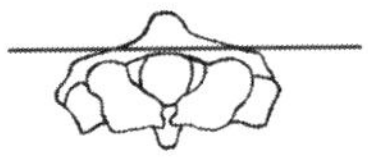	medial	20	chop	chopped through
21	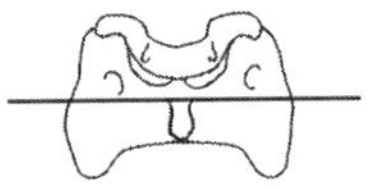	ventral	21	chop	chopped through
22	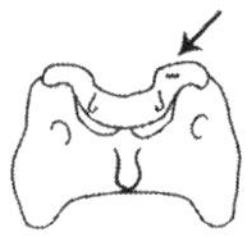	ventral	22	cut	on cranial surface

VERTEBRAE

20 21	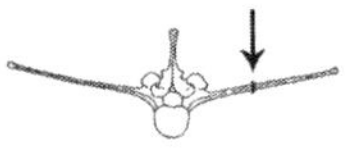	cranial	20 21	chop cut	on processus transversus as 20, but cut
22 23	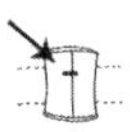	ventral	22 23	cut chop	on ventral corpus vertebrae as 22, but chop
24	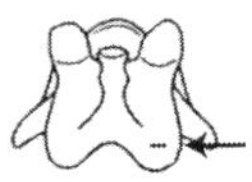	dorsal	24	chop	as 8, but not chopped through
25		lateral	25	cut	on lateral side of the corpus
26 27	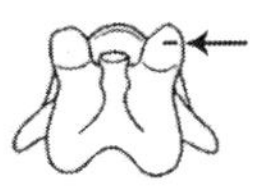	dorsal	26 27	cut chop	on processus articularis cranialis as 26, but chop

SCAPULA

41	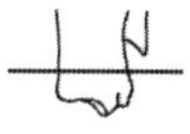	distal	41	chop	as 29, but chopped through

HUMERUS

37		distal cranial	37	chop	as 18, but chopped through

ULNA

16	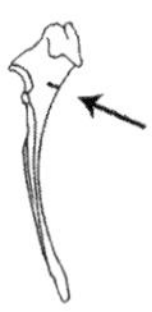	lateral	16	cut	on volar-lateral side of the proximal end
17		lateral	17	chop	on lateral side of the distal part
18	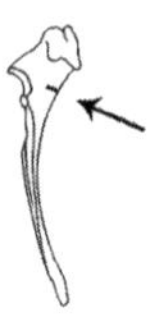	medial	18	chop	as 16, but chop
19	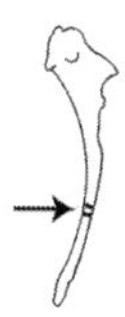	medial	19	cut	on medial side of the distal part

METAPODIA

31		plantar	31		shaving mark

PHALANGES

13		distal	13	chop	on distal surface

PELVIS

30	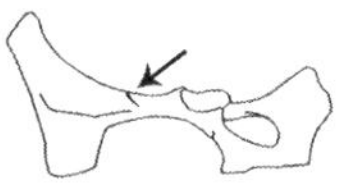	medial	30	cut	on medial side of the ischium

SACRUM

1 2	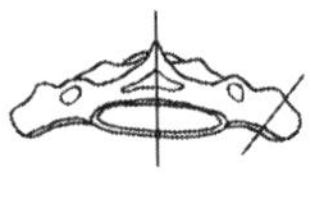	cranial	1 2	chop chop	corpus chopped through processus articularis chopped through

CENTROTARSALE

1	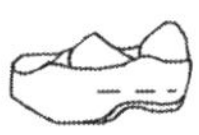	dorsal	1	cut	on corpus

ASTRAGALUS

14		plantar	14	cut	on middle part of the plantar side
15 16		dorsal	15 16	chop	on proximal part of the dorsal side as 15, but carved out

Appendix II - Aurochs atlas measurements

Table 1. Measurements of finds of atlases from aurochs (*Bos primigenius*) in Europe.

Find	Sex	Age	GB	GL	BFcr	BFcd	GLF	H	Index*
Dongjum 98 III 513	?	?	184.2	116.6	108.4	106.8	95.4	-	158.0
Potsdam-Schlaatz 19	Male	7-8 years	252.0	150.0	134.0	132.0	119.0	117.0	168.0
Seeberg Burgäschisee-Süd (several individuals)	?	?	-	-	116-140	108-130	-	-	-
Jura (Ain, France)	Male	Juvenile	202.8	121.5	120.0	129.7	106.0	-	166.9
Friedland, Kreis Neubrandenburg	Male	>9 years	234.0	133.0	127.0	119.0	(106.0)	104.0	175.9
Neumark-Nord	Male	>4 years	(>244)	165	157	(145)	136	126	(>148)
Bedburg-Köningshoven	?	?	240	133	131	127	95	113	180.5
Kanaalwerken Buinen-Schoonoord (49)	Male	Subadult	175.5	120.2	118.4	110.5	99.9	93.1	146.0
Zwolle (74)	?	?	265.2	159.6	152.3	141.1	117.9	129.4	166.2
Koerhuisbeek Deventeer (75)	?	?	218.0	141.8	137.2	131.5	108.0	105.1	153.7
Unknown location (76) (Bison priscus ?)	?	?	-	-	143.0	146.0	117.7	-	-
Zwarte Water (77)	?	?	190.0	125.0	114.0	106.0	100.9	100.9	152.0

* (GB*100)/GL

Table 2. References to the finds in table 1.

Find	Reference
Potsdam-Schlaatz 19, Seeberg Burgäschisee-Süd	Teichert, L. 1987. Knochenfunde com Ur (Bos primigenius Bojanus 1827) am Schlaatz bei Potsdam. *Veröffentlichungen des Museums für Ur- und Frügeschichte Potsdam. Bd. 21/1987.* pp. 37-45.
Jura (Ain, France)	Chaix, L. & Valton, B. 1984. Note sur un aurochs (Bos primigenius Bojanus) subatlantique de Jura Gessien (Ain, France). *Revue de paléobiologie. Vol. 3. N° 2.* Genéve. pp. 185-190.
Friedland, Kreis Neubrandenburg	Lehmkuhl, U. 1988. Funde vom Ur (Bos primigenius Bojanus, 1827) bei Friedland, Kreis Neubrandenburg. *Bodendenkmalpflege in Mecklenburg. Jb. 1987.* Berlin. pp. 9-28.
Neumark-Nord	Döhle. H-J. 1990. Osteologische Untersuchungen am Ur (Bos primigenius Bojanus, 1827) von Neumark-Nord. *Veröffentlichungen des Landesmuseums in Halle. Bd. 43.* pp. 177-192 (Tafeln pp. 300-308).
Bedburg-Köningshoven	Street. M. 1999. Remains of aurochs (Bos primigenius) from the early Mesolithic site Bedburg-Köningshoven (Rhineland, Germany). *Archäologie und Biologie des Auerochsen. Wissenschaftliche Schriften des Neanderthal Museums. Bd. 1.* Neanderthal Museum. pp. 173-194.
Kanaalwerken Buinen-Schoonoord (49), Zwolle (74), Koerhuisbeek Deventeer (75), Unknown location (76) (Bison priscus ?), Zwarte Water (77)	Es, L.J.M. van. 1990. *De oeros in noord en midden Nederland.* Doctoraalonderzoek. Biologisch-Archaeologisch Instituut. Groningen.

References

Abbott, J. 1932. *Keys of Power: A Study of Indian Ritual and Belief.* Delhi.

Alekshin, V.A. 1983. Burial Customs as an Archaeological Source. *Current Anthropology. Vol. 24. No. 2.* pp. 137-149.

Andersen, N.H. 2004. Sarup. In Bogucki, P. & Crabtree, P.J. (eds.). *Ancient Europe 8000 B.C.-A.D. 1000 - Encyclopedia of the Barbarian World. Volume I. The Mesolithic to Copper Age (C. 8000-2000 B.C.).* New York. pp. 301-304.

Andrén, A. 1989. State and towns in the Middle Ages. The Scandinavian experience. *Theory and Society 18.* Dordrecht. pp. 585-609.

- 1998. En centralort utan textbelägg? - Uppåkra som ett historiskt-arkeologiskt problem. In Larsson, L. & Hårdh, B. (eds.). *Centrala Platser - Centrala Frågor. Samhällsstrukturen under Järnåldern.* Uppåkrastudier 1. Acta Archaeologica Lundensia Series In 8°, No. 28. Lund. pp. 137-146.

Anthony, D.W. 2004 Domestication of the Horse. In Bogucki, P. & Crabtree, P.J. (eds.). *Ancient Europe 8000 B.C.-A.D. 1000. Encyclopedia of the Barbarian World. Volume I. The Mesolithic to Copper Age (C. 8000-2000 B.C.).* New York. pp. 363-367.

Appelgren, K. 2007. *Gravar och begravningar? Utbyggnad av Väg 73, delen Hammarbacken-Nyfors Södermanland, Ösmo socken, Valla 1:1 och 2:1, RAÄ 266.* UV Mitt, dokumentation av fältarbetsfasen 2007:1. Arkeologisk undersökning. Hägersten.

Arjaans, J. 1991. Terpafgravingen in Friesland. *Jaarverslagen van de Vereniging voor Terpenonderzoek 75.* pp. 45-55.

Artelius, T. & Nordin, P. 2007. *Stavkyrka, gravar och skeppssättning - Halland, Tölö socken, Tölö 8:12, Tölö 121 och 212.* UV Väst Rapport 2007:11. Arkeologisk förundersökning och undersökning. Riksantikvarieämbetet. Mölndal.

Azier, I.J.J. 2010. *Uitwerking steilkandonderzoek van de terp Dongjum.* Masterscriptie. Archeologie. Rijksuniversiteit Groningen. Groningen.

Baal, J. van. 1976. Offering, Sacrifice and Gift. *Numen, Vol 23. Fasc. 3.* pp. 161-178.

Baaren, T.P. van. 1964. Theoretical Speculations on Sacrifice. *Numen, Vol 11. Fasc. 1.* pp. 1-12.

Backe, M., Edgren, B. & Herschend, F. 1993. Bones Thrown into a Water-Hole. *Pact 38 - III. 6.* pp. 327-342.

Bagge, S. 1998. *Mennesket i middelalderens Norge. Tanker, tro og holdninger 1000-1300.* Oslo.

Bailey, D.W. 2000. *Balkan Prehistory: Exclusion, Incorporation and Identity.* London.

Bar-Oz, G. 2001. An Inscribed Astragalus with a Dedication to Hermes. *Near Eastern Archaeology Vol. 64, No. 4.* pp. 215-217.

Bazelmans, J. 1991. Conceptualising early Germanic political structure: a review of the use of concept of Gefolgschaft. In Roymans, N. & Theuws, F. (eds.). *Images of the past. Studies on Ancient Societies in Northwestern Europe.* Studies in Pre- en Protohistorie 7. Amsterdam. pp. 91-129.

- 1999. *By Weapons Made Worthy. Lords, retainers and their relationship in Beowulf.* Archaeological Studies 5. Amsterdam.

- 2000. Een Laat-Romeins Bewoningshiaat in het Nederlandse Kustgebiet en het Voortbestaan van de Friezennaam. In Taayke, E. & Kramer, E. (eds.). *Jaarverslagen van de Vereniging voor Terpenonderzoek 76-82. 1992-1998.* Groningen. pp. 14-75.

- 2005. Die Wurten von Dongjum-Heringa, Peins-Oost und Wijnaldum-Tjitsma: kleinmaßstäblicher Deichbau in ur- und frühgeschichtlicher Zeit des nördlichen Westergo. *Küste und Marsch.* Oldenburg. pp. 68-84.

Bazelmans, J., Gerrets, D., de Koning, J. & Vos, P. 1999. Zoden aan de dijk. Kleinschalige dijkbow in de late prehistorie en protohistorie van noordelijk Westergo. *De Vrije Fries. Vol. 79.* Leeuwarden. pp. 7-73.

Becker, C.J. 1972. "Mosepotter" fra Danmarks jernealder - Problemer omkring mosefundne lerkar og deres tolkning. *Aarbøger for Nordisk Oldkyndighed og Historie. 1971.* pp. 5-60.

Behrensmeyer, A.K. 1978. Taphonomic and Ecologic Information from Bone Weathering. *Paleobiology 4.* pp. 150-162.

Bermann, M. & Estévez Castillo, J. 1995. Domestic Artifact Assemblages and Ritual Activities in the Bolivian Formative. *Journal of Field Archaeology, Vol. 22, No. 4 (Winter, 1995).* pp. 389-398.

Berryman, H.E. 2002. Disarticulation Pattern and Tooth Mark Artifacts Associated with Pig Scavenging of Human Remains: A Case Study. In Haglund, W.D. & Sorg, M.H. (eds.). *Advances in Forensic Taphonomy. Method, Theory, and Archaeological Perspectives.* Boca Raton. pp. 487-495.

Besteman, J.C., Bos, J.M., Gerrets, D.A., Heidinga, H.A. & de Koning, J. 1999. *The Excavations at Wijnaldum. Repports on Frisia in Roman and Medieval Times.* Volume 1. Rotterdam.

Binford, L. 1967. Smudge Pits and Hide Smoking: The Use of Analogy in Archaeological Reasoning. *American Antiquity, Vol. 32, No. 1.* pp. 1-12.

- 1981. *Bones - Ancient Men and Modern Myths.* Studies in Archaeology. Academic press. New York.

Bloch, M. 1996. Ancestors. In Barnard, A. & Spencer, J. (eds.). *Encyclopedia of Social and Cultural Anthropology.* London. pp. 66-67.

Bolyanatz. A.H. 2000. *Mortuary feasting on New Ireland: the activation of matriliny among the Sursurunga.* Westport.

Bonnefoy, Y. 1991. *Asian mythologies.* Translated by Doniger, W. Chicago.

Bonsall, C. 2004. Iron Gates Mesolithic. In Bogucki, P. & Crabtree, P. J. (eds.). *Ancient Europe 8000 B.C.-A.D. 1000 - Encyclopedia of the Barbarian World. Volume I. The Mesolithic to Copper Age (C. 8000–2000 B.C.).* New York. pp. 175-179.

Bookidis, N. & Stroud R.S. 1997. *The sanctuary of Demeter and Kore, Topography and architecture.* Corinth - Results of excavations conducted by the American school of classical studies at Athens volume xviii, part iii. New Jersey.

Branca, A., Helgesson, B., Hårdh, B. & Tegnér, M. 1999. Detektorfunna föremål från järnåldern. Översikt av materialet vid årsskiftet 1998/1999. In Hårdh, B. (ed.). *Fynden i centrum. Keramik, glas och metall från Uppåkra.* Uppåkrastudier 2. Acta Archaeologica Lundensia Series In 8°, No. 30. Lund. pp. 59-65.

Brothwell, D. 1976. Further Evidence of Bone Chewing by Ungulates: the sheep of North Ronaldsay, Orkney. *Journal of Archaeological Science 3.* pp. 179-182.

Campbell, E. 1991. Excavations of a wheelhouse and other Iron Age structures at Sollas, North Uist, by R J C Atkinson in 1957. *Proceedings of the Society of Antiquaries of Scotland, 121 (1991).* pp. 117-173.

Capelle, T. 1987. Eisenzeitliche Bauopfer. *Frühmittelalterliche Studien. Jahrbuch des Instituts für Frühmittelalterforschung der Universität Münster. Band 21.* Berlin. pp. 182-205.

Carlie, A. 2004. *Forntida byggnadskult. Tradition och regionalitet i södra Skandinavien.* Riksantikvarieämbetet. Arkeologiska undersökningar, Skrifter No 57. Stockholm.

Carlie, L. 1992. *Brogård – ett brons- och järnålderskomplex i södra Halland. Dess kronologi och struktur.* Hallands Länsmuseers skriftserie N° 6. Varberg.

Chacon, R.J. & Dye, D.H. 2007. 1. Introduction to Human Trophy Taking: An Ancient and Widespread Practice. In Chacon, R.J. & Dye, D.H. (eds.). *The Taking and Displaying of Human Body Parts as Trophies by Amerindians.* Interdisciplinary contributions to archaeology. New York. pp. 5-31.

Chidester, D., Kwenda, C., Petty, R., Tobler, J. & Wratten, D. 1997. *African traditional religion in South Africa : an annotated bibliography.* Westport.

Clunies Ross, M. 2002. Närvaron och frånvaron av ritual i norröna medeltida texter. In Jennbert, K., Andrén, A. & Raudevere, C. (eds.). *Plats och praxis. Studier av nordisk förkristen ritual.* Vägar till Midgård 2. Lund. pp. 13-30.

Conkey, M.W. & Spector, J.D. 1984. Archaeology and the Study of Gender. In Schiffer, M.B. (ed.). *Advances in Archaeological Method and Theory, Volume 7.* Orlando. pp. 1-38.

Cope, C. 2004. The butchering patterns of Gamla and Yodefat: beginning the search for *kosher* practices. In O'Day, S.J., Neer, W. van & Ervynck, A. (eds.). *Behaviour Behind Bones. The zooarchaeology of ritual, religion, status and identity.* Proceedings of the 9th ICAZ Conference, Durham 2002. Oxford. pp. 25-33.

Crawford, S. Votive Deposition, Religion and the Anglo-Saxon Furnished Burial Ritual. *World Archaeology. Vol. 36. No. 1.* pp. 87-102.

Crump, T. 1996. Household activities. In Barnard, A. & Spencer, J. (eds.). *Encyclopedia of Social and Cultural Anthropology.* London. pp. 433-465.

Cuijpers, A.G.F.M., Haverkort, C.M., Pasveer, J.M. & Prummel, W. 1999. The human burials. In Besteman, J.C., Bos, J.M., Gerrts, D.A., Heidinga, H.A. & de Koning, J. (eds.). *The excavations at Wijnaldum: Reports on Frisia in Roman and Medieval Times. Volume 1.* Rotterdam. pp. 305-322.

Cunliffe, B. 1995. *Danebury: an Iron Age hillfort in Hampshire. Volume 6. A hillfort community in perspective.* Research Report 102. Council for British Archaeology. London.

Dart, R.A. 1956. The Myth of the Bone-Accumulating Hyena. *American Anthropologist 58.* pp. 40-62.

Davis, S.J.M. 1987. *The archaeology of animals.* London.

Dickmann, E. 1997. *Erin. Archäologie in Castrop-Rauxel.* Ausstellungskatalog. Castrop-Rauxel.

Dijkstra, J. & Nicolay, J.A.W. 2006. Een terp op de schop: Het archeologisch onderzoek op het Oldehoofsterkerkhof in Leeuwarden (Fr.). *Paleo-Aktueel 17.* pp. 135-141.

- 2008. *Een terp op de schop. Archeologisch ondezoek op het Oldehoofsterkerkhof te Leeuwarden.* ADC Monografie 3/ADC Rapport 1227. Amersfoort.

Dijkstra, J., Gerrets, D.A. & Nicolay, J.A.W. 2008. 15. Synthesis: origin and development of the Oldehove terp at Leeuwarden. In Dijkstra, J.& Nicolay, J.A.W. (eds.). *Een terp op de schop. Archeologisch ondezoek op het Oldehoofsterkerkhof te Leeuwarden.* ADC Monografie 3/ADC Rapport 1227. Amersfoort. pp. 307-342.

Driesch, A. von den 1976. *A guide to the measurement of animal bones from archaeological sites.* Peabody Museum Bullentin 1. Harvard University. Cambridge.

Driesch, A. von den & Boessneck, J. 1975. Schnittspuren an neolitischen Tierknochen. Ein Beitrag zur

Schlachttierzerlegung in vorgeschichtlicher Zeit. *Germania. Anzeiger der Römisch-Germanischen Kommission des Deutschen Archäologischen Instituts.* Jahrgang 53. Berlin.

Driver, H.E. & Massey, W.C. 1957. Comparative Studies of North American Indians. *Transactions of the American Philosophical Society. New Series. Vol. 74. No. 2.* pp. 165-456.

Ekman, J. 1973. *Early Medieval Lund – the fauna and the landscape.* Archaeologica Lundensia. Investigationes de antiqvitatibus urbis Lundae V. Museum of Cultural History. Lund.

Erikson, M. 2001. En väg till Uppåkra. In Larsson, L. (ed.). *Uppåkra. Centrum i analys och rapport.* Uppåkrastudier 4. Acta Archaeologica Lundensia Series In 8°, No. 36. Lund. pp. 167-176.

Eriksson, M. & Iregren, E. 2002. *Bildkompendium Historisk Osteologi.* University of Lund. Department of Archaeology and Ancient History. Report series No. 85. Lund.

Eriksson, M. & Magnell, O. 2001. Det djuriska Tågerup. In Karstens, P. & Knarrström, B. (eds.). *Tågerup specialstudier. Skånska spår – arkeologi längs Västkustbanan.* Riksantikvarieämbetet. Lund. pp. 156-237.

Ersgård, L. 2002. Västannortjärn - En rituell plats från högmedeltid. In Jennbert, K., Andrén, A. & Raudevere, C. (eds.). *Plats och praxis. Studier av nordisk förkristen ritual.* Vägar till Midgård 2. Lund. pp. 287-297.

Ervynck, A. 2005. Detecting the Seasonal Slaughtering of Domestic Mammals: Inferences from the Detailed Recording of Tooth Eruption and Wear. *Environmental Archaeology 10 (2).* pp. 153-169.

Fabech, C. 1998. Kult og Samfund i yngre jernalder – Ravlunda som eksempel. In Larsson, L. & Hårdh, B. (eds.). *Centrala Platser – Centrala Frågor. Samhällsstrukturen under Järnåldern.* Uppåkrastudier 1. Acta Archaeologica Lundensia Series In 8°, No. 28. Lund. pp. 147-163.

Falk, A-B. 2008. *En grundläggande handling: byggnadsoffer och dagligt liv i medeltid.* Vägar till Midgård 12. Lund.

Fendin, T. 2006. Grinding processes and reproductive metaphors. In Andrén, A., Jennbert, K. & Raudvere, C. (eds.). *Old Norse religion in long-term perspectives: origins, changes, and interactions.* Vägar till Midgård. Lund. pp. 159-163

Ferdinand, J. & Ferdinand, K. 1962. Jernalderofferfund i Valmose ved Rislev. *KUML. Årbog for Jysk Arkæologisk Selskab.* Aarhus. pp. 47-90.

Foster, G.V. 1984. The Bones from the Altar West of the Painted Stoa. *Hesperia, Vol. 53, No. 1.* pp. 73-82.

Frölich, A. 2007. Kirurgiska instrument. In Hårdh, B. & Larsson, L. (eds.). *Uppåkra – Lund före Lund.* Föreningen Gamla Lund Årsbok 2007. Lund. pp. 59-60.

Gabriel, I & Heinrich, D. 1976. Spendegefäss und Brandopfer der älteren vorrömishchen Eisenzeit aus Oldenburg in Holstein. *Offa. Berichte und Mitteilungen zur Urgeschichte, Frügeschichte und Mittelalterarchäologie. Band 33.* Neumünster. pp. 123-131.

Gerritsen, F. 2003. *Local Identities. Landscape and community in the late prehistoric Meuse-Demer-Scheldt region.* Archaeological Studies 9. Amsterdam.

Giffen, A. E. van. 1936. Der Warf in Ezinge, Provinz Groningen, Holland, und seine westgermanischen Häuser. *Germania 20.* pp. 40-47.

- 1963. Het bouwoffer uit de oudste hoeve te Ezinge. *Helinium 3.* pp. 246-253.

Gifford-Gonzales, D.P., Damrosch, D.B., Damrosch, D.R., Pryor, J. & Thunen, R.L. 1985. The Third Dimension on Site Structure: An Experiment in Trampling and Vertical Dispersal. *American Antiquity 50.* pp. 803-818.

Gijn, A. van & Raemaekers, D. C. M. 1999. Tool Use and Society in the Dutch Neolithic: The Inevitability of Ethnographic Analogies. In Owen, L. R. & Porr, M. (eds.). *Ethno-Analogy and the Reconstruction of Prehistoric Artefact Use and Production.* Tübingen. pp. 43-52.

Gilbert, M. 1987. 8. The person of the king: Ritual power in a Ghanaian state. In Cannadine, D. & Price, S. (eds.). *Rituals of royalty: power and ceremonial in traditional societies.* Cambridge. pp. 298-330.

Gilchrist, J. 1994. *Gender and Material Culture. The Archaeology of Religious Women.* London.

Gilmour, G. H. 1997. The Nature and Function of Astragalus Bones from Archaeological Contexts in the Levant and Eastern Mediterranean. *Oxford Journal of Archaeology 16(2).* pp. 167-175.

Girard, J. 1810. *Anatomie der Hausthiere Bd 1.* Fransösischen überseßt. München.

Graham-Campbell, J. & Kidd, D. 1980. *The Vikings.* British Museum Publications. London.

Grane, T. 2003. Romerske kilder til Germaniens geografi og etnografi. In Jørgensen, L., Storgaard, B. & Gebauer Thomsen, L. (eds.). *Sejrens triumf – Norden i skyggen af det romerske Imperium.* Nationalmuseet. København. pp. 126-147.

Grant, A. 1982. The use of tooth wear as a guide to the age of domestic ungulates. In Wilson, B., Grigson, C. & Payne, S. (eds.). *Ageing and Sexing Animal Bones from Archaeological Sites.* BAR British Series 109. Oxford. pp. 91-107.

- 1984a. Animal husbandry. In Cunliffe, B. (ed.). *Danebury: an Iron Age hillfort in Hampshire. Volume 2. The excavations. 1969-1978: the finds.* Research Report 52. Council for British Archaeology. London. pp. 496-543.

- 1984b. Survival or sacrifice? A critical appraisal of animal burials in Britain in the Iron Age. In Griegson, C. & Clutton-Brock (eds.). *Animals and Archaeology: 4 Husbandry in Europe.* BAR International Series 227. Oxford. pp. 221-227.

- 1989. Animals and ritual in early Britain: The visible and the invisible. *Anthropozoologica. L'animal dans les pratiques religieuse: Les manifestation materielles. Troisième numéro spécial*. pp. 79-86.

- 1991. Economic or Symbolic? Animals and Ritual Behaviour. In Garwood, P., Jennings, D., Skeates, R. & Toms, J. (eds.). *Sacred and Profane. Proceedings of a Conference on Archaeology, Ritual and Religion. Oxford, 1989*. Oxford University Committee for Archaeology Monograph. Oxford. pp. 109-114.

Grant, J., Gorin, S. & Fleming, N. 2008. *The Archaeology Coursebook - An introduction to themes, sites, methods and skills*. Third Edition. London.

Green, M. 1992. *Animals in Celtic Life and Myth*. London.

Greenfield, H.J. 1988. Bone Consumptions by Pigs in a Contemporary Serbian Village: Implications for the Interpretation of Prehistoric Faunal Assemblages. *Journal of Field Archeology, 15*. pp. 473-479.

Groenewoudt, B., Groothedde, M. & van der Velde, H. 2006. De Romeinse tijd, Middeleeuwen en Vroegmoderne tijd in het Midden- en Oost-Nederlands zandgebied. *NOaA chapter 20 (version 1.0), (www.noaa.nl)*. pp. 1-59.

Groot, M. 2008. *Animals in Ritual and Economy in a Frontier Community. Excavations in Tiel-Passewaaij*. Amsterdam Archaeological Studies 12. Amsterdam University Press. Amsterdam.

Grunfeld, F.V. 1975. *Games of the world: how to make them, how to play them, how they came to be*. New York.

Gräslund, A-S. 2004. Dogs in graves – A question of symbolism? In Santillo Frizell, B. (ed.). *PECUS. Man and animal in antiquity*. Proceedings of the conference at the Swedish Institute in Rome, September 9-12, 2002. The Swedish Institute in Rome. Projects and Seminars, 1, Rome 2004. Rome. pp. 168-176.

Grønnegaard, T. 1997. *Yngre jernalders centralpladser – infaldsvinkler til et helhedsbillede*. Hovedfagsspeciale. København.

Göckenjan, H. & Zimonyi, I. 2001. *Orientalische Berichte über die Völker Osteuropas und Zentralasiens im Mittelalter. Die Ǧayhānī-Tradition*. Veröffentlichungen der Societas Uralo-Altica. Band 54. Wiesbaden.

Göthberg, H., Schütz, B. & Jonsson, L. 2006. *I närheten av E4 2005 - Arkeologiska schaktningsövervakningar och undersökningar. Gamla Uppsala och Vaksala socknar, Uppsala kommun, Uppland*. Rapport 2006:14, Arkeologiska avdelningen, Upplandsmuseet. Uppsala.

Haarnagel, W. 1979. *Die Grabung Feddersen Wierde. Methode, Hausbau, Siedlungs- und Wirtschaftsformen sowie Sozialstruktur*. Textband. Wiesbaden.

Habermehl, K. H. 1975. *Die Altersbestimmung bei Haus- und Labortieren*. 2nd ed. Parey. Berlin.

Hagberg, U.E. 1967. *The archaeology of Skedemosse. 2, The votive deposits in the Skedemosse fen and their relation to the Iron-Age settlement on Öland, Sweden*. Stockholm.

Halbertsma, H. 1984. Het heidendom waar Liudger onder de Friezen mee te maken kreeg. In Duindam, H.F.J. (ed.). *Luidger 742-809. De confrontatie tussen heidendom en christendom in de Lage Landen*. De Bataafsche Leeuw. Muiderberg. pp. 21-43.

Hallowell, I. 1926. Bear Ceremonialism in the Northern Hemisphere. *American Anthropologist New Series, Vol. 28 January-March, No. 1*. pp. 1-175.

Hamerow, H. 2006. 'Special Deposits' in Anglo-Saxon Settlements. *Medieval Archaeology 50*. pp. 1-30.

Harsema, O.H. 1980. *Drents boerenleven van de bronstijd tot de middeleeuwen*. Museumfonds. Publicatie nummer 6. Assen.

Hatt, G. 1938. Jernalderens bopladser i Himmerland. *Aarbøger for nordisk Oldkyndighed og Historie*. pp. 119-266.

- 1960. The Roman Iron Age dwelling Site at Mariesminde, Vestervig. *Acta Archaeologica XXVI. 1960*. pp. 63-83.

Hatting, T. 1993. Oksefund fra Bukkerup og Turup. *Fynske Minder, Odense Bys Museer, årbok 1993*. pp. 93-98.

Haynes, G. 1980. Evidence of carnivore gnawing on Pleistocene and Recent mammalian bones. *Paleobiology* 6. pp. 341-351.

- 1983a. A guide for differentiating mammalian carnivore taxa responsible for gnawing damage to herbivore limb bones. *Paleobiology* 9. pp. 164-172.

- 1983b. Frequencies of spiral and green-bone fractures on ungulate limb bones in modern surface assemblages. *American Antiquity 48. No 1*. pp. 102-114.

Haverkort, C.M., Hopman, M., Pasveer, J.M. & Prummel, W. 1993. De Jongste Bewoners van Wijnaldum. *Paleoaktueel, Vol. 4*. pp. 123-126.

Hawkes, C. 1954. Wenner-Green Foundation Supper Conference: Archaeological Theory and Method: Some Suggestions from the Old World. *American Anthropologist. Vol 56 (2)*. pp. 155-168.

Headrick, A. 2007. *The Teotihuacan Trinity. The sociopolitical structure of an ancient Mesoamerican city*. Austin.

Hedeager, L. 2007 Scandinavia and the Huns: An Interdisciplinary Approach to the Migration Era. *Norwegian Archaeological Review, Vol. 40, No. 1*. pp. 42-58.

Heidinga, H.A. 1997. *Frisia in the First Millennium. An outline*. Utrecht.

Henriksen, M.B. 1998. Guden under gulvet - ofringer under fynske huse fra ældre jernalder. *Fynske Minder. 1998*. pp. 191-212.

Helgesson, B. 1998. Vad är centralt? – fenomen och funktion; lokalisering och person. In Larsson, L. & Hårdh, B. (eds.). *Centrala Platser – Centrala Frågor. Samhällsstrukturen under Järnåldern*. Uppåkrastudier 1. Acta Archaeologica Lundensia Series In 8°, No. 28. Lund. pp. 39-46.

- 2004. Tributes to be Spoken of. Sacrifice and Warriors in Uppåkra. In Larsson, L. (ed.). *Contiuity for centuries – A ceremonial building and its context at Uppåkra, Southern Sweden*. Uppåkrastudier 10. Acta Archaeologica Lundensia Series In 8°, No. 48. Lund. pp. 223-239.

Hertz, R. 2004. A Contribution to the Study of the Collective Representation of Death. In Robben, A.C.G.M. (ed.). *Death, Mourning, and Burial: A Cross-Cultural Reader.* Malden. pp. 197-212.

Hedman, S-D. 2003. *Boplatser och offerplatser. Ekonomisk strategi och boplatsmönster bland skogssamer 700-1600 AD.* Studia Archaeologica Universitatis Umensis 17. Institutionen för arkeologi och samiska studier. Umeå.

Hill, A. 1976. On Carnivore and Weathering Damage to Bone. *Current Anthropology 17. No 2.* pp. 335-336.

Hill, J.D. 1995. *Ritual and Rubbish in the Iron Age of Wessex – A study on the formation of a specific archaeological record.* BAR British Series 242. Oxford.

Hodder, I. 1982a. *Symbols in action: Ethnoarchaeological studies of material culture.* Cambridge University Press. Cambridge.

- 1982b. *The Present Past. An introduction to anthropology for archaeologists.* London.

- 1986. *Reading the past. Current approaches to interpretation in archaeology.* Cambridge.

Hodgson, G.W.I. & Jones, A. 1982. The animal bone. In Murray, J.C. (ed.). *Excavations in the medieval burgh of Aberdeen 1973-81.* Monographs of the Society of Antiquaries of Scotland. Monograph Series Number 2. Edinburgh. pp. 229-238.

Holl, J. 2005. *Een dierlijk crematiekuil in Midlaren. Thematisch Bronnenonderzoek.* Studentscriptie Archaeologie. Rijksuniversiteit Groningen.

Holmgren, R. 2004. "Money on the hoof" The astragalus bone – religion, gaming and primitive money. In Santillo Frizell, B. (ed.). *PECUS. Man and animal in antiquity. Proceedings of the conference at the Swedish Institute in Rome, September 9-12, 2002.* The Swedish Institute of Rome. Projects and Seminars 1. Rome. pp. 212-220.

Holmqvist, W. 1961. Introduction. In Holmqvist, W. (ed.). *Excavations at Helgö I. Report for 1954-1956.* Kungl. vitterhets historie och anmtikvitets akademien. Stockholm. pp. 22-44.

Hüster-Plogmann, H. 1993. Eine Experimentelle Schweineschlachtung nach Studien an Frümittelalterlichem Knochenmateriale aus dem Haithabu-Hafen. *Zeitschrift für Archäologie 27.* pp. 225-234.

Hårdh, B. 1998. Preliminära notiser kring detektorfynden från Uppåkra. In Larsson, L. & Hårdh, B. (eds.). *Centrala Platser – Centrala Frågor. Samhällsstrukturen under Järnåldern.* Uppåkrastudier 1. Acta Archaeologica Lundensia Series In 8°, No. 28. Lund. pp. 113-127.

- 2003. Förord. In Hårdh, B. (ed.). *Fler Fynd i centrum. Materialstudier i och kring Uppåkra.* Uppåkrastudier 9. Acta Archaeologica Lundensia Series In 8°, No. 45. Lund. pp. 1.

- 2004. A Metal Beaker with Embossed Foil Bands. In Larsson, L. (ed.). *Contiuity for centuries – A ceremonial building and its context at Uppåkra, Southern Sweden.* Uppåkrastudier 10. Acta Archaeologica Lundensia Series In 8°, No. 48. Lund. pp. 49-91.

Hårdh, B. & Larsson, L. 2007. *Uppåkra – Lund före Lund.* Föreningen Gamla Lund Årsbok 2007. Lund.

Hänninen, K., den Ouden, P., van Rijn, P. & Vorst, Y. 2008. 9. Constructiehout en houten voorwerpen. In Dijkstra, J.& Nicolay, J.A.W. (eds.). *Een terp op de schop. Archeologisch ondezoek op het Oldehoofsterkerkhof te Leeuwarden.* ADC Monografie 3/ADC Rapport 1227. Amersfoort. pp. 191-210.

Ilkjær, J. 2002. Den beviste ødeleggelse i krigsbytteofringerne. In Jennbert, K., Andrén, A. & Raudevere, C. (eds.). *Plats och praxis. Studier av nordisk förkristen ritual.* Vägar till Midgård 2. Lund. pp. 203-214.

- 2003. Danske krigsbytteofringer. In Jørgensen, L., Storgaard, B. & Gebauer Thomsen, L. (eds.). *Sejrens triumf – Norden i skyggen af det romerske Imperium.* Nationalmuseet. København. pp. 44-64.

Insoll, T. 2004. *Archaeology, Ritual, Religion.* London.

Iregren, E. 1997. Why animal bones in human graves – an attempt to interpret animals present in Iron age cremations in Sweden. In Smits, E., Iregren, E. & Drusini, A.G. (eds.). *Proceedings of the Symposium "Cremation studies in Archaeology", Amsterdam, 26-27 October 1995.* Padova. pp. 9-32.

Jameson, R. & Shaw, I. 1999. *A Dictionary of Archaeology.* Oxford.

Jans, J. 2005. *Rituelen in Midlaren? Een vroegmiddeleeuws dierlijk crematiegraf.* Bijvakscriptie Archaeologie. Rijksuniversiteit Groningen. Groningen.

Jansson, I. 1991. Gotland och omvärlden. In Jansson, I. (ed.). *Gutar och vikingar.* Statens Historiska Museum. Stockholm. pp. 207-247.

Jebb, R.C. 1898. *Homer: An Introduction to the Iliad and the Odyssey.* Glasgow.

Jensen, J. 1995. *The Prehistory of Denmark.* London.

Jensen, S. 1991. Dankirke-Ribe. Fra handelsgård til handelsplads. In Mortensen, P. (ed.). *Fra Stamme til stat i Danmark 2. Høvdingesamfund og Kongemagt.* Jydske Arkæologiske Selskab. Skr. XXVII. Århus. pp. 73-88.

Jensen, S. & Watt, M. 1993. Handelscentre og centralpladser. In Hvass, S. & Storgaard, B. (eds.). *Da klinger i muld... 25 års arkæologi i Danmark.* Det Kgl. Nordiske Oldskriftsselskab. Jysk Arkæologisk Selskab. Aarhus. pp. 195-199.

Jensen, X.P. 2003. Vimosefundet. In Jørgensen, L., Storgaard, B. & Gebauer Thomsen, L. (eds.). *Sejrens triumf - Norden i skyggen af det romerske Imperium.* Nationalmuseet. København. pp. 224-238.

Jeppesen, J. & Madsen, H.J. 1997. Trækirke og stormandshal i Lisbjerg. *KUML 1995-96. Årbog for Jysk Arkæologisk Selskab.* Århus. pp. 149-171.

Johnson, E. 1985. Current Developments in Bone Technology. In Schiffer, M.B. (ed.). *Advances in Archaeological Method and Theory 8.* Orlando. pp. 157-235.

Jonuks, T. & Konsa, M. 2007. The Revival of Prehistoric Burial Practices: Three Archaeological Experiments. *Folklore. Issue 37.* pp. 91-110.

Jonsson, R. 1972. Osteologisk analys av brandgravsmaterial från bronsåldershögar och bronsåldersgravar i södra Halland. In Lundborg, L. (ed.). *Undersökningar av bronsåldershögar och bronsåldersgravar i södra Halland. Höks, Tönnersjö och Halmstads härader under åren 1854-1970.* Hallands Museum, 2. Halmstad. pp. 131-142.

Junker, L.L. 2000. *Raiding, trading, and feasting: the political economy of Philippine chiefdoms.* Manila.

Jørgensen, L. 1998. En storgård fra vikingetid ved Tissø, Sjælland – en foreløbig præsentation. In Larsson, L. & Hårdh, B. (eds.). *Centrala Platser – Centrala Frågor. Samhällsstrukturen under Järnåldern.* Uppåkrastudier 1. Acta Archaeologica Lundensia Series In 8°, No. 28. Lund. pp. 39-46.

- 2002. Kongsgård - kultsted – marked. Overvejelser omkring Tissøkompleksets struktur of function. In Jennbert, K., Andrén, A. & Raudevere, C. (eds.). *Plats och praxis. Studier av nordisk förkristen ritual.* Vägar till Midgård 2. Lund. pp. 215-247.

- 2003. Sejerens triumf – Norden i skyggen af det romerske Imperium. In Jørgensen, L., Storgaard, B. & Gebauer Thomsen, L. (eds.). *Sejrens triumf – Norden i skyggen af det romerske Imperium.* Nationalmuseet. København. pp. 12-16.

Karsten, P. 1994. *Att kasta yxan i sjön - En studie över rituell tradition och förändring utifrån skånska neolitiska offerfynd.* Acta Archeologica Lundensia-Series in 8° No 23. Lund.

Kaul, F. 2003. Hjortspringsfundet – Det ældste af de store nordiske krigsbytteofre. In Jørgensen, L., Storgaard, B. & Gebauer Thomsen, L. (eds.). *Sejrens triumf - Norden i skyggen af det romerske Imperium.* Nationalmuseet. København. pp. 212-223.

Kelly, R.L. & Thomas, D.H. 2006. *Archaeology.* Belmont.

Kimura, T. 1999. Bearing the 'Bare Facts' of Ritual. A Critique of Jonathan Z. Smiths Study of the Bear Ceremony Based on a study of the Ainu Iyomante. *Numen, Vol. 46. No. 1.* pp. 88-114.

Kirsch, A.T. 1997. *Feasting and Social Oscillation: Religion and Society in Upland Southeast Asia.* Southeast Asia Program. Data paper, 92. New York.

Kjær, H. 1928. Oldtidshuse ved Ginderup i Thy. *Nationalmuseets Arbejdsmark 1928.* pp. 7-20.

Klindt-Jensen, O. 1957. *Bornholm i Folkevandringstiden.* Nationalmuseet. København.

Knol, E. 1987. Knucklebones in urns: Playful grave-goods in Early Medieval Friesland. *Helinium XXVII.* pp. 280-288.

Knol, E., Prummel, W., Uytterschaut, H.T., Hoogland, M.L.P., Casparie, W.A., de Langen, G.J., Kramer, E. & Schelvis, J. 1996. The early medieval cemetery of Oosterbeintum (Friesland). *Palaeohistoria 37/38.* pp. 245-416.

Krol, T. & Bos, J. 2005. Een verklaring voor de hoeveelheid Angelsaksisch aardewerk in Drenthe. *Paleo-aktueel 16.* pp. 101-104.

Kyle, D.G. 1998. *Spectacles of Death in Ancient Rome.* London.

Lambert, M. 1993. Ancient Greek and Zulu Sacrificial Ritual: A Comparative Analysis. *Numen, Vol. 40, No. 3.* pp. 293-318.

Larsson, L. 1998. Gjort och ogjort i Uppåkra. In Larsson, L. & Hårdh, B. (eds.). *Centrala Platser – Centrala Frågor. Samhällsstrukturen under Järnåldern.* Uppåkrastudier 1. Acta Archaeologica Lundensia Series In 8°, No. 28. Lund. pp. 95-112.

- 2003. The Uppåkra Project. Preconditions, Performance and Prospects. In Larsson, L. & Hårdh, B. (eds.). *Centrality – Regionality. The social structure of Southern Sweden during the Iron Age.* Uppåkrastudier 7. Acta Archaeologica Lundensia Series In 8°, No. 40. Lund. pp. 3-26.

- 2007. The Iron Age ritual building at Uppåkra, southern Sweden. *Antiquity. Vol. 81, No. 311.* pp. 11-25.

- 2009. Mordutredning pågår i Uppåkra. *Populär Arkeologi 1/2009.* pp. 14-15.

Larsson, L. & Hårdh, B. 1998. Förord. In Larsson, L. & Hårdh, B. (eds.). *Centrala Platser – Centrala Frågor. Samhällsstrukturen under Järnåldern.* Uppåkrastudier 1. Acta Archaeologica Lundensia Series In 8°, No. 28. Lund. pp. V.

Larsson, L. & Lenntorp, K-M. 2004. The Enigmatic House. In Larsson, L. (ed.). *Contiuity for centuries – A ceremonial building and its context at Uppåkra, Southern Sweden.* Uppåkrastudier 10. Acta Archaeologica Lundensia Series In 8°, No. 48. Lund. pp. 3-48.

Larsson, L. & Wyszomirska, B. 1989. *Arkeologi och Religion. Rapport från arkeologidagarna 16-18 januari 1989.* University of Lund. Institute of Archaeology Report Series No. 34. Lund.

Last, R. 2000. Inscribed Astragalus from Sha'ar Ha-'Amaqim. *Zeitschrift für Papyrologie und Epigraphik. Bd. 130.* pp. 248

Lauwerier, R.C.G.M. 1988. *Animals in Roman Times in the Dutch Eastern River Area.* Nederlandse Oudheden 12. Project Oostelijk Rivierengebied 1. ROB. Amersfoort.

- 2002. Animals as food for the soul. In Dobney, K. & O'Connor, T. (eds.). *Bones and the Man. Studies in honour of Don Brothwell.* Oxford. pp.63-71.

Lauwerier, R.C.G.M., Groenewoudt, B.J., Brinkkemper, O. & Laarman, F.J. 1999. Between ritual and economics: animals and plants in a fourth-century native settlement at Heeten, the Netherlands. *Berichten van de Rijksdienst voor het Oudheidkundig Bodemonderzoek 43, 1998-99.* pp. 155-198.

Legge, A., Williams, J. & Williams, P. 2000. Lambs to the slaughter: sacrifice at two Roman temples in southern England. In Rowley-Conwy, P. (ed.). *Animal Bones, Human Societies.* Oxford. pp. 152-157.

Lenntorp, K-M. & Lindell, M. 2000. *Rapport. Arkeologisk*

Förundersökning Lilla Uppåkra 4:1 & 8:1, fornlämning nr 5, Uppåkra socken, Staffanstorps kommun, Skåne Län. Lunds Universitet. Uppåkra.

Lentacker, A., Ervynck, A. & van Neer, W. 2004. Gastronomy or religion? the animal remains from the *mithraeum* at Tienen (Belgium). In O'Day, S.J., van Neer, W. & Ervynck, A. (eds.). *Behaviour Behind Bones. The zooarchaeology of ritual, religion, status and identity.* Proceedings of the 9th ICAZ Conference, Durham 2002. Oxford. pp. 77-94.

Lillie, M. 2004a. Late Neolithic/Copper Age Eastern Europe. In Bogucki, P. & Crabtree, P.J. (eds.). *Ancient Europe 8000 B.C.-A.D. 1000. Encyclopedia of the Barbarian World. Volume 2. The Bronze Age to Early Middle Ages (C. 3000 B.C.-A.D. 1000).* New York. pp. 354-363.

- 2004b. Late Neolithic/Copper Age Eastern Europe. In Bogucki, P. & Crabtree, P.J. (eds.). *Ancient Europe 8000 B.C.-A.D. 1000. Encyclopedia of the Barbarian World. Volume 2. The Bronze Age to Early Middle Ages (C. 3000 B.C.-A.D. 1000).* New York. pp. 368-370.

Lindeblad, K. & Nielsen, A-L. 1997. *Kungens Gods i Borg - om utgrävningarna vid Borgs säteri.* Arkeologisk slutundersökning. Borgs säteri 6702, RAÄ 276. Riksantikvarieämbetet Avdelning för arkeologiska undersökningar. Rapporter UV Linköping 1997:12. Linköping.

Lindell, M. 2001. Utgrävningar i Uppåkra – en översiktlig beskrivning av de inledande utgrävningarna utförda under åren 1996-1998. In Larsson, L. (ed.). *Uppåkra – Centrum i analys och rapport.* Uppåkrastudier 4. Acta Archaeologica Lundensia Series In 8°, No. 36. Lund. pp. 3-22.

Looijenga, J.H. 1997. *Runes around the North Sea and on the Continent AD 150-700.* Proefschrift Rijksuniversiteit Groningen. Groningen.

Lucas, G. & McGovern, T. 2008. Bloody slaughter: Ritual decapitation at the Viking settlement of Hofstaðir, Iceland. *European Journal of Archaeology. Vol. 10 (1).* pp. 7-30.

Luff, R-M. 1982. *A Zooarchaeological Study of the Roman North-western Provinces.* BAR International Series 137. Oxford.

Lyman, R.L. 1994. *Vertebrate taphonomy.* Cambridge manuals in archaeology. Cambridge.

Magnell, O. 2003. Butchering of Wild Boar (*Sus scrofa*) in the Mesolithic. In Larsson, L., Kindgren, H., Knutsson, K., Loeffler, D. & Åkerlund, A. (eds.). *Mesolithic on the Move. Papers presented at the Sixth International Conference on the Mesolithic in Europe, Stockholm 2000.* Oxford. pp. 671-679.

- 2006. Tracking Taphonomy. Analysis of Skeletal Remains of Wild Boar (Sus scrofa) and Red Deer (Cervus elaphus) from Mesolithic Sites in Scania, Southern Sweden. In *Tracking Wild Boar and Hunters. Osteology of Wild Boar in Mesolithic South Scandinavia.* Studies in Osteology 1. Acta Archaeologica Lundensia Series in 8°, No 51. Stockholm. pp. 1-14.

Magner, L.N. 2005. *A History of Medicine.* Second Edition. Boca Raton.

Maltby, M. 1989. Urban-rural variation in the butchering of cattle in Romano-British Hampshire. In Serjeantson, D. & Waldron, T. (eds.). *Diets and Crafts in Towns – The evidence of animal remains from the Roman to the Post-Medieval periods.* BAR British Series 199. Oxford. pp. 75-106.

Marcus, J. & Flannery, K.V. 2004. The coevolution of ritual and society: New 14C dates from ancient Mexico. *Proceedings of the National Academy of Sciences of the United States of America. Vol. 101, No. 52.* pp. 18257–18261.

Marshall, L.G. 1989. Bone Modification and "The Laws of Burial". In Bonnichsen, R. & Sorg, M.H. (eds.). *Bone modification.* Peopling of the Americas Publication. Center for the Study of the First Americans. University of Maine. Orono. pp. 7-24.

Martin, H. 1998. Analysis of Dental Cementum Rings as an Approach to Azilian Hunting Strategies. *Environmental Archaeology 3.* pp. 13-22.

McKinley, I. 1993. Bone Fragment Size in British Cremation Burials and its Implications for Pyre Technology and Ritual. *Journal of Archaeological Science 21.* pp. 339-342.

Mellars, P. 1996. *The Neanderthal legacy: an archaeological perspective from western Europe.* Princeton University. New Jersey.

Merrifield, R. 1987. *The Archaeology of Ritual and Magic.* London.

Metcalf, P. & Huntington, R. 1991. *Celebrations of death: the anthropology of mortuary ritual.* Second edition. Cambridge.

Monks, G.G. 1981. Seasonality Studies. In Schiffer, M.B. (ed.). *Advances in Archaeological Method and Theory. Volume 4.* New York. pp. 177-240.

Montelius, O. 1888. *The Civilisation of Sweden in Heathen Times.* London.

Montgomery, J. E. 2000. Ibn Fadlān and the Rūsiyyah. *Journal of Arabic and Islamic Studies 3.* pp. 1-25.

Müller, H-H. 1974. Die Tierreste aus der slawischen Burganlage von Arkona auf der Insel Rügen. *ZFA : Zeitschrift für Archäologie. Vol. 8.* pp. 255-295.

Müller-Wille, M. 1971. Pferdegrab und Pferdeopfer im frühen Mittelalter. *Berichten van de Rijksdienst voor het Oudheidkundig Bodemonderzoek 20-2. 1970-71.* pp. 119-248.

- 1999. *Opferkulte der Germanen und Slawen.* Archäologie in Deutschland. Sonderheft 1999. Stuttgart.

- 2002. Offerplatser på kontinenten. Några exempel från förkristen tid. In Jennbert, K., Andrén, A. & Raudevere, C. (eds.). *Plats och praxis. Studier av nordisk förkristen ritual.* Vägar till Midgård 2. Lund. pp. 135-166.

Møhl, U. 1962. Rislevsfundets dyreknogler. *KUML. Årbog for Jysk Arkæologisk Selskab.* Aarhus. pp. 96-106.

- 1972. Oksehorn fra førromersk Jernalder Hornskederne fra »Nørrekær« og »Grøntoft«, Vestjylland. *Aarbøger for Nordisk Oldkyndighed og Historie. 1971.* pp. 61-71.

Naumann-Steckner, F. 1997. Death on the Rhine: changing burial customs in Cologne, 3rd-7th century. In Webster, L. & Brown, M. (eds.). *The transformation of the Roman world AD 400-900*. Berkeley. pp. 143-159.

Nickel, R., Schummer, A. & Seiferle, E. 1968. *Lehrbuch der Anatomie der Haustiere. Bd. 1. Bewegungsapparat*. Berlin.

Nicolay, J.A.W. 2005. Midlaren – 'De Bloemert': een archeologisch paradijs aan de oevers van het Zuidlaardermeer (Dr.). *Paleo-aktueel 16*. pp. 57-70.

- 2008a. 3. Sporen van gebouwen en woonerven uit de Romeinse tijd, de Volksverhuisingstijd and de Vroege-Middleeuwen. In Dijkstra, J.& Nicolay, J.A.W. (eds.). *Een terp op de schop. Archeologisch ondezoek op het Oldehoofsterkerkhof te Leeuwarden*. ADC Monografie 3/ADC Rapport 1227. Amersfoort. pp. 43-98.

- 2008b. 7. Ovrige structuren: hutcommen, schuren, spiekers, roedenbergen en veekralen. In Nicolay, J.A.W. (ed.). *Opgravingen bij Midlaren. 5000 jaar wonen tussen Hondsrug en Hunzedal. Deel 1*. Groningen Archaeological Studies. Vol. 7/1. Groningen. pp. 127-150.

- 2008c. 8. Waterputten, waterkuilen en natuurlijke laagtes: de watervoorziening in en direct rondom de woonplats. In Nicolay, J.A.W. (ed.). *Opgravingen bij Midlaren. 5000 jaar wonen tussen Hondsrug en Hunzedal. Deel 1*. Groningen Archaeological Studies. Vol. 7/1. Groningen. pp. 151-179.

- 2008d. 9. Kuilen met brandlagen: het lokaliseren van metaalwerkplaatsen. In Nicolay, J.A.W. (ed.). *Opgravingen bij Midlaren. 5000 jaar wonen tussen Hondsrug en Hunzedal. Deel 1*. Groningen Archaeological Studies. Vol. 7/1. Groningen. pp. 182-190.

- 2008e. Samenvatting. In Dijkstra, J.& Nicolay, J.A.W. (eds.). *Een terp op de schop. Archeologisch ondezoek op het Oldehoofsterkerkhof te Leeuwarden*. ADC Monografie 3/ADC Rapport 1227. Amersfoort. pp. 1-3.

- 2010. *Terpbewoning in oostelijk Friesland. Twee opgravingen in het voormalige kweldergebied van Oostergo*. Groningen Archaeological Studies 10. Groningen.

Nicolay, J.A.W. & Den Hengst, P. 2008. 29. Synthese: De Bloemert in een regionale context. In Nicolay, J.A.W. (ed.). *Opgravingen bij Midlaren. 5000 jaar wonen tussen Hondsrug en Hunzedal. Deel 2*. Groningen Archaeological Studies 7/1. Groningen. pp. 577-622.

Nicolay, J.A.W. & Waterbolk, H. T. 2008. 6. Huisplattegronden: de ontwikkeling van de bronstijd tot in de middeleeuwen. In Nicolay, J.A.W. (ed.). *Opgravingen bij Midlaren. 5000 jaar wonen tussen Hondsrug en Hunzedal. Deel 1*. Groningen Archaeological Studies 7/1. Groningen. pp. 91-126.

Nieuwhof, A. 2006. 1. Inleiding. In Nieuwhof, A. (ed.). *De wierde Wierum (provincie Groningen). En archeologisch steilkantonderzoek*. Groningen Archaeological Studies 3. Groningen. pp. 1-8.

- 2008. Restanten van rituelen. In Nieuwhof, A. (ed.). *De Leege Wier van Englum. Archeologisch onderzoek in het Reitdiepgebiet*. Jaarverslagen van de Vereniging voor Terpenonderzoek 91. Groningen. pp. 187-248.

Nieweg, D.C. 2009. Archeozoölogie. In Siemons, H. & Lanzing, J.J. (eds.). *Bewoningssporen uit de Romeinse tijd in het Wateringse Veld, Den Haag*. Den Haag. pp. 301-315.

Nilsson, L. 1995. The Bones from Hindby Votive Fen. Delicate Problems of Interpretation. *Lund Archaeological Review 1, 1995*. pp. 65-74.

- 2001. Benmaterialet från Uppåkra 98:2 - preliminära resultat. In Larsson, L. (ed.). *Uppåkra. Centrum i analys och rapport*. Uppåkrastudier 4. Acta Archaeologica Lundensia. 8°, No. 36. Almqvist & Wiksell International. Lund. pp. 87-96.

Noddle, B. 1977. Animal and Plant Remains. In Clark, H. & Carter, A. (eds.). *Excavations in King's Lynn 1963-1970*. The Society for Medieval Archaeology Monograph Series: No. 7. Vol II. London. pp. 378-410.

Noe-Nygaard, N. 1995. Ecological, Sedimentary, and Geochemical Evolution of the Late-Glacial to Postglacial Åmose Lacustrine Basin, Denmark. *Fossils and Strata 37*. Scandinavian University Press. Oslo.

Noe-Nygaard, N. & Richter, J. 1990. Seventeen Wild Boar Mandibles from Slutegårds Sømose - Offal or Sacrifice? In Robinson, E.D. (ed.). *Experimentation and Reconstruction in Environmental Archaeology, Symposia of the Association for Environmental Archaeology No. 9 Roskilde, Denmark 1988*. Oxford.

Norr, S. 1998. *The Rede and to Rown. Expression of Early Scandinavian Kingship in Written Sources*. OPIA 17. Uppsala.

Näsman, U. 1988. Analogislutning i Nordisk Jernalderarkæologi. Et bidrag til udviklingen af en nordisk historisk etnografi. In Mortensen, P & Rasmussen, B.M. (eds.). *Fra stamme til Stat i Danmark 1. Jernalderens stammesamfund*. Jysk Arkæologisk Selskabs Skrifter XXII, 1988. Aarhus. pp. 123-140.

Näsström, B-M. 2001. *Fornskandinavisk religion. En grundbok*. Lund.

O'Connor, T.P. 1996. A Critical Overview of Archaeological Animal Bone Studies. *World Archaeology, Vol. 28, No. 1, Zooarchaeology: New Approaches and Theory*. pp. 5-19.

- 1998. On the Difficulty of Detecting Seasonal Slaughtering of Sheep. *Environmental Archaeology 3*. pp. 5-11.

- 2000. *The archaeology of animal bones*. Sutton.

Oestigaard, T. 2000. Sacrifices of Raw, Cooked and Burnt Humans. *Norwegian Archaeological Review, Vol. 33, No. 1*. pp. 41-58.

Olsen, O. 1992. Kristendommen og Kirkerne. In Roesdahl, E. (ed.). *Viking of Hvidekrist. Norden og Europa 800-1200. København*. pp. 152-161.

Outram, A. 2001. A New Approach to Identify Bone Marrow and Grease Exploitation: Why the

"Indeterminate" Fragments should not be Ignored. *Journal of Archaeological Science 28.* pp. 401-410.

Page, R.I. 1995. *Runes and Runic Inscriptions: Collected Essays on Anglo-Saxon and Viking Runes.* Woodbridge.

Pager, H. 1982. The Ritual Hunt: Parallels Between Ethnological and Archaeological Data. *The South African Archaeological Bulletin. 38.* pp. 80-87.

Paulsson, T. 1993. *Huset och lyckan: en studie i byggnadsoffer från nordisk järnålder och medeltid.* C-uppsats. Lunds universitets arkeologiska institution vt 1993. Lund.

Paulsson-Holmberg, T. 1997. Iron Age building offerings. A contribution to the analysis of a die-hard phenomenon in Swedish preindustrial agrarian society. *Fornvännen. Vol. 96. 3-4.* pp. 163-175.

Pegg, C. 2001. *Mongolian music, dance, & oral narrative.* Washington.

Peralta, J.T. & Scott, W.H. 2001. *Reflections on Philippine culture and society: festschrift in honor of William Henry Scott.* Manila.

Petrén, T. 1984. *Lärobok i anatomi. Del I. Rörelseapparaten.* Stockholm.

Pendleton, C. 2001. 16. Firstly, let's get rid of ritual. In Brück, J. (ed.). *Bronze Age Landscapes. Tradition and Transformation.* Oxford. pp. 170-178.

Piggot, S. 1962. Heads and Hoofs. *Antiquity. Vol. 36.* pp. 110-118.

Prummel, W. 2006. 5. Dierlijk bot. In Nieuwhof, A. (ed.). *De wierde Wierum (provincie Groningen). En archeologisch steilkantonderzoek.* Groningen Archaeological Studies 3. Groningen. pp. 31-45.

- 2008. Dieren op de wierde Englum. In Nieuwhof, A. (ed.). *De Leege Wier van Englum. Archeologisch onderzoek in het Reitdiepgebiet.* Jaarverslagen van de Vereniging voor Terpenonderzoek 91. Groningen. pp. 116-159.

Prummel, W., Halici, H., & Verbaas, A. 2011. The bone and antler tools from the Wijnaldum Tjitsma *terp. Journal of Archaeology in the Low Countries 3-1/2.* pp. 65-106.

Prummel, W., Holl, J., Jans, J.E.A., Huisman, N. & Thilderqvist, J.G.M. 2008. 13. Het dierlijk botmateriaal: voedselvoorziening en rituele praktijken. In Nicolay, J.A.W. (ed.). *Opgravingen bij Midlaren. 5000 jaar wonen tussen Hondsrug en Hunzedal. Deel 1.* Groningen Archaeological Studies 7/1. Groningen. pp. 235-259.

Raemaekers, D.C.M., van der Sanden, W.A.B. & Nicolay, J.A.W. 2008. 1. Voorgeschiedenis en onderzoeksstrategie. In Nicolay, J.A.W. (ed.). *Opgravingen bij Midlaren. 5000 jaar wonen tussen Hondsrug en Hunzedal. Deel 1.* Groningen Archaeological Studies 7/1. Groningen. pp. 1-15.

Raudevere, C. 2002. Mellan liv och text. Ordets makt som litterärt motiv och rituell praktik I norrön litteratur. In Jennbert, K., Andrén, A. & Raudevere, C. (eds.). *Plats och praxis. Studier av nordisk förkristen ritual.* Vägar till Midgård 2. Lund. pp. 31-68.

Rech, M. 2006. Pferd und Wagen in Nordwestdeutschland. In Rech, M. (ed.). *Pferdeopfer - Reiterkrieger. Fahren und Reiten durch die Jahrtausende.* Bremer Archäologische Blätter. Beiheft 4. Bremen. pp. 1-62.

Reitz, E.J. & Wing, E.S. 1999. *Zooarchaeology.* Cambridge manuals in archaeology. Cambridge.

Reichstein, H. 1991. *Die Fauna des Germanischen Dorfes Feddersen Wierde.* Teil 1: Text. Stuttgart.

Renfrew, C. 1989. Comments on Archaeology into the 1990s. *Norwegian Archaeological Review. Vol 22. No 1.* pp. 33-41.

Riddersporre, M. 1998. Ravlunda och Uppåkra. Två exempel på försvunna storgårdar? In Larsson, L. & Hårdh, B. (eds.). *Centrala Platser - Centrala Frågor. Samhällsstrukturen under Järnåldern.* Uppåkrastudier 1. Acta Archaeologica Lundensia Series In 8°, No. 28. Lund. pp. 165-177.

Rixson, D. 1989. Butchery evidence on animal bones. *Circaea - The Bulletin of the Association for Environmental Archaeology. Vol 6. Issue 1.* pp. 49-62.

Roundtree, H. C. 1998. Powhatan Indian Women: The People Captain John Smith Barely Saw. *Ethnohistory. Vol. 45. No. 1.* pp. 1-29.

Ross, A. 1968. Shafts, pits, wells - sanctuaries of the Belgic Britons? In *Studies in Ancient Europe.* Coles, J. M. & Simpson, D. D. A. (eds.). Leicester University. Leicester. pp. 255-285.

Rydving, H. & Kristoffersson, R. 1993. Några samiska offerplatser. *Fornvännen 88.* pp. 195-210.

Sanday, P.R. 1986. *Divine Hunger: Cannibalism as a Cultural System.* Cambridge University. Cambridge.

Sawyer, B. & Sawyer, P. 1993. *Medieval Scandinavia. From Conversion to Reformation circa 800-1500.* The Nordic Series, Volume 17. Minneapolis.

Scheuer, L. & Black, S. 2004. *The Juvenile Skeleton.* Oxford.

Scott, E. 1991. Animal and Infant Burials in Romano-British Villas: A Revitalization Movement. In Garwood, P., Jennings, D., Skeates, R. and Toms, J. (eds.). *Sacred and Profane. Proceedings of a Conference on Archaeology, Ritual and Religion. Oxford, 1989.* Oxford University Committee for Archaeology Monograph. Oxford. pp. 115-121.

Schmid, B. 1965. Opferplatz und Gräberfeld des 6. Jahrhunderts bei Oberwerschen, Kr. Hohenmölsen. *Ausgrabung und Funde - Nachrichtenblatt für Vor- und Frühgeschichte 10.* pp. 39-42.

Schmid, E. 1972. *Atlas of Animal bones: For Prehistorians, Archaeologists and Quaternal Geologists.* Amsterdam.

Schwab, D.K.L. 1833. *Lehrbuch der Anatomie der Hausthiere.* 2nd ed. München.

Selinsky, P. 2005. A Preliminary Report on the Human Skeletal Material from Gordion's Lower Town Area. In Kealhofer, L. (ed.). *The archaeology of Midas and the Phrygians: recent work at Gordion.* Philadelphia. pp. 117-123.

Serjantson, D. 1998. Birds: a Seasonal Resource. *Environmental Archaeology 3.* pp. 23-33.

Shanks, M. & Tilley, C. 1989. Archaeology into the 1990s. *Norwegian Archaeological Review. Vol 22. No 1.* pp. 1-12.

Shipman, P., Bosler, W. & Davis, K. L. 1981. Butchering of giant galadas at an Acheulian site. *Current Anthropology, Vol. 22, No. 3.* pp. 257-268.

Skibsted Klæsøe, I. 1995. 7. Det trefligede belag fra Trabjerg Bakker og andre detektorfund. In Bender Jørgensen, L. & Eriksen, P. *Trabjerg. En Vestjysk landsby fra Vikingetiden.* Holstebro Museums række. Jysk Arkæologisk Selskabs Skrifter XXXI:1. Aarhus Universitetsforlag. Højbjerg. pp. 107-119.

Skinner, A. 1913. Notes on the Florida Seminole. *American Anthropologist. New Series. Vol. 15. No. 1.* pp. 63-77.

Sigvallius, B. 1994. *Funeral Pyre. Iron Age Cremations in North Spånga.* Theses and papers in osteology 1. Osteological Research Laboratory Stockholm University. Stockholm.

Silver, A. 1969. The Ageing of Domestic Animals. In Brothwell, D. & Higgs, E. (eds.). *Science in Archaeology: a survey of progress and research.* London. pp. 283-302.

Smith, W. 1859. *A dictionary of Greek and Roman antiquities.* Second edition. Boston.

Solberg, B. 2002. Courtyard Sites north of the Polar Circle – Reflections of Power in the Late Roman and Migration Period. In Hårdh, B. & Larsson, L. (eds.). *Central Places in the Migration and Merovingian Periods – Papers from the 52nd Sachsensymposium.* Uppåkrastudier 6. Acta Archaeologica Lundensia Series In 8°, No. 39. Lund. pp. 219-229.

Spence, L. 1945. *The Magic Arts in Celtic Britain.* London.

Stalsberg, A. 1991. Tradeswomen during the Viking Age. Nordic TAG. *Archeology and Environment 11.* Umeå. pp. 45-52.

Steinsland, G. 1992. Nordisk hedendom. In Roesdahl, E. (ed.). *Viking og Hvidekrist. Norden og Europa 800-1200.* København. pp. 144-151.

- 2005. *Norrøn religion. Myter, riter, samfunn.* Oslo.

Sten, S. & Vretemark, M. 1988. Storgravsprojektet – osteologiska analyser av yngre järnålderns benrika brandgravar. *Fornvännen 83.* pp. 145-156.

Stjernquist, B. 1997. *The Röekillorna spring : spring-cults in Scandinavian prehistory.* Contributions by Tage Nilsson and Ulrik Møhl. Acta Regiae Societatis humaniorum litterarum Lundensis 82. Stockholm.

- 2004. A Magnificent Glass Bowl from Uppåkra. In Larsson, L. (ed.). *Contiuity for centuries – A ceremonial building and its context at Uppåkra, Southern Sweden.* Uppåkrastudier 10. Acta Archaeologica Lundensia Series In 8°, No. 48. Lund. pp. 103-151.

Straten, F.T. van 1995. *Hierà kalá: images of animal sacrifice in archaic and classical Greece.* Religions in the Graeco-Roman World, 127. Leiden.

Sugiyama, S. 2005. *Human Sacrifice, Militarism, and Rulership: Materialization of State Ideology at the Feathered Serpent Pyramid, Teotihuacan.* New Studies in Archaeology. Cambridge.

Sulimirski, T. & Taylor, T. 2006. 33a The Scythians. In Boardman, J., Edwards, I.E.S., Hammond, N.G.L. & Sollberger, E. (eds.). *The Cambridge Ancient History. Second Edition, Volume III, Part 2. The Assyrian and Babylonian Empires and other States of the Near East, from the Eighth to the Sixth Centuries B.C.* Cambridge. pp. 547-590.

Taayke, E. 2005. 8. Het noorderlijke kustgebied in de ijzertijd en Romeinse tijd. In Knol, E., Bardet, A.C. & Prummel, W. (eds.). *Professor van Giffen en het geheim van de wierden.* Groninger Museum. Veendam. pp. 152-165.

- 2008. In het spoor van Wijnaldum. De Vrije Fries als eigentijds platform voor archeologische interpretaties. *De Vrije Fries. Vol. 88.* Leeuwarden. pp. 189-211.

Tegnér, G. 1992. En båtgrav i Vendel. In Roesdahl, E. (ed.). *Viking og Hvidekrist. Norden og Europa 800-1200.* København. pp. 239.

Therkorn, L.L. 2004. *Landscaping the Powers of Darkness & Light. 600 BC – 350 AD settlement concerns of Noord-Holland in wider perspective.* Proefschrift Universiteit van Amsterdam. Amsterdam.

Thilderqvist, J. 2001. *Experimentella försök med djurpåverkan på ben.* D-uppsats Ht 2001. Institutionen för Arkeologi och Antikens historia. Lund.

Thomsen, C.J., Rafn, C.C. & Petersen, N.M. 1836. *Ledetraad til nordisk oldkyndighed.* Det Kongelige Nordiske Oldskrifts Selskab. København.

Thrane, H. 1991. Gudmeundersøgelserne. In Mortensen, P. & Rasmussen, B.M. (eds.). *Fra Stamme til Stat i Danmark 2. Høvdingesamfund og Kongemagt.* Jysk Arkæologisk Selskabs Skrifter XXII: 2. Aarhus. pp. 67-70.

Todd, M. 1994. Barbarian Europe, AD 300-700. In Cunliffe, B. (ed.). *The Oxford Illustrated Prehistory of Europe.* Oxford. pp. 447-482

Tooker, E. 1991. *An ethnography of the Huron Indians, 1615-1649.* New York.

Torsetnes, E. 2004. *Analogislutningers rolle i kunnskapsproduksjonen. En kritisk studie eksemplifisert ved arkeologiske analyser av kjønn og rom.* Hovedfagsoppgave i arkeologi. Universitet i Tromsø. Tromsø.

Trigger, B.G. 1989. *A history of archaeological thought.* Cambridge.

Tringham, R. 1978. Experimentation, Ethnoarchaeology, and Leapfrogs in Archaeological Methodology. In Gould, R.A. (ed.). *Explorations in Ethnoarchaeology.* Albuquerque. pp.169-199.

Vallet, F. & Roesdahl, E. 1992. Vikingegraven fra Ille de Groix. In Roesdahl, E. (ed.). *Viking og Hvidekrist. Norden og Europa 800-1200.* København. pp. 322.

Vifot, B-M. 1936. *Järnåldersboplatsen vid Uppåkra.* Meddelande från Lunds Universitets Historiska Museum. K. Humanistiska Vetenskapssamfundets i Lund Årsberättelse 1935-1936, VI. Lund. pp. 97-142.

Vretemark, M. 1997. *Från ben till boskap. Kosthåll och djurhållning med utgångspunkt i medeltida benmaterial från Skara*. Skrifter från Länsmuseet Skara, nr 25. Skara.

Wait, G.A. 1985. *Ritual and Religion in Iron Age Britain*. Part I. BAR British Series 149 (i). Oxford.

Walker, W.H. 1998. Where Are the Witches of Prehistory? *Journal of Archaeological Method and Theory, Vol. 5, No. 3*. pp. 245-308.

Walter, R. & Sheppard, P. 2006. 7. Archaeology in Melanesia: A Case Study from the Western Province of the Solomon Islands. In Lilley, I. (ed.). *Archaeology of Oceania: Australia and the Pacific Islands*. Blackwell Studies in Global Archaeology. Malden. pp. 137-159.

Waterbolk, H. T. 2009. *Getimmerd verleden. Sporen van voor- en vroeghistorische houtbouw op de zand- en kleigronden tussen Eems en Ijssel*. Groningen Archaeological Studies Volume 10. Groningen.

Watt, M. 1991. Sorte Muld. Høvdingesæde og kultcentrum fra Bornholms yngre jernalder. In Mortensen, P. & Rasmussen, B.M. (eds.). *Fra Stamme til Stat i Danmark 2. Høvdingesamfund og Kongemagt*. Jysk Arkæologisk Selskabs Skrifter XXII: 2. Aarhus. pp. 89-106.

- 1999. Guldgubber og patricer til guldgubber fra Uppåkra. In Hårdh, B. (ed.). *Fynden i centrum. Keramik, glas och metall från Uppåkra*. Uppåkrastudier 2. Acta Archaeologica Lundensia Series In 8°, No. 30. Lund. pp. 177-190.

- 2004. The Gold-Figure Foils ("Guldgubbar") from Uppåkra. In Larsson, L. (ed.). *Contiuity for centuries – A ceremonial building and its context at Uppåkra, Southern Sweden*. Uppåkrastudier 10. Acta Archaeologica Lundensia Series In 8°, No. 48. Lund. pp. 167-221.

Wells, P.S. 2004. Hjortspring. In Bogucki, P. & Crabtree, P.J. (eds.). *Ancient Europe 8000 B.C.–A.D. 1000. Encyclopedia of the Barbarian World. Volume 2. The Bronze Age to Early Middle Ages (C. 3000 B.C.–A.D. 1000)*. New York. pp. 354-363.

White, E. M. & Hannus, L. A. 1983. Chemical Weathering of Bone in Archaeological Soils. *American Antiquity 48*. pp. 316-322.

Widht, T. 1997. Innledning. In *Agricola og Germania*. Thorleif Dhals Kulturbibliotek. Oslo. pp. 7-19.

Wigh, B. 2001. *Animal Husbandry in the Viking Age Town of Birka and its Hinterland*. Excavations in the Black Earth 1990-95. Stockholm.

Wigham, C.F.W. 2004. *Encyclopedia of Ancient Asian Civilizations*. New York.

Wijngaarden-Bakker, L. Van. 1990. Replication of Butchering Marks on Pig Mandibles. In Robinson, D. E. (ed.). *Experimentation and Reconstruction in Environmental Archaeology. Symposia of the Association for Environmental Archaeology No. 9 Roskilde, Denmark, 1988*. Oxford. pp. 167-174.

Wilkens, B. 2002. Roman souvitaurilia and its predecessors. In O'Day, S.J., Van Neer, W. & Ervynck, A. (eds.). *Behaviour behind bones: the zooarchaeology of ritual, religion, status and identity*. Proceedings of the 9th Conference of the International Council of Archaeozoology, Durham, August 2002. Oxford. pp. 73-76.

Wilson, B. 1992. Considerations for the Identification of Ritual Deposits of Animal Bones in Iron Age Pits. *International Journal of Osteoarchaeology, Vol. 2*. pp. 341-349.

- 1996. *Spatial Patterning among Animal Bones in Settlement Archaeology. An English regional exploration*. BAR British Series 251. Oxford.

- 1999. Displayed or concealed? Cross cultural evidence for symbolic and ritual activity depositing Iron Age animal bones. *Oxford Journal of archaeology, Vol. 18, No. 3*. pp. 297-305.

Worsaae, J.J.A. 1886. *The Pre-history of the North*. London.

Wylie, A. 2002. *Thinking from Things. Essays in the Philosophy of Archaeology*. Berkeley.

Zeist, W. van 1974. Palaeobotanical Studies of Settlement Sites in the Costal Area of the Netherlands. *Palaeohistoria. XVI*. Bussum. pp. 226-371.

Ørsnes, M. & Ilkjær, J. 1993. Offerfund. In Hvass, S. & Storgaard, B. (eds.). *Da klinger i muld... 25 års arkæologi i Danmark*. Det Kgl. Nordiske Oldskriftsselskab. Jysk Arkæologisk Selskab. Aarhus. pp. 215-219.

Personal comments

Nicolay, J.A.W. (24-02-2009).

Magnell, O. (01-06-2009).

de Winter, J. (10-02-2010).

Classical and medieval sources

Annales

1942. The Annals By Tacitus. In *The Complete Works of Tacitus*. The Modern Library. Translated into English by A.J. Church & W.J. Brodribb.

Argonautica

1889. *"The Argonautica" of Apollonius Rhodius'. Translated into English prose from the text of R. Merkel*. London. Translated into English by Coleridge, E.P.

Beowulf

Garmonsway, G.N. & Simpson, J. 1968. *Beowulf and its analogues*. London.

De Bello Gallico

1915. *"De bello gallico" & other commentaries of Caius Julius Caesar*. Rhys, E. (ed.). Veryman's library. Translated into English by Macdevitt, W.A.

1995. *The Celtic Heroic Age: Literary Sources for Ancient Celtic Europe & Early Ireland & Wales*. Celtic Studies Publications. Malden. Translated into English by Koch, J.T. & Carey, J.

De Divinatione
1923. *Cicero on Divination.* Loeb Classical Library, Harvard University Press, vol. XX. English translation by Falconer, W.A.

Eddukvæði
1906. *The Elder Eddas of Saemund Sigfusson.* Norroena society, London. Translated into English by Thorpe, B.
1992. *Eddan. De nordiska guda- och hjältesagorna.* Uddevalla. Translated into Swedish by Brate, E.

Egils saga Skallagrímssonar
1893. *The story of Egil Skallagrimsson: being an Icelandic family history of the ninth and tenth centuries.* London. Translated into English by Green, W.C.
1976. *Egil's saga.* Penguin Classics. London. Translated into English by Pálsson, H. & Edwards, P.
1992. *Egil Skallagrimssons och Gunnlaug Ormstungas Sagor.* Stockholm. Translated into Swedish by Alving, H.

Epigrams
1897. *Martial, Epigrams.* Book 4. Bohn's Classical Library. Anonymous translator.

Lex Frisionum
1982. *Lex Frisionum.* Monumenta Germaniae historica. Fontes iuris germanici antiqui in usum scholarum separatim editi. Hanover. Translated into German by Eckhardt, K.A. & Eckhardt, A.

Lex Saxonum
1993. *Die Capitulatio de partibus Saxoniae.* Translated into German by Schubert, E.

Geographica
1995. *The Celtic Heroic Age: Literary Sources for Ancient Celtic Europe & Early Ireland & Wales.* Celtic Studies Publications. Malden. Translated into English by Koch, J.T. & Carey, J.

Germania
1737. Tacitus: Germania. In Ellot C.W. (ed.). 1910. *Voyages and travels : ancient and modern, with introductions, notes and illustrations.* The Harvard classics, [vol. XXXIII]. New York. Translated into English by Gordon, T.
1997. *Agricola og Germania.* Thorleif Dhals Kulturbibliotek. Oslo. Translated into Norwegian by Widht, T.

Gesta Hammaburgensis Ecclesiae Pontificum
1893. *Adam von Bremen Hamburgische Kirchengeschichte.* Geschichtschr. d. deutschen Vorzeit. 2. Gesammtausg. Bd. XLIV. Leipzig. Translated into German by Laurent, J.C.M.
1993. *Adam of Bremen. Beretningen om Hamburg stift, erkebiskopenes bedrifter og øyrikene i Norden.* Thorleif Dahls Kulturbibliotek. Oslo. Translated into Norwegian by Danielsen, B.T. & Frihagen, A.K.
2002. *History of the Archbishops of Hamburg-Bremen.* Records of western civilization. New York. Translated into English by Tschan, F.J.

Gísla saga Súrssonar
1996. *Soga om Gisle Sursson.* LNUs Litteraturserie. Landslaget för Norskundervisning (LNU). Gjøvik. Translated into Norwegian by Haglund, J.R.

Gutasaga
2004. *The History of the Gotlanders.* Translated into English by Tunstall, P.

Historiae Adversum Paganos
1964. *Paulus Orosius: The Seven Books of History against the Pagans. Deferrari, R.J. (ed.).* The fathers of the Church vol. 50, Washington DC. Translated into English by Deferrari, R.J.

The Histories
1858. *The History of Herodotus: a new English version.* In four volumes. Volume 1. Translated into English by Rawlingson, G.

Hrafnkel's Saga
1882. *The Story of Hrafnkell, Frey's Priest.* Translated into English by John Coles, J.

Kitāb
Montgomery, J.E. 2000. Ibn Fadlān and the Rūsiyyah. *Journal of Arabic and Islamic Studies 3.* pp. 1-25.

Kitāb al-alāq an-nafīsa
Göckenjan, H. & Zimonyi, I. 2001. *Orientalische Berichte über die Völker Osteuropas und Zentralasiens im Mittelalter. Die Ǧayhānī-Tradition.* Veröffentlichungen der Societas Uralo-Altaica. Band 54. Wiesbaden. Translation into German by Kmoskó, M.

The Library of History
1814. *The Historical Library of Diodorus the Sicilian, in fifteen books to which are added the Fragments of Diodorus and those published by H. Valesius, I. Rhodomannus, and F. Ursinus.* Volume 1. London. Translated into English by Booth, G.

The New Testament
1611. *The Authorized King James Version.* The 1987 printing.

Njáls saga
1960. *Njal's Saga.* Penguin Classics. London. Translated into English by Pálsson, H. & Edwards, P.
1991. *Njals Saga.* Göteborg. Translated into Swedish by Alving, H.

The Origin and Deeds of the Goths
1915. *The Gothic History of Jordanes: In English with an Introduction and a Commentary.* Translated into English by Mierow, C.C.

Povest vremennych let
1998. *Nestorskrönikan.* Kulturhistoriskt bibliotek Stockholm. Translated into Swedish by Oxenstierna, G.

The Saga of Håkon Góði
1844. *Heimskringla. The Chronicle of the Kings of Norway.* London. Translated into English by Laing, S.

Skáldskaparmál
1880. *The younger Edda, also called Snorre's Edda of the prose Edda. Skáldskaparmál.* Translated into English by Anderson, R.B.
1928. *Edda-Myterne II - Snorris Eddasagn. Af Skjaldskablæren.* Translated into Danish by Larsen, T.

Vita Anskarii
1921. *Anskar, The Apostle of the North, 801-865, translated from the Vita Anskarii by Bishop Rimbwert his fellow missionary and sucessor.* London. Translated into English by Robinson, C.H.
Vita Willibrordii, Vita Bonifatii & Bonifatii Epistulae
Talbot. C.H. 1954. *The Anglo-Saxon Missionaries in Germany, Being the Lives of SS. Willibrord, Boniface, Leoba and Lebuin together with the Hooepericon of St. Willibald and a selection of the correspondence of St.Boniface.* London.
1968. *Briefe des Bonifatius – Willibalds Leben des Bonifatius. Nebst Einigen Zeitgenøssischen Dokumenten.* Darmstadt. Translated into German by Tangl & Külb, H. Text further edited by Rau, R.

Ynglinga saga
1844. *Heimskringla. The Chronicle of the Kings of Norway.* London. Translated into English by Laing, S.
1995. *Norges kongesagaer.* Bind I. Translated into Norwegian by Storm, G. Text further edited by Hansen, K-W., Larsen, J. & Vaa, D. Andresen, K.M. & Fonnes, Ø. (eds.).

English summary

Introduction

This thesis deals with the problems involved in interpreting the human actions behind deposits of mass materials in general, and specifically bones. It is a known fact that even at archaeological sites where the preservation conditions for organic material are good, only a tiny fraction of the bones that once existed is ever found. Moreover most studies dealing with mass materials have broad scopes and seem to assume that everything we find is normal refuse. This in combination with a common lack of interest for taphonomy result in two major issues. First, the reader will receive little information about specific human action and instead sweeping indications about diet and economy. Secondly, even these indications will be uncertain since they, however broad in scope, are still only based on a very small percentage of the material that once existed.

With this in mind, the study of human actions behind individual deposits should be a priority, both as an interesting subject in itself, as well as a vital part to produce reliable large scale studies. To be able to interpret individual deposits we need a good methodology and an array of methods suited for the task.

In order to limit the scope of the research, the specific subject of identifying what are frequently called special or ritual deposits of animal bones was chosen. The problem of separating refuse from other activities is recurring within archaeozoology, and ritual is a label that is often carelessly used for any unusual deposit found, for the most part without much effort being spent on finding out more specifically what actions were behind the find. The main research question was therefore expressed as follows:

> **The main aim of the thesis is to test and evaluate different methods in order to formulate a broad methodological framework for approaching the identification of ritual deposits of bones in settlements.**

To test, evaluate and possibly develop methods, four excavated settlements were chosen as case study-sites. In order to limit the extent of the research, to avoid the direct influence from the Roman cultural sphere and to ensure that transition of ritual customs in connection with the introduction of Christianity would be minimal, only Early Medieval Period sites were considered. The sites should furthermore contain well dated bone materials, excavated and recovered with modern archaeological methods. It was also essential that both the bones and the field documentation from the excavations were accessible for first hand examination.

Three bone materials from the Northern Netherlands and one material from Southern Sweden were selected. The four sites represent three very different types of settlements. Dongjum and Leeuwarden (Oldehoofsterkerkhof) were artificial dwelling mounds (*terps*) situated in the then undiked salt marches of the Northern Netherlands, while Midlaren was an inland settlement located on the sandy soils in the province of Drenthe. The final site, Uppåkra was a central place in the Southern plains of the province of Skåne in Sweden.

The bone fragments from the four materials were counted, weighed and identified with the help of bone reference collections and the results were registered on paper and in Access databases. A number of taphonomic processes such as weathering, trampling, heating, bone fragmentation, human processing marks and gnawing were registered. In addition the animals' sex and age were determined and measurements were taken when possible. Finally the minimum number of individuals and a basic element distribution were calculated for every material.

Identifying and interpreting deposits

The interpretation of special and ritual deposits is not a new phenomenon in archaeological research and many methods have been used for this purpose. In order to find such deposits the most common characteristics to look for are symbolic meaning, irregularity, non functionality and repetition of the type of deposit.

There are however several problems both in connection with these characteristics, and with analyses of unusual deposits in general. These problems include a tendency to generalize and apply categories used successfully at a certain place over wide areas and time periods. Often there is an ignorance of the normal material, despite the fact that a search for remains of explicit rituals or symbolic activities often is a search for the irregular, the result of which can thus only be found in contrast to normality. Another common problem is the lack of source criticism, or at least the lack of a thorough discussion of the limitations of the specific material. Furthermore there is sometimes an over-emphasis of modern symbolism.

Finally there exists a paradox regarding the repetition of deposits. If a certain type of deposit is not repeated it will usually not be interpreted as ritual but as a coincidence, but if it is repeated too often it will be considered a normality.

To structure the analysis and interpretation of the deposits in the case-study materials, a five step process was devised. This process include the steps definition, description, identification, interpretation and explanation. The definition of what is sought for in the analysis is undoubtedly the most important step of the process and without a concise definition, the aim of the research becomes unclear. That a detailed and methodical description should be produced for the studied material both in general and for every deposit might sound obvious, but this step is far from always performed in a satisfactory way. The third step in the process is to identify which deposits differ from the normality of the material that is investigated. It is at this point not yet necessary to interpret what type of activity the special deposits are the result of, but this step should instead be seen as a selection process in which everything out of the ordinary is put aside for further study.

Unfortunately many researchers never go beyond this step in their analysis, either because it is too time consuming, too difficult, or because it would in their opinion only produce vague or speculative results. It is however important to realise that the identification of special deposits is not a very interesting result in itself. The fourth step is therefore to produce a broad interpretation of the special deposits according to the definition in step one. The final and most difficult step in the process is to produce more detailed explanations of the already interpreted deposits with the help of analogies, written sources and iconography.

As can be seen from the description of the process, reliable methods are required to identify the special deposits and various analogies as well as written sources, if available, are needed to interpret and explain the meaning behind these. A number of methods that could be used to identify special deposits were tested for the four materials. These methods can be divided into archaeozoological and archaeological ones. The first included special patterns for species selection, age structure, seasonal use, sex selection, element distribution and taphonomic history while the second constituted special patterns for artefact context, feature type, stratigraphical position, spatial context and general find context.

The final two parts of the five step process require either the use of analogies or of historical sources. Analogy is the transferring of information from one particular situation or period to another, and is assumed to be reliable if there exists a relevant relation of some kind between the two. One of the most important aspects of the use of analogies in archaeology is their quality, e.g. their suitability for a certain situation or place. An analogy can never be verified as being correct or incorrect, but it can be strengthened or weakened in three main ways:

- Relevance (how are the similarities in the analogy actually of interest in the specific case).
- Generality (how much cross-cultural viability does the analogy seem to have).
- Goodness-of-fit (how many similarities exist between the particulars).

It is important to note that analogies are used on all levels of interpretation, and as such permeate virtually all stages of archaeological research. The subconscious use of analogies is a serious problem when interpreting human behaviour. We as researchers possess an almost endless number of scraps of information from different sources and how we use them is furthermore affected by our political attitudes and social preconceptions. To avoid or at least lessen this problem it is vital that we as far as possible are aware of which analogies we use when we make interpretations and what their sources are. If we do not use conscious analogies but instead rely on "common sense", the risk is great that we subconsciously use an analogy that might be of little relevance for the specific situation.

Traditionally the historical sources are the domain of historians while archaeologists deal with material culture and human interaction with the environment. In reality however written sources have been used by archaeologists since the very beginning of the discipline. Historical sources are important within archaeology in three principal ways. First they are used in historical and protohistorical periods to give a general understanding of the technological, cultural and political situation in the area that is studied. Secondly the written sources can give direct information about human behaviour, which in turn can be used to interpret archaeological finds. The third way an historical text can be used by an archaeologist, and the most common way in protohistorical and early historical periods, is as an analogy. A text might not deal with exactly the right subject, but may show some resemblance to it, similar to how ethnographic or contemporary observations are used.

The period and area of my research, roughly 400-1000 AD in Northern Europe, is really on the border between protohistory and history. Some indigenous written texts exist, but they are few compared with later periods. Most of the sources available were written either later or by foreigners that only in a few cases had first-hand experience of the places and subjects they wrote about. The various available sources include runic inscriptions, early Scandinavian Christian manuscripts, Eddic and scaldic poetry, the prose Edda, classical authors that wrote about Northern Europe, Arabic travellers that described Rūs in what is now modern day Russia, Ukraine and Belarus, texts dealing with Christian missionaries in Northern Europe, Early Medieval laws and Old English poetry.

As with all written sources there are various problems involved in using these texts. Source criticism regarding the use of written documents plays a major part in any historical research and if we as archaeologists are to use written sources in our research, this cannot be neglected. The problems involved include errors that appear after repeated copying of the text, errors in translation between languages, the transformation of information between an event and when it was written down, the agenda and personal perspective of the author, the reliability of the author's sources and the fact that the texts were written in a specific literary genre that require certain elements to be presented in a certain way regardless of the actual historic events described.

If we as researchers are to use the written sources we have to address these problems and the first step towards this is being aware of them. The next step is to learn as much as possible about the written sources and especially about the context in which they were written. Only by knowing the individual situation and circumstances of a written text can we hope to address the problems of interpretation. Finally if it is at all possible the researcher should study the original text, or otherwise analyses and translations made by researchers that did so.

Dongjum

The partially destroyed artificial dwelling mound (*terp*) Heringa in Dongjum was excavated as part of the Frisia project and only covered a small portion of the settlement. The *terp* was first inhabited in the 2nd century AD when a single at least 15 meter long house-podium was raised. During the 3rd century this initial *terp* was expanded at least five times and eventually reached a diameter of more than 30 meters. At the end of the 3rd century the site was abandoned. In the early 5th century the *terp* was resettled and traces of habitation have been found until the 11th or 12th centuries.

The animal bone material from Dongjum dated to the selected time period is well preserved with only a small degree of modification by taphonomic processes such as weathering and trampling. The material is dominated by cattle and sheep/goat while a smaller part has been identified as pig and horse. This distribution does not deviate from what could be expected for a *terp*-material. The vast majority of the animals were fully-grown, with only less than 20 percent of the fragments of the major species coming from young individuals.

Three deposits from Dongjum were interpreted as special. One was categorised as articulated body parts and included the left ankle of an adult horse found in a ditch. The only traces of taphonomic processes shown on the bones are clear marks of gnawing. The bones fit together and the placement of the gnawing marks on the protruding parts of the joint and their absence from the articular surfaces show that the ankle was articulated when it was placed in the ditch.

The second deposit belong to the category selected elements and come from the same ditch system as the horse ankle. The fact that makes this deposit special is the element distribution of the cattle and sheep/goat bones that show a substantially higher percentage of mandible fragments than for the remaining Dongjum material.

The third and final special deposit from Dongjum belongs to the age structure category, but was also determined as special due to the archaeological finds. The beehive-shaped feature was interpreted as a well with three distinct fillings of which two contained bone material. The topmost filling contained some bone fragments as well as a grinding stone. As a contrast the thin middle filling was empty, while the bottommost one contained a few fragments of animal bone and a sharpening stone deposited at the very bottom of the feature. The only aspect that clearly differs from the rest of the material is the age distribution and then especially for the topmost filling that has an unusually large number of bones from young animals. This, together with the deposited grinding and sharpening stones is enough to qualify the deposit as special.

Leeuwarden

The archaeological project at Leeuwarden-Oldehove was carried out between 2004 and 2006 by the company ADC ArchaeoProjecten and the Groningen Institute of Archaeology of the University of Groningen.

The earliest excavated remains of the site are from the Early Roman Period during which the settlement constituted of a core podium with a single farm. Around 300 AD the site was abandoned. During the middle of the 5th century the *terp* was resettled and gradually developed into a larger settlement with two to four farms in the Merovingian and Carolingian periods. The *terp*-settlement of Oldehoofsterkerkhof can generally be described as a normal agricultural community that lacks high status finds directly connected with the ruling classes of society as well as indications of specialized craftsmen.

The Early Medieval animal bone material from the Leeuwarden excavation is well preserved with cattle and sheep/goat dominating the species distribution.

There are 12 deposits in the bone material that have been interpreted as special. The first three belong to the category articulated body parts and consisted of bone from the front parts of sheep/goats found in similar type layers.

The fourth deposit included four complete left astragali, three from cattle and one from sheep. One of the cattle astragali shows traces of carving on its proximal end. That four astragali from the left side were found alone in a feature and only show superficial surface deterioration is definitely out of the ordinary and enough to consider the deposit special.

The fifth deposit was found in a sunken hut and was categorised as selected species. In addition to a few pieces of ceramics, a handful of comb fragments and an unidentified iron object the finds from this feature are dominated by animal bones found in two dark ash stained fillings separated by cleaner layers. The fillings contained 400 fragments that could be identified as wild bird bones. These bones also had showed a remarkably high MNI with at least 25 individuals for the older deposit and 8 individuals for the younger one. Since this deposit constitutes a focused selection of birds that will have produced a quantity of meat far beyond the normal daily consumption need of a settlement of this size, it qualifies as special.

The following two deposits were made up of an undamaged horse cranium found in a layer and a complete sheep skull found in a ditch. Since cranial fragments from the site normally show processing marks, the cranium and skull qualify as special.

The last five special deposits from Leeuwarden contained one or more human bones in what was not interpreted as grave contexts. A well dated as late as 828 ± 6 AD was the most spectacular of these. At the very bottom of the feature three human skulls and several other large human bones was placed directly underneath an upside down turned wooden trough made of ash. Unfortunately the human bones from the well were initially thought to come from a later Christian grave context and were therefore reburied. Despite this it is clear that the bones were parts of skeletons from at least three different individuals deposited in an unusual way.

Midlaren

The inland settlement of Midlaren was excavated in 2003-2004 by Groningen Institute of Archaeology in collaboration with the company ARC. The permanent settlement had its origin during the Iron Age, probably in the form of a single farm. During the Middle Roman Period the settlement consisted of two or three farmsteads and had long distance contacts with the Roman areas to the south, with the *terp*-region to the north, and with Scandinavia. In addition to imported objects the site also shows signs of local craft production, including metalworking. In most aspects such as pottery forms and house layout the village showed strong connections with the coastal *terp*-area. In the Late Roman Period and Early Medieval Period the settlement decreased somewhat in size and was during its continued existence made up by one or two farms. The investigated area was finally abandoned in the middle of the 11th century.

Due to the sandy soils the bone material is poorly preserved and highly fragmented. The material is nevertheless predictably dominated by cattle followed by sheep/goat and pig.

Three different Early Medieval deposits were identified as special. Two of these contained unprocessed skulls and crania. In an irregular shaped pit the remains of at least four cattle skulls were found. Due to the preservation conditions they were fragmented but had been deposited as complete skulls.

In a well several teeth from a cattle maxilla were found and this is most likely another example of the same type of deposit as the other unprocessed skulls/crania.

The third deposit is a unique feature in the form of an oval shaped pit dated to the Carolingian Period that contained the cremated remains of several animals. The soil was sieved resulting in an assemblage of almost half a million bone fragments. At first the feature was thought to be a human cremation but during the analysis only animal bones from cattle, sheep/goat, pig and dog were found.

Cattle dominated the material while dog was only represented by four identified fragments. The material was very fragmented due to the high temperature at which the bones were burned and a very large proportion of the material could not be determined to species. The bone elements come from all parts of the animals' skeletons. At least nine adult domestic animals, three cattle, two sheep/goats, two pigs and two dogs were cremated. The shape and compactness of the bone concentration strongly indicate that the bones were collected from the cremation place and deposited wrapped in cloth or leather. This deposit is considered special for a number of reasons. Even if old animals were mainly chosen for the event, it would still be a substantial economic investment for the settlement. The cremated remains were furthermore collected and neatly wrapped in a piece of cloth or leather. This package was finally placed in a pit together with a bronze object and covered with soil.

Uppåkra

The site Uppåkra is a central place located five kilometres south of Lund (Sweden). It has been subject to a few small rescue excavations throughout the 1900s, and to yearly campaigns of archaeological research investigations performed by the Department of Archaeology and Ancient History at Lund University since the mid 1990s.

The settlement was first established during the Late Iron Age on a low elevation in the otherwise flat landscape of South-western Skåne. Corings, detector-surveys and excavations have revealed a settlement that formed a roughly oval shape, 1100 meter from north to south and 600 meters from east to west. Numerous structures, including houses dating to the first millennium AD have been found. The settlement seems to have functioned as a unit until the second half of the 10th century when the amount of finds is declining.

During its existence Uppåkra was a supra-regional centre with economic, religious and administrative functions, and had a settlement structure that can best be described as proto-urban with a number of farms situated close to each other.

Much of the recent excavation campaigns have been concentrated to the central parts of the northern half of the settlement where several interesting buildings were found, among these a number of large longhouses and a smaller house of an unusual construction. This smaller house has been interpreted as a temple or ceremonial house and inside it several unique finds were excavated, most notably a copper beaker with silver and gold decorations, a finely crafted glass bowl and as many as 111 gold-figure foils, a type of artefact that is closely associated with ritual activities.

Situated only 30 meters north of the unusual house numerous consciously destroyed weapons, mainly lance and spearheads, and other military equipment were found together with large amounts of animal bones and three human bones in an area covering roughly 86 m^2. This area was interpreted as being the stage for ritual activity in the form of weapon sacrifices, but without any attention to the animal bones and their function in these rituals.

As such it was an interesting material for analysis in the thesis, but also one that differed somewhat from the other three sites. Here the material had already been given a ritual interpretation. Since an analysis of the complete animal bone material from the settlement would be an impossible task, a comparative bone assemblage representing normality had to be found. For this purpose a bone material from the southern part of the settlement was chosen.

The animal bone assemblage from the weapon deposit area in Uppåkra consists of almost 190 kg of bones and 26 of the 86 excavated square meters were chosen for analysis. The material is relatively fragmented and shows a lot of weathering, although mainly in the lighter stages. Since the preservation condition of bone material is generally good in Uppåkra this fragmentation could be explained by that the bones were left in the open for some time or that later weapon deposits were dug into the bone layer, exposing and fragmenting bones in the process.

The bone assemblage is dominated by cattle, even to a larger degree than the materials from Dongjum and Leeuwarden. There is slightly more pig in the material than sheep/goat, also something that differs from Dongjum and Leeuwarden, but that is to be expected since the *terp*-sites usually contain little pig. Fragments from the head dominate the element distribution of the assemblage and especially for the cattle bones. The general age at death distribution in the weapon deposit material shows a majority of fully-grown animals. For cattle this tendency is even more clear with very few young cattle identified and almost 20 percent of the cattle being killed at old age.

The comparative assemblage from the southern area differs from the weapon deposit area in that sheep/goat are represented by more fragments than cattle and that very few horse bones were found. The element distribution, like in the weapon deposit area, show many loose teeth, but few skull and mandible fragments. The age at death distribution in the southern assemblage also shows a lot more younger individuals than is the case for the weapon deposit area.

The conclusion is that the bone material from the weapon deposit area is special in itself due to the unusually large amount of cattle bones, mainly in the form of cranial and mandible fragments from fully-grown or old individuals. These deposits of animals were substantial with at least 31 individual cattle, in addition to 18 other domestic animals identified in the analysed assemblage, which included only 20 percent of the total material.

Interpretation and explanation of activity

The methods described above are meant to be universal in the sense that they are built on patterns of recognition basically independent from the contexts of the observed deposits. This is however only one step in a process that should not end with the identification of special deposits, but continue with the aim of interpreting and understanding the reasons behind these deposits.

Of the 20 special deposits identified in the case study materials, 3 came from Dongjum, 13 from Leeuwarden, 3 from Midlaren and 1 from Uppåkra. Of these 3 were considered normal refuse and 17 were interpreted as ritual.

The three deposits of articulated parts of sheep/goats found in layers in Leeuwarden are most likely examples of partial offerings. There are few sources that describe this type of phenomenon regarding sheep or goats, but plenty of examples of partial sacrifice of other species in which only a small part of an animal is given to the supernatural power and the remainder of the body is divided up and eaten. One source that describes the use of sheep in a similar ritual is Ibn Fadlān's *Kitāb's Rūsiyyah*, according to which the Rūs tied the skulls of cows and sheep to poles as an offering to the gods after a successful trade transaction. The rest of the meat from the animals was shared and eaten.

Four left cattle and sheep astragali found at the bottom of a pit in Leeuwarden, constitute a deposit that brought the use of bones as game pieces and divination tools to mind. The use of astragali for games and ritual purposes such as divination has been recorded by ethnographers at various places and is also well documented in classical sources. Games using astragali, for example "knucklebones", are still played today but are usually games of dexterity played with the smaller astragali from sheep, goats or calves. Other types of games more resembling boules or some marbles games, using both sheep and cattle astragali however exist in various parts of Asia, for example in Turkey, Turkmenistan and Mongolia. Ethnographic observations of the Mongolian style game also show that it has a strong ritual component. The archaeological record corresponds well with the ethnographic sources and shows a long and widespread use of astragali in games and divination. With this tradition in mind it is likely that the bones from this feature was a cache of game pieces or fortune telling bones that were discarded, hidden, or offered to supernatural beings.

The special deposits made up of complete skulls and crania are represented by four features at the sites of Leeuwarden and Midlaren. These include seven or eight different skulls and crania, five or six from cattle, one from horse and one from sheep. The use of skulls and crania in rituals is a widespread phenomenon and the strong symbolism of the head of animals can be seen in cultures all over the world. One interesting ethnographic example comes from the Zulu people in Southern Africa who in connection with the sacrifice of cattle to the ancestor spirits used to place the skull or horns of the animal on display in the village. Some of the remaining parts of the animal were offered to the spirits while the rest was eaten by the people of the community during a feast.

Archaeological parallels to this type of deposit exist in various places, for example in the *terp*-settlement Feddersen Wierde in Germany where two horse skulls and one cattle skull were found in two pits and a posthole, or in Elisenhof, another German *terp*-settlement where a cattle skull was found near the southern wall of a house. A historical source that describes the use of animal skulls is the already mentioned passage from Ibn Fadlān about the skulls of cows and sheep tied to poles. The skulls and crania in Leeuwarden and Midlaren were in line with this interpreted as partial offerings where the heads were given to supernatural powers while most of the animals were eaten.

In a sunken hut in Leeuwarden a large number of bones from wild birds were deposited at two separate occasions. During the first instance the bones from most parts of the bodies of at least 25 individual birds were deposited while at the second occasion a smaller number of bones from at least 9 individual birds were laid down. The element distribution and the indication of roasting suggest that the bird bones are cooking or consumption refuse. The fact that the birds from the first depositi at the very least could supply meat for forty people indicates that the feature represents the remains of a large feast, possibly of a ritual character and probably connected to bird hunting since only wild birds are represented.

A number of human bones were found in five different features dated to the Early Medieval Period in Leeuwarden. Various types of ritual handling of human remains have been recorded by ethnographers in societies around the world. These include exhumation and reburial, some types of ancestor worship, cannibalism, trophy hunting, human sacrifice and various magical practices. Finds of stray human bones outside of grave contexts are not very frequent in the *terp*-region during the Early Middle Ages, but during earlier periods some interesting finds have been made. Historical sources covering Northern Europe do occasionally talk about human sacrifice or other forms of ritual handling of human remains, but these rarely shed any light on the existence of stray bones in settlements. Taken the various analogies into consideration it is likely that at least some of the five deposits are the result of trophy hunting, ancestor worship or were disposed religious or magical tools.

The most intriguing deposit discussed in the thesis is the Carolingian Period cremation pit from Midlaren that contained nearly half a million bone fragments thoroughly burned at high temperatures. Archaeological finds comparable to the cremation pit in Midlaren are few and far between. Burnt animal bones in human cremation graves are a common feature in cemeteries from the Bronze Age and onwards in Northern Europe, but in most cases the amounts of burnt animal bones are small. Occasionally pits have been found in Early Medieval cemeteries containing cremated animal bones and totally lacking human bones, but also these deposits contain a comparably small amount of bones. The find that is most comparable with the Midlaren pit comes from the Slavic fortress in Oldenburg, Germany. In an oval shaped pit dated to the Iron Age, potsherds from a large vessel of early Slavic type was found. This pot has been deposited whole but later collapsed on top of a concentration of burnt bones. The pile of bones was so dense that it was interpreted as being deposited wrapped in a piece of cloth or leather and contained 3.1 kg burnt bones from at least two cattle and one pig.

The burning of animals or of portions of animals as part of an offering is known from various cultures in different time periods. A large number of Roman and Greek sources show that it was regularly done around the Mediterranean in classical times. The fact that the Midlaren animals were most likely burned as complete bodies leads to the interpretation of the deposit as an offering to supernatural beings, possibly connected to a ritual of the dead.

The interpretative circumstances for the studied bone assemblage from Uppåkra are slightly different from that of the other three materials. Here the deposition area in question had already been identified as a special feature and been given a ritual interpretation as the place for repeated weapon sacrifices. There is a large ethnographic record that deals with the two main aspects of the feature, sacrifice and feasting. The very term sacrifice implies an offering of something that is ritually killed, either in practice or at a symbolic level. This type of offering is known from countless societies around the world and while a sacrifice is often associated with the killing of an animal, or more rarely a human, it is also frequently linked to the destruction of non-living things. The custom of feasting, the gathering together of many people that consume large quantities of food and drink for a specific reason, is also a common feature in ethnographic descriptions, and is something that still exists in contemporary societies. Feasting can be performed for various reasons beyond the official, and often symbolic, motivation. Such reasons include gaining prestige (by the sponsor of the feast), strengthening social bonds, keeping traditions and upholding social hierarchies.

The artefacts from the weapon deposit area show some similarities with the large Roman and Migration Period weapon offering sites from Southern Scandinavia. The finds from Uppåkra does however differ from these sites in a few respects, mainly in the relatively small number of artefacts, the composition of many small deposits rather than a few large events, and the continuation of the weapon deposits well into the 8th century. Another difference is the character of the bone material. Animal bones have been retrieved from several of the large weapon sacrifices, but in these cases the composition is quite different from that of the bone layer in Uppåkra. As a contrast the animal bones from more mundane offering sites such as Eketorp, Röekillorna and Valmose show similarities with those at Uppåkra.

The practice among the people of Northern Europe of sacrificing the weapons of defeated enemies after a battle has been described by several classic authors. Regarding feasting, the importance of this custom as a social and political event has long been an established idea in the study of Iron Age, Roman Period and Early Medieval Scandinavia and have its roots in the many references made, especially in the Sagas where feasting often had a religious content, but was also manifested power structures and loyalties. Taking the different aspects of the material and the relevant analogies into consideration, the bone layer is interpreted as the result of repeated feasting in combination with partial offerings of animals. During such an event one or several animals (usually cattle) were butchered. Some parts of the animal, especially skulls or mandibles, were offered to the supernatural entities and deposited either together with, or at least in the same place where weapons were sacrificed.

Conclusions

A number of methods were applied for identifying special deposits among bone materials. The main conclusion is that no single, or even a set of methods stood out as a flawless tool for identifying ritual deposits. Even if similarities did exist between different human societies in the past, either because of direct connections between them, or as a result of the influences and constraints of human biology and the surrounding physical world, the variation in human behaviour is too great for any method to be generally applicable.

When a material from an area, period or culture is to be analysed, a suitable set of methods of identification can only be put together through experimentation and evaluation. In the same way relevant analogies to be used for interpretation have to be established for each type of deposit within such a society.

Of the various methods used, the archaeozoological ones turned out to have a high potential for identifying special deposits and as many as 19 of the 20 identified deposits showed anomalies within species selection, age structure, seasonal use, sexual selection or element distribution, or a combination of these.

The archaeological methods of artefact context, feature type, stratigraphical position, spatial context and general find context showed some potential. With the exception of the bone layer from Uppåkra these methods did not provide the primary signs of special deposits, something that however is natural considering the archaeozoological focus of the research, but they nevertheless gave supporting indications for most of the identified deposits.

The interpretation of the meaning of human behaviour in the past requires either access to direct historical sources, or the use of analogies. Since direct historical sources dealing with the exact places and time periods in question were in effect absent, analogies had to be used in the process of interpretation. Historical analogies were used to interpret several deposits as were ethnographical analogies although relevant detailed ethnographical examples of ritual behaviour including animal bones were less common than expected. Contemporary analogies were most certainly used in the interpretation of the deposits since these are integrated in the general human thought process. I did however try to limit unconscious and generic inference from modern society as far as this was possible.

Even though the methods for the interpretation of deposits are the main focus of this thesis, it was interesting to analyse for what purpose and in which contexts the special deposits were made. The most common ritual interpretation was partial offering, something that is not surprising considering what we know from historical and ethnographic sources. It seems to have been common practice not to spend complete animals as offerings to supernatural powers, but to consume most of the meat. Only if the ritual also functioned as an indicator of wealth and status, were complete animals given to the supernatural powers. The interpretations in general show that we have to carefully observe the composition of deposits. It is always important to take a step back and think about what was actually found and why it ended up in that place.

The results of the analysis and interpretations lead to a methodological framework for understanding individual deposits based on a holistic perspective where all information should be regarded as potentially valuable and every possible scrap of data and every idea have to be assessed. This does not imply that the researcher should produce incredibly detailed studies and get bogged down in a never ending search for information. Rather it suggests openness in the approach, to take various methods and ideas into consideration and to avoid formalisation, simplification and stereotyping.

For a holistic approach some responsibilities of the researchers involved in the archaeological process have to be fulfilled. The excavating archaeologist is the most important person for making the approach suggested in this chapter possible. The information and the material that are not recorded in the field and not recovered will be irretrievably lost regardless of how thorough and methodical later analyses are. The experience from the four materials analysed in this thesis is that there is a direct correlation between the amount of information recorded about a feature during the excavation, and the extent and reliability of later interpretations. The following points are the most important responsibilities of the excavator in order to make detailed intra-site analyses of deposits possible:

- A thorough archaeological methodology and knowledge of the theoretical possibilities and limitations of the discipline.
- An appreciation of the importance of individual features for understanding human behaviour.
- An attentiveness for find circumstances, also when large materials are excavated.
- A good field documentation that is made available to the expert as early as possible.
- An openness towards, and integration of, the expert early in a project, preferably already in the planning stages, and a willingness to make use of his or her skills both in the field and in the archaeological interpretation phase.

In addition to the duties of the field archaeologist, the expert analysing the material has corresponding responsibilities. The following points are the most important duties of the expert, both as a potential participant in the archaeological field work, as well as during the later stages of analysis and interpretation:

- A willingness to learn the theory, methodology and limitations of archaeology as a discipline, and not only the specifics of the expert's own sub-discipline.
- A readiness to study the area, time period and society from which the material originates, also outside of the expert's sub-discipline.
- A willingness to participate both in the planning phase and in the field part of an excavation in order to support the field archaeologists
- The delivery of a report of the material analysis focusing on the archaeological problems, including interpretations based on the expert's knowledge, and not only a technical report.

Dutch summary

Inleiding

Dit proefschrift behandelt de problemen rond het interpreteren van de menselijke handelingen die verborgen gaan achter de deposities van afval en andere materialen in het algemeen, en die van beenderen in het bijzonder. Het is een bekend feit dat zelfs op archeologische vindplaatsen waar de conserveringsomstandigheden voor organisch materiaal goed zijn, slechts een gering deel van de botten die ooit aanwezig waren, worden teruggevonden. Bovendien richten de meeste studies die zich bezighouden met deze afvalmaterialen zich op een breed gebied en lijken zij te veronderstellen dat alles wat op een opgraving gevonden wordt, normaal afval is. Dit punt en een algemeen gebrek aan interesse voor de tafonomie resulteren in twee belangrijke kwesties. Ten eerste, ontvangt de lezer weinig informatie over de specifieke menselijke acties en in plaats daarvan algemene informatie over dieet en economie. Ten tweede, zelfs de aanduidingen over dieet en economie zijn, hoe algemeen ook gesteld, toch onzeker, omdat zij slechts gebaseerd zijn op een zeer klein percentage van het materiaal dat eens aanwezig was.

Met dit in het achterhoofd moet de studie van menselijke handelingen achter individuele deposities van het hoogste belang worden geacht, omdat het zowel een interessant onderwerp op zichzelf is, als een vitaal onderdeel van betrouwbare grootschalige studies. Om individuele deposities te kunnen interpreteren hebben wij een goede methodologie en een reeks van methoden nodig, die geschikt zijn voor deze taak.

Om het onderwerp van dit onderzoek te beperken, koos ik het specifieke onderwerp van de bijzondere of rituele deposities van dierlijke beenderen. Het probleem van het scheiden van normaal afval van dat van andere activiteiten komt steeds terug binnen de archeozoölogie, terwijl het label 'ritueel' vaak achteloos op ongewone deposities wordt geplakt, doorgaans zonder veel moeite te doen te zoeken naar de specifieke handelingen die aan de depositie ten grondslag lagen. De belangrijkste onderzoeksvraag van dit proefschrift is daarom:

> **Het belangrijkste doel van de thesis is de verschillende methoden te testen en te evalueren om een breed methodologisch kader te formuleren voor het herkennen van rituele afzettingen van beenderen in nederzettingen.**

Om deze methoden te testen, te evalueren en eventueel te ontwikkelen werden vier nederzettingen waar opgravingen hadden plaatsgevonden, gekozen als casestudies. Om de omvang van het onderzoek te beperken, de directe invloed van de Romeinse cultuur te vermijden en om er eveneens voor te zorgen dat de overgang van rituele gebruiken ten gevolge van de invoering van het christendom minimale invloed zouden hebben, werden alleen vindplaatsen uit de Vroege Middeleeuwen bij het onderzoek betrokken. De vindplaatsen moesten bovendien goed gedateerde botmaterialen hebben opgeleverd, die waren opgegraven en verzameld met moderne archeologische methoden. Het was ook belangrijk dat zowel de beenderen als de originele velddocumentatie van de opgravingen voor de onderzoeker toegankelijk waren.

Drie botmaterialen uit Noord-Nederland en een uit Zuid-Zweden werden geselecteerd voor het onderzoek. De vier vindplaatsen vertegenwoordigen drie zeer verschillende soorten nederzettingen. Dongjum en Leeuwarden (Oldehoofsterkerkhof) zijn kunstmatig opgeworpen woonheuvels (terpen) gelegen in het toen onbedijkte kweldergebied van Noord-Nederland (provincie Friesland), terwijl Midlaren in het binnenland gelegen was, op de zandgronden van de provincie Drenthe. De laatste vindplaats, Uppåkra, was een centrale plaats in de zuidelijke vlakten van de provincie Skåne in Zweden.

De botfragmenten van de vier materialen werden geteld, gewogen en gedetermineerd met behulp van archeozoölogische vergelijkingscollecties. De gegevens werden op papier en in Access-databases geregistreerd. De sporen van een aantal tafonomische processen zoals verwering, betreding, verhitten, fragmentatie en slacht-, snij- en knaagsporen werden geregistreerd. Verder werden de sekse en de leeftijd van de dieren bepaald en werden botmetingen genomen wanneer dat mogelijk was. Ten slotte werden voor elk materiaal de minimum aantallen individuen en de verdeling van de botfragmenten over het skelet vastgesteld.

Het identificeren en het interpreteren van deposities

Het achterhalen van de betekenis van bijzondere en rituele deposities is geen nieuw verschijnsel binnen het archeologisch onderzoek en vele methoden werden reeds gebruikt voor dit doel. Om dergelijke deposities te vinden is het zaak de meest voorkomende kenmerken op te sporen op het gebied van symbolische betekenissen, onregelmatigheden, non-functionaliteit en herhaling van het type depositie.

Er zijn echter verschillende problemen met deze kenmerken en met analyses van ongebruikelijke deposities in het algemeen. Deze problemen houden onder meer een neiging tot generaliseren in en het toepassen van categorieën die op een bepaalde plaats met succes zijn gebruikt, op velerlei gebieden en perioden. Vaak is er een gebrek aan kennis van het normale materiaal, ondanks het feit dat een zoeken naar overblijfselen van expliciete rituelen of symbolische activiteiten vaak een zoektocht naar het onregelmatige is, waarvan de resultaten dus alleen als contrast met het normale gevonden kunnen worden. Een ander veelvoorkomend probleem is het gebrek aan bronnenkritiek, of althans het ontbreken van een grondige bespreking van de beperkingen van het specifieke materiaal. Daarnaast is er soms een overdreven nadruk op moderne symboliek. Ten slotte bestaat er een paradox over de herhaling van deposities. Als een bepaald type depositie niet wordt herhaald, wordt het meestal niet als ritueel maar als toevallig geïnterpreteerd, maar als het te vaak wordt herhaald, wordt het beschouwd als een een normaliteit.

Om de analyse en de interpretatie van de deposities in de casestudies te structureren, werd een proces in vijf stappen ontworpen. Dit proces omvat de stappen definitie, beschrijving, identificatie, interpretatie en verklaring. De definitie van wat in deze analyse wordt nagestreefd, is zonder twijfel de belangrijkste stap van het proces en zonder een exacte definitie, is het doel van het onderzoek onduidelijk. Dat een gedetailleerde en methodische beschrijving moet worden geleverd voor het bestudeerde materiaal, zowel in het algemeen als voor elke depositie, klinkt misschien als voor de hand liggend, maar deze stap wordt lang niet altijd op een bevredigende manier uitgevoerd. De derde stap in het proces is het identificeren van deposities die afwijken van de normaliteit van het materiaal dat wordt onderzocht. Het is op dit moment nog niet nodig om te interpreteren welk type activiteit ten grondslag lag aan welke speciale deposities, maar deze stap moet gezien worden als een selectieproces waarin alles wat buitengewoon is apart wordt gehouden voor verdere studie.

Helaas gaan vele onderzoekers nooit verder dan deze stap in hun analyse, hetzij omdat het te tijdrovend of te moeilijk is, hetzij omdat het naar hun mening alleen vage of speculatieve resultaten zou produceren. Het is echter belangrijk te beseffen dat de identificatie van speciale deposities op zichzelf niet een zeer interessant resultaat is. De vierde stap is dan ook om een brede interpretatie te geven van de speciale deposities volgens de definitie in stap 1. De laatste en moeilijkste stap in het proces is om een meer gedetailleerde verklaring van de reeds geïnterpreteerde deposities te geven met behulp van analogieën, geschreven bronnen en iconografie.

Zoals uit de beschrijving van het proces blijkt, zijn betrouwbare methoden nodig voor de identificatie van speciale deposities, terwijl diverse analogieën en schriftelijke bronnen, indien beschikbaar, nodig zijn om de betekenis achter deze deposities te interpreteren en uit te leggen. Een aantal methoden die gebruikt konden worden om speciale deposities te identificeren, werden op de vier materialen getest. Deze methoden kunnen worden onderverdeeld in archeozoölogische en archeologische methoden. De archeozoölogische methoden zoeken naar bijzondere patronen qua soortselectie, leeftijdsopbouw, seizoensgebonden gebruik, selectie naar sekse, verdeling van de skeletelementen en tafonomische geschiedenis, terwijl de archeologische methoden zoeken naar speciale patronen in de bijbehorende artefacten, spoortype, stratigrafische positie, ruimtelijke context en de algemene vondstcontext.

De twee laatste van de vijf stappen vereisen het gebruik van ofwel analogieën ofwel historische bronnen. Analogie is het overbrengen van informatie van de ene specifieke situatie of periode naar een andere. Het wordt betrouwbaar geacht indien er op de een of andere manier een relevante relatie wordt verondersteld tussen de twee. Een van de belangrijkste aspecten van het gebruik van analogieën in de archeologie is hun kwaliteit, zoals bijvoorbeeld hun geschiktheid voor een bepaalde situatie of plaats. Een analogie kan nooit geverifieerd worden als juist of onjuist, maar kan versterkt of verzwakt zijn volgens drie belangrijke criteria:

- Relevantie (zijn de overeenkomsten in de analogie werkelijk van belang in het specifieke geval).
- Algemeenheid (hoeveel interculturele levensvatbaarheid heeft de analogie).
- Goodness-of-fit (hoeveel overeenkomsten in detail zijn er aanwezig).

Het is belangrijk op te merken dat analogieën gebruikt worden op alle interpretatieniveaus en als zodanig doordringen in vrijwel alle stadia van archeologisch onderzoek. Het onderbewuste gebruik van analogieën is een ernstig probleem bij het interpreteren van menselijk gedrag. Wij als onderzoekers beschikken over een bijna eindeloos aantal kruimels informatie uit verschillende bronnen en hoe we ze gebruiken wordt bovendien beïnvloed door onze politieke attitudes en sociale vooroordelen. Om dit probleem te voorkomen of ten minste te beperken is het belangrijk dat we ons zoveel mogelijk bewust zijn welke analogieën wij gebruiken wanneer we interpreteren, en waar deze analogieën vandaan komen. Als wij geen bewuste analogieën gebruiken, maar in plaats daarvan vertrouwen op het 'gezonde verstand', is de kans groot dat we in het onderbewuste een analogie gebruiken die weinig relevant is voor de specifieke situatie.

Traditioneel zijn de historische bronnen het domein van historici, terwijl archeologen de materiële cultuur en de menselijke interactie met de omgeving behandelen. In werkelijkheid echter zijn geschreven bronnen gebruikt door archeologen sinds het begin van deze discipline. Historische bronnen zijn op drie verschillende manieren belangrijk binnen de archeologie. Ten eerste geven ze voor de historische en vroeghistorische perioden een algemeen begrip van de technologische, culturele en politieke situatie in het gebied dat wordt bestudeerd. Ten tweede kunnen de geschreven bronnen directe informatie over menselijk gedrag geven, die vervolgens kan worden gebruikt om archeologische vondsten te interpreteren. De derde manier waarop een historische tekst door een archeoloog kan worden gebruikt, is de analogie; dit is tevens de meest voorkomende manier voor de vroeghistorische perioden. Een tekst gaat wellicht niet precies over hetzelfde onderwerp, maar kan hiermee toch enige overeenkomst vertonen, vergelijkbaar met hoe etnografische of hedendaagse waarnemingen worden gebruikt.

De periode en het gebied van mijn onderzoek, Noord-Europa tussen circa 400 en 1000 na Christus, liggen precies op de grens tussen de protohistorie en de geschiedenis. Er bestaan enkele inheemse geschreven teksten, maar het zijn er weinig in vergelijking met die uit latere perioden. De meeste beschikbare bronnen zijn ofwel later geschreven of zijn geschreven door vreemdelingen waarvan er slechts enkelen aanwezig zijn geweest op de plaatsen waarover en ervaring hebben gehad met de onderwerpen waarover zij schreven. De verschillende beschikbare bronnen omvatten runeninscripties, vroege Scandinavische christelijke manuscripten, Edda- en skaldenpoëzie, de proza Edda, klassieke auteurs die over Noord-Europa schreven, Arabisch reizigers die de Rūs beschreven in wat nu hedendaags Rusland, Oekraïne en Wit-Rusland is, teksten over de christelijke missionarissen in Noord-Europa, vroegmiddeleeuwse wetten en Oud-Engelse poëzie.

Zoals met alle schriftelijke bronnen zijn er verschillende problemen met het gebruik van deze teksten. Bronnenkritiek met betrekking tot het gebruik van teksten speelt een belangrijke rol in elk historisch onderzoek en als wij als archeologen geschreven bronnen in ons onderzoek willen gebruiken, kan bronnenkritiek niet worden verwaarloosd. De problemen zijn onder andere fouten die ontstaan na het herhaald kopiëren van de tekst, fouten in de vertaling van taal naar taal, het omzetten van informatie tussen een gebeurtenis en het opschrijven ervan, de agenda en de persoonlijke visie van de auteur, de betrouwbaarheid van de bronnen van de auteur en het feit dat de teksten in een specifiek literair genre werden geschreven, dat vereiste dat bepaalde elementen op een bepaalde manier werden gepresenteerd, ongeacht de feitelijke historische gebeurtenissen.

Als wij als onderzoekers de geschreven bronnen willen gebruiken, moeten wij deze problemen oplossen. De eerste stap is dat we ons bewust zijn van deze problemen. De volgende stap is dat we zo veel mogelijk te weten komen over de schriftelijke bronnen en vooral over de context waarin ze werden geschreven. Alleen door het kennen van de individuele ontstaansgeschiedenis van een geschreven tekst kunnen we hopen de problemen van de interpretatie op te lossen. Tot slot: de onderzoeker moet, als het maar enigszins mogelijk is, de oorspronkelijke tekst bestuderen, of anders in elk geval analyses en vertalingen bestuderen, die gemaakt werden door onderzoekers dat dit deden.

Dongjum

Een klein gedeelte van de reeds eerder gedeeltelijk afgegraven kunstmatige woonheuvel (terp) Heringa in Dongjum werd opgegraven in het kader van het Frisiaproject. De terp werd voor het eerst in de 2e eeuw na Chr. bewoond, nadat een ten minste 15 m lang huispodium was opgeworpen. Tijdens de 3e eeuw werd de terp minstens vijf keer uitgebreid en uiteindelijk bereikte deze een diameter van meer dan 30 m. Aan het einde van de 3e eeuw werd de terp verlaten. In de vroege 5e eeuw werd de terp weer in gebruik genomen. De laatste sporen van bewoning dateren in de 11e of 12e eeuw.

De dierlijke beenderen uit Dongjum die dateren uit de geselecteerde periode, zijn goed geconserveerd en slechts in geringe mate aangetast door tafonomische processen zoals verwering en betreding. Het materiaal wordt gedomineerd door resten van runderen en schapen en/of geiten, terwijl een kleiner deel geïdentificeerd is als resten van varken en paard. Deze verdeling wijkt niet af van die in ander terpenmateriaal. De overgrote meerderheid van de dieren was volgroeid op het moment van de slacht; slechts minder dan 20 procent van de fragmenten van de belangrijkste soorten is afkomstig van jonge dieren.

Drie vondstgroepen uit Dongjum werden als speciale deposities opgevat. Een was beschreven als een groep lichaamsdelen in articulatie en omvatte de botten uit de linkerenkel van een volwassen paard, die waren gevonden in een greppel. De enige sporen op de beenderen van tafonomische processen zijn duidelijke knaagsporen. De beenderen passen in elkaar en de plaatsing van de knaagsporen op de uitstekende delen van de gewrichten en hun afwezigheid op de gewrichtsvlakken zelf tonen aan dat de enkel compleet in elkaar zat, toen hij in de greppel werd geplaatst.

De tweede depositie behoort tot de categorie van de geselecteerde skeletelementen. Deze is afkomstig uit hetzelfde greppelsysteem als de paardenenkel. Wat deze depositie speciaal maakt is de verdeling van de runder- en schaap/geitbotten over het skelet: een aanzienlijk hoger percentage onderkaakfragmenten is in deze depositie aangetroffen dan het resterende Dongjummateriaal laat zien.

De derde en laatste speciale depositie van Dongjum valt in categorie met een opvallende leeftijdsstructuur. Maar deze depositie is ook speciaal door de archeologische vondsten bij deze depositie. Het bijenkorfvormige spoor werd geïnterpreteerd als een waterput met drie afzonderlijke vullingen, waarvan er twee botmateriaal bevatten. De bovenste vulling bevatte enkele botfragmenten en een slijpsteen. Daarentegen was de dunne middelste vulling leeg, terwijl de onderste vulling helemaal onderin enkele fragmenten van dierlijke beenderen en een slijpsteen bevatte. Het enige aspect waardoor deze dierlijke resten duidelijk verschillen van de rest van het materiaal is de leeftijdsverdeling. Dit geldt vooral voor de bovenste vulling, die een ongewoon groot aantal beenderen van jonge dieren bevatte. Dit, samen met de gedeponeerde slijpstenen, is genoeg om deze hele depositie als speciaal aan te merken.

Leeuwarden

Het archeologische project Leeuwarden-Oldehove werd uitgevoerd tussen 2004 en 2006 door de firma ADC ArchaeoProjecten en het Groninger Instituut voor Archeologie van de Rijksuniversiteit Groningen.

De vroegste opgegraven resten van deze vindplaats dateren uit de Vroeg-Romeinse tijd. De nederzetting bestond toen uit een kernpodium met een boerderij. Rond 300 na Chr. werd de vindplaats verlaten. Tijdens het midden van de 5e eeuw werd de terp opnieuw in gebruik genomen. Deze ontwikkelde zich in de Merovingische en de Karolingische perioden geleidelijk tot een grotere nederzetting, met twee tot vier boerderijen. De terpnederzetting Oldehoofsterkerkhof kan over het algemeen als een normale agrarische gemeenschap worden aangeduid. Er zijn geen vondsten gedaan die zouden wijzen op een hoge status van de bewoners als heersende klasse van de maatschappij. Evenmin zijn er aanwijzingen voor gespecialiseerde ambachtslieden.

Het vroegmiddeleeuwse dierlijk botmateriaal van deze Leeuwarder opgraving is goed geconserveerd. Runderen en schapen/geiten domineren de soortenverdeling.

Er bevinden zich 12 deposities in het botmateriaal, die als speciaal worden beschouwd. De eerste drie behoren tot de categorie gearticuleerde lichaamsdelen (deelskeletten) en bestaan uit beenderen uit de voorste delen van schapen en/of geiten. Zij werden in overeenkomstige vondstlagen aangetroffen.

De vierde depositie bestaat uit vier complete linkersprongbenen (astragali), drie van runderen en een van een schaap. Een van de rundersprongbenen toont op het proximale uiteinde sporen van bewerking. Dat vier linkersprongbenen als enige vondsten in een grondspoor zijn gevonden en alleen oppervlakkige verwering vertonen, is zeker uitzonderlijk genoeg om deze resten als speciale depositie te beschouwen.

De vijfde depositie werd in een hutkom gevonden en werd als bijzonder gecatalogiseerd vanwege de geselecteerde diersoorten. Naast een paar stukjes aardewerk, een handvol kamfragmenten en een onbekend ijzeren voorwerp bestaan de vondsten uit dit spoor voornamelijk uit dierenbotten die gevonden werden in twee donkere, door as gekleurde lagen, die gescheiden werden door schonere lagen. De vullingen bevatten 400 fragmenten van wilde vogels. Voor deze botten werden opmerkelijk hoge MNI's berekend, minstens 25 individuele vogels voor de onderste vondstlaag en 8 individuele vogels in de jongere vondstlaag. Aangezien deze depositie een gerichte selectie van vogels laat zien, die een hoeveelheid vlees vertegenwoordigen die ver buiten de normale dagelijkse consumptie van een nederzetting van deze grootte lag, werd deze vogelbotten als speciaal aangemerkt.

De volgende twee deposities bestonden uit een gave paardenschedel gevonden in een terplaag, en een complete schapenschedel gevonden in een greppel. Aangezien de meeste schedelfragmenten uit deze vindplaats gebroken zijn en slachtsporen bevatten, werden deze complete paarden- en schapenschedels als speciaal aangemerkt.

De laatste vijf speciale deposities uit Leeuwarden bevatten een of meer menselijke beenderen in sporen die niet als graven waren geïnterpreteerd. Een late waterput gedateerd als 828 ± 6 na Chr. was de meest spectaculaire van deze deposities. Op de bodem van dit spoor waren onder een ondersteboven gekeerde essenhouten trog drie menselijke schedels en verscheidene andere grote menselijke beenderen geplaatst. Helaas werd tijdens de opgraving gedacht dat deze menselijke beenderen afkomstig waren van een later graf, uit de christelijke periode. Daarom werden deze resten herbegraven op een moderne begraafplaats. Desondanks is het duidelijk dat deze botten skeletdelen van ten minste drie mensen zijn, die aan het begin van de 9e eeuw na Chr. op een ongebruikelijke manier in deze waterput waren neergelegd.

Midlaren

De landinwaarts gelegen nederzetting Midlaren werd in 2003-2004 door Groninger Instituut voor Archeologie in samenwerking met het archeologische bedrijf ARC opgegraven. Deze permanente nederzetting had zijn oorsprong in de ijzertijd, waarschijnlijk als een enkele boerderij. In de Romeinse tijd bestond de nederzetting uit twee of drie boerderijen. De nederzetting had langeafstandscontacten met de Romeinse gebieden in het zuiden, de terpenregio in het noorden en met Scandinavië. De bewoners voerden bepaalde objecten in, maar maakten ook zelf, in eigen ambachten, allerlei producten, waaronder metalen voorwerpen. Naar de meeste aspecten, zoals aardewerkvormen en de indeling van de huizen, vertoonde het dorp sterke verbindingen met het terpengebied. In de Laat-Romeinse tijd en de Vroege Middeleeuwen nam de nederzetting enigszins in grootte af. Hij bestond toen nog maar uit een of twee boerderijen. Het onderzochte gebied werd uiteindelijk in het midden van de 11e eeuw verlaten.

Als gevolg van de zandgrond ter plaatse is het botmateriaal slecht geconserveerd en zeer gefragmenteerd.

Het materiaal bestaat desondanks, zoals te verwachten was, in de eerste plaats uit resten van runderen, gevolgd door die van schapen en/of geiten en varkens.

Drie verschillende vroegmiddeleeuwse deposities werden als bijzonder gekwalificeerd. Twee hiervan bestaan uit schedels zonder slachtsporen of andere sporen. In een onregelmatig gevormde kuil werden de overblijfselen van ten minste vier runderschedels gevonden. Als gevolg van de slechte conserveringsomstandigheden waren er nog slechts fragmenten van over. Duidelijk is wel dat zij waren gedeponeerd als complete schedels.

In een waterput werd een aantal runderkiezen gevonden; deze kiezen komen waarschijnlijk van eenzelfde type depositie als de hiervoor genoemde schedels zonder slacht- of bewerkingssporen.

De derde depositie is een uniek spoor. Het heeft de vorm van een ovale kuil en wordt gedateerd in de Karolingische periode. In de kuil bevonden zich de verbrande overblijfselen van verschillende dieren. De grond van dit spoor werd gezeefd, wat resulteerde in bijna een half miljoen botfragmenten. In eerste instantie werd gedacht aan de restanten van een menselijke crematie, maar tijdens de analyse werden uitsluitend fragmenten van dierlijke beenderen aangetroffen, namelijk van runderen, schapen/geiten, varkens en honden.

Het rund domineert het materiaal, terwijl de hond slechts door vier gedetermineerde fragmenten is vertegenwoordigd. Het materiaal is zeer gefragmenteerd ten gevolge van de hoge temperatuur waarbij de botten werden verbrand. Een zeer groot deel van het materiaal kon daarom niet op soort worden gebracht. De botfragmenten komen uit alle delen van de skeletten van de dieren. Ten minste negen volwassen huisdieren werden gecremeerd, namelijk drie koeien, twee schapen of geiten, twee varkens en twee honden. De vorm en compactheid van de botconcentratie geven zeer sterk de indruk dat de botten werden verzameld van de brandstapel en in deze kuil werden gedeponeerd, gewikkeld in doek of een stuk leer. Deze depositie wordt als speciaal beschouwd om een aantal redenen. Zelfs als voornamelijk oude dieren voor deze gebeurtenis werden gekozen, was het nog altijd een substantiële economische investering voor de nederzetting. De gecremeerde resten werden bovendien verzameld en netjes verpakt in een stuk stof of leer. Dit pakket werd uiteindelijk geplaatst in een kuil samen met een bronzen voorwerp en werd vervolgens bedekt met aarde.

Uppåkra

De vindplaats Uppåkra was een centrale plaats 5 km ten zuiden van Lund (Zweden). Het was het doel van een aantal kleine noodopgravingen in de hele 20e eeuw en van jaarlijkse campagnes van archeologisch veldonderzoek uitgevoerd door het Instituut voor Archeologie en Oude Geschiedenis van de Universiteit van Lund vanaf halverwege de jaren 1990.

De nederzetting werd voor het eerst in gebruik genomen in de Late IJzertijd op een geringe verhoging in het overigens zo vlakke landschap van zuidwestelijk Skåne. Boringen, metaaldetectoronderzoek en opgravingen hebben een nederzetting aan het licht gebracht met een min of meer ovale vorm, 1100 m van noord naar zuid en 600 m van oost naar west. Talrijke structuren werden gevonden, waaronder huizen uit het eerste millennium na Chr. De nederzetting functioneerde vermoedelijk als een eenheid tot de tweede helft van de 10e eeuw, toen de hoeveelheid vondsten begon af te nemen.

Tijdens haar bestaan was Uppåkra een bovenregionaal centrum met economische, religieuze en bestuurlijke functies. De structuur van de nederzetting kan het beste worden beschreven als vroegstedelijk met een aantal boerderijen dicht op elkaar.

Vele van de recente opgravingscampagnes waren geconcentreerd op het centrale gedeelte van de noordelijke helft van de nederzetting, waar verschillende interessante gebouwen werden gevonden, waaronder een aantal grote langhuizen en een kleiner huis van ongewone bouwwijze. Dit kleinere huis wordt geïnterpreteerd als een tempel of ceremonieel huis en daarin werden verschillende unieke vondsten gedaan, waarvan de opmerkelijkste een bronzen beker met zilveren en gouden versieringen, een fijn vervaardigde glazen kom en maar liefst 111 goudfiguren (guldgubbar, gold-figure foils), een soort artefact dat nauw met rituele activiteiten verbonden was.

Slechts 30 m ten noorden van dit ongewone huis werden in een gebied dat ongeveer 86 m^2 groot was, vele bewust vernietigde wapens, voornamelijk lans- en speerpunten en ander militair materieel, gevonden, samen met grote hoeveelheden dierlijke beenderen en drie menselijke beenderen. Dit gebied werd geïnterpreteerd als de plaats waar rituele activiteiten in de vorm van wapenoffers werden uitgevoerd, zonder enige aandacht te besteden aan de dierlijke beenderen en aan hun functie bij deze rituelen.

Als zodanig is het een interessant analysemateriaal voor dit proefschrift, maar ook een dat enigszins verschilt van de materialen van de andere drie locaties. Het dierlijk materiaal van deze vindplaats had van meet af aan een rituele duiding gekregen. Aangezien analyse van het volledige dierlijk botmateriaal van de nederzetting een onmogelijke taak zou zijn, moest een bottenassemblage worden geselecteerd dat de normale situatie zou representeren. Voor dit doel werd een botmateriaal uit het zuidelijke deel van de nederzetting gekozen.

De dierlijke beenderen uit de wapenopslagplaats in Uppåkra wegen samen bijna 190 kg. De beenderen uit 26 van de 86 opgegraven vierkante meters werden voor analyse uitgekozen. Het materiaal is tamelijk sterk gefragmenteerd en toont veel sporen van verwering, maar vooral de lichtere verweringsstadia. Aangezien het botmateriaal van Uppåkra in het algemeen goed geconserveerd is, kan deze fragmentatie worden verklaard doordat de botten gedurende enige tijd onafgedekt lagen of doordat latere wapendeposities werden ingegraven in de bottenlaag, waardoor de beenderen open en bloot kwamen te liggen en verder gefragmenteerd raakten.

De botmateriaal wordt gedomineerd door runderen, in zelfs nog sterkere mate dan in de materialen uit Dongjum en Leeuwarden. Er zijn iets meer resten van varkens dan van schapen en/of geiten in het materiaal, in tegenstelling tot de situatie in Dongjum en Leeuwarden, maar dat was te verwachten aangezien de terpen meestal weinig resten van varkens bevatten. Schedelfragmenten overheersen in het assemblage en dit geldt vooral voor de runderbeenderen. Het slachtleeftijdpatroon van het wapendepotmateriaal laat zien dat de meeste dieren werden gedood als volwassen dieren. Dit geldt het meest voor het rundvee: slechts zeer weinig jonge runderen werden geïdentificeerd, terwijl bijna 20 procent van de runderen op hoge leeftijd werd gedood.

Het vergelijkingsassemblage uit het zuidelijke deel van Uppåkra verschilt van het materiaal uit het wapendepot doordat schapen en/of geiten door meer fragmenten zijn vertegenwoordigd dan de runderen en doordat er maar heel weinig paardenbeenderen werden gevonden. De skeletelementverdeling laat net als die van het wapendepot, vele losse tanden zien en maar weinig schedel- en onderkaakfragmenten. Het slachtleeftijdpatroon voor het zuidelijke assemblage wijst op het slachten van runderen op veel jongere leeftijden dan voor het wapendepot het geval was.

De conclusie is dat het botmateriaal uit het wapendepot ook op zichzelf speciaal is dankzij de ongewoon grote hoeveelheid runderbeenderen, hoofdzakelijk in de vorm van schedel- en onderkaakfragmenten van volwassen of oude runderen. Deze dierendeposities waren van aanzienlijke omvang: onder de 20 procent van het totale materiaal dat werd onderzocht, werden minstens 31 individuele runderen en 18 andere huisdieren geteld.

Interpretatie en verklaring van de activiteit

De hierboven beschreven methoden zijn bedoeld als universele methoden in de zin dat zij gebaseerd zijn op herkenningspatronen die in principe onafhankelijk zijn van de context van de onderzochte deposities. Dit is echter slechts één stap in een proces dat niet zou behoren te stoppen na het aanwijzen van speciale deposities, maar verder moet gaan en de interpretatie en het begrijpen van de redenen achter deze deposities als doel moet hebben.

Van de 20 bijzondere deposities die zijn geïdentificeerd in de casestudy-materialen, komen drie uit Dongjum, dertien uit Leeuwarden, drie uit Midlaren en een uit Uppåkra. Hiervan werden drie geïnterpreteerd als normaal afval en zeventien als rituele deposities.

De drie deposities bestaande uit deelskeletten van schapen en/of geiten in de vondstlagen van Leeuwarden zijn hoogstwaarschijnlijk voorbeelden van offers van delen van dieren. Er zijn maar weinig bronnen die dit fenomeen voor schapen of geiten beschrijven. Daarentegen zijn er genoeg bronnen die offers van delen van andere diersoorten beschrijven. Van deze dieren krijgt de bovennatuurlijke macht slechts een klein gedeelte, terwijl de rest van het lichaam onder de mensen wordt verdeeld en wordt opgegeten. Een bron die het gebruik van schapen in een soortgelijk ritueel beschrijft is *Kitāb Rūsiyyah* van de hand van Ibn Fadlān, die beschrijft dat de Rūs de schedels van koeien en schapen op palen bonden als een offer aan de goden na een succesvolle handelstransactie. De rest van het vlees van de dieren werd verdeeld en opgegeten.

Vier linker sprongbeenderen (astragali, bikkels) van runderen en schapen die werden gevonden op de bodem van een kuil in Leeuwarden, vormen een depositie die doet denken aan het gebruik van beenderen in spelen en waarzeggerij. Het gebruik van sprongbenen in spelen en voor rituele doeleinden zoals voorspellingen is op verschillende plaatsen beschreven door etnografen en is ook goed gedocumenteerd in klassieke bronnen. Er wordt nog steeds gespeeld met deze sprongbeenderen, bijvoorbeeld in het bikkelspel, maar dit zijn meestal behendigheidsspelen met de kleine sprongbeenderen van schapen, geiten of kalveren. Andere soorten spelen, die meer gelijken op jeu de boules of op knikkeren, maar waarvoor zowel schapen- als rundersprongbenen worden gebruikt, bestaan echter in verschillende delen van Azië, bijvoorbeeld in Turkije, Turkmenistan en Mongolië. Etnografische waarnemingen van de Mongoolse vorm van dit spel tonen eveneens aan dat dit spel een sterk rituele component heeft. De archeologische data komen goed overeen met de etnografische bronnen en laten een lang en het wijdverspreid gebruik van sprongbeenderen in spelen en bij voorspellingen zien. Met deze traditie in het achterhoofd is het waarschijnlijk dat de botten uit dit grondspoor in Leeuwarden een hoeveelheid speelstenen of botten vertegenwoordigen, waarmee de toekomst werd voorspeld, en die werden weggegooid, werden verborgen of werden geofferd aan bovennatuurlijke wezens.

De speciale deposities van complete schedels en schedels zonder onderkaken zijn vertegenwoordigd door vier sporen uit Leeuwarden en Midlaren. Zij omvatten zeven of acht afzonderlijke schedels, vijf of zes van runderen, één van een paard en één van een schaap. Het gebruik van schedels met of zonder onderkaken bij rituelen is een wijdverbreid fenomeen en de sterke symboliek van het hoofd van de dieren is te zien in culturen over de hele wereld. Een interessant etnografisch voorbeeld komt van de Zulu-bevolking in Zuid-Afrika. Deze mensen plaatsten de schedel of de horens van het rund dat werd geofferd aan de vooroudergeesten, in het dorp ter bezichtiging. Een deel van het karkas van het rund werd aan de geesten geofferd, terwijl de rest van het dier door de gemeenschap tijdens een feestmaal werd opgegeten.

Archeologische parallellen voor dit soort deposities zijn op verschillende plaatsen aangetroffen, zoals bijvoorbeeld in de terp Feddersen Wierde in Duitsland, waar twee paardenschedels en een runderschedel werden gevonden in twee kuilen en in een paalgat, en in Elisenhof, een andere Duitse terp, waar een runderschedel werd gevonden nabij de zuidelijke muur van een huis. Een historische bron die het gebruik van dierlijke schedels beschrijft, is de reeds genoemde passage van Ibn Fadlān over de schedels van koeien en schapen, die waren gebonden aan palen. De schedels uit Leeuwarden en Midlaren werden in overeenstemming hiermee geïnterpreteerd als deeloffers, waarbij de hoofden van de dieren geschonken werden aan bovennatuurlijke krachten, terwijl het meeste vlees van de dieren door de mensen werd genuttigd.

In een hutkom in Leeuwarden werd een groot aantal botten van wilde vogels gedeponeerd bij twee afzonderlijke gelegenheden. Tijdens de eerste fase werden vrijwel alle beenderen van ten minste 25 vogels neergelegd, terwijl in de tweede fase een geringer aantal beenderen van minstens 9 individuele vogels werd gedeponeerd. De verspreiding over het skelet en de brandsporen op de botten suggereren dat de vogelbeenderen afval zijn van voedselbereiding of consumptie. Dat de vogels uit de eerste depositie vlees voor ten minste veertig personen kunnen hebben geleverd, geeft aan dat het grondspoor de overblijfselen van een groot festijn representeren, dat eventueel een ritueel karakter had en waarschijnlijk verbonden was met de vogeljacht, aangezien alleen vogelbotten zijn aangetroffen.

Een aantal menselijke beenderen werd gevonden in vijf verschillende sporen uit de Vroege Middeleeuwen in Leeuwarden. Verschillende types van ritueel handelen met menselijke resten zijn geregistreerd door etnografen in samenlevingen over de gehele wereld. Deze omvatten het weer opgraven en herbegraven, verschillende typen van voorouderverering, kannibalisme, trofeejacht, mensenoffers en verschillende magische praktijken. Verspreide menselijke beenderen buiten grafcontext zijn niet zeer frequent in het terpengebied wat de Vroege Middeleeuwen betreft, maar er zijn enkele interessante vondsten uit eerdere perioden. Historische bronnen voor Noord-Europa spreken soms over mensenoffers of andere vormen van ritueel handelen met menselijke resten, maar deze werpen zelden een licht op de aanwezigheid van losse beenderen in nederzettingen. De verschillende analogieën in aanmerking genomen is het waarschijnlijk dat ten minste een deel van de vijf deposities het resultaat waren van trofeejacht of van voorouderverering, of van deposities van religieuze of magische voorwerpen bestaande uit mensenbotten.

De meest intrigerende depositie die besproken is in dit proefschrift is de kuil met verbrande botfragmenten uit de Karolingische periode uit Midlaren. Deze bevatte bijna een half miljoen botfragmenten die bij hoge temperaturen waren verbrand. Er bestaan maar zeer weinig archeologische vondsten die vergelijkbaar zijn met deze crematiekuil uit Midlaren. Verbrande dierlijke beenderen in menselijke crematiegraven zijn een algemeen verschijnsel op grafvelden vanaf de Bronstijd in Noord-Europa, maar in de meeste gevallen gaat het om kleine hoeveelheden verbrande dierlijke beenderen. Hier en daar zijn kuilen gevonden op vroegmiddeleeuwse grafvelden met gecremeerde dierlijke beenderen, maar zonder enige menselijke beenderen, maar ook deze deposities bevatten steeds betrekkelijk geringe hoeveelheid botten. De vondst die het meest vergelijkbaar is met de kuil uit Midlaren is afkomstig uit de Slavische burcht in Oldenburg (Sleeswijk-Holstein, Duitsland). In de ovale kuil die dateert uit de IJzertijd, werden scherven van een grote aardewerken pot van vroeg-Slavisch type gevonden. Deze pot werd in zijn geheel gedeponeerd, maar brak later boven op een concentratie van verbrande beenderen. De botten lagen zo dicht opeen dat geconcludeerd werd dat zij waren gewikkeld in een stuk stof of leer en zo in de kuil werden geplaatst. De depositie bevatte 3,1 kg verbrande beenderen van ten minste twee runderen en één varken.

Het verbranden van dieren of van gedeelten van dieren als onderdeel van een offer is bekend voor verschillende culturen in verschillende perioden. Een groot aantal Romeinse en Griekse bronnen laat zien dat deze offers rond de Middellandse Zee in klassieke tijden regelmatig werden gebracht. Het feit dat de dieren uit Midlaren waarschijnlijk als complete lichamen waren verbrand, leidt tot de interpretatie dat de depositie een offer aan bovennatuurlijke wezens was, dat eventueel gekoppeld was aan een ritueel rond de dood.

De verklarende omstandigheden voor het bestudeerde botmateriaal uit Uppåkra verschillen enigszins van die voor de andere drie materialen. Hier was het gebied waar de dierlijke resten waren gedeponeerd reeds geïdentificeerd als een bijzonder spoor met een rituele betekenis als de plaats van herhaalde wapenoffers. Er bestaat een groot etnografisch bronnenbestand rond de twee belangrijkste aspecten van dit spoor: offers en feesten. De term offer impliceert het aanbieden van iets dat ritueel is gedood, hetzij in werkelijkheid of symbolisch. Dit type offer is bekend uit talloze samenlevingen uit de hele wereld, en terwijl een offer vaak het doden van een dier, of zeldener van een mens, inhoudt, wordt de term ook vaak gebruikt voor de vernietiging van niet-levende dingen. Het feesten, het samenkomen van veel mensen die grote hoeveelheden voedsel en drank consumeren voor een specifieke reden, is ook een gemeenschappelijk kenmerk in etnografische beschrijvingen, en is iets dat nog steeds in hedendaagse samenlevingen bestaat. Er kan gefeest worden voor diverse redenen buiten de officiële, vaak symbolische motivatie. Deze andere redenen zijn bijvoorbeeld het verkrijgen van prestige (door de sponsor van het feest), het versterken van sociale banden, het in stand houden van tradities en het handhaven van sociale hiërarchieën.

De artefacten van de wapendepositie vertonen enige gelijkenis met de grote wapendepositievindplaatsen uit de Romeinse tijd en de Volksverhuizingstijd in zuidelijk Scandinavië. De vondsten uit Uppåkra, echter, verschillen van deze vindplaatsen in enkele opzichten, vooral door het relatief kleine aantal artefacten, de samenstelling van vele kleine deposities in plaats van enkele grote en de voortzetting van de wapendeposities tot diep in de 8e eeuw. Een ander verschil is de aard van het botmateriaal. De samenstelling van de dierlijke beenderen uit verschillende van de grote wapenofferplaatsen verschilt nogal van die van de beenderlaag in Uppåkra. Daarentegen vertonen de dierlijke beenderen van meer alledaagse offervindplaatsen zoals Eketorp, Röekillorna en Valmose meer overenkomsten met die van Uppåkra.

Het gebruik van Noord-Europese volkeren de wapens van vijanden te offeren na een gevecht is beschreven door diverse klassieke auteurs. Dat deze feesten in Scandinavië in de IJzertijd, de Romeinse tijd en de Vroege Middeleeuwen belangrijke sociale en politieke gebeurtenissen waren, is sinds lang een vaststaande opvatting. De wortels voor deze opvatting liggen in de vele verwijzingen, met name in de Sagas, naar de religieuze betekenis van feesten. Maar het feesten was ook een manier om uiting te geven aan machtsstructuren en loyaliteiten. Rekening houdend met de verschillende aspecten van het materiaal en de relevante analogieën in aanmerking genomen, wordt de bottenlaag van Uppåkra in dit proefschrift geïnterpreteerd als het resultaat van herhaalde feesten in combinatie met offers van delen van dieren. Tijdens een dergelijk evenement werden één of meerdere dieren (meestal runderen) geslacht. Sommige delen van het dier, met name de schedels of de onderkaken, werden geofferd aan de bovennatuurlijke machten en gedeponeerd ofwel samen met wapens, of althans op dezelfde plaats waar ook wapens werden geofferd.

Conclusies

Een aantal methoden werd in dit proefschrift toegepast voor het identificeren van speciale deposities in de onderzochte beenmaterialen. De belangrijkste conclusie is dat niet één enkele, of zelfs niet één set van methoden zich onderscheidde als hét perfecte hulpmiddel voor het identificeren rituele deposities. Zelfs als er overeenkomsten tussen verschillende menselijke samenlevingen in het verleden bestaan, hetzij vanwege directe verbindingen, of als gevolg van de invloeden en de beperkingen van de menselijke biologie en van de omringende fysieke wereld, is de variatie in menselijk gedrag te groot om elke methode algemeen toepasbaar te doen zijn.

Wanneer een materiaal uit een gebied, periode of cultuur moet worden geanalyseerd, kan een adequate set van identificatiemiddelen alleen worden samengesteld via experimenten en evaluatie. Op dezelfde wijze moeten relevante analogieën worden vastgesteld om deze te kunnen gebruiken voor de interpretatie van elk type depositie binnen een dergelijke samenleving.

Van de verschillende gebruikte methoden, bleken de archeozoölogische methoden een groot potentieel te hebben om speciale deposities te kunnen identificeren. Maar liefst 19 van de 20 geïdentificeerde bijzondere deposities vertoonden anomalieën wat betreft de soortenselectie, de leeftijdsopbouw, seizoensgebonden gebruik, selectie naar sekse, de keuze van de skeletelementen, of een combinatie van deze aspecten.

De archeologische methoden zoals de samen met de dierlijke resten gevonden artefacten, het spoortype, de stratigrafische positie, de ruimtelijke context en de algemene vondstcontext bleken enige mogelijkheden te geven bijzondere deposities aan te tonen. Met uitzondering van het bottenlaag uit Uppåkra, verschaften deze methoden niet de belangrijkste aanwijzingen voor speciale deposities. Dit was wellicht te verwachten omdat de nadruk bij het onderzoek op de archeozoölogie lag. Wel gaven de archeologische methoden ondersteunende indicaties voor de meeste geïdentificeerde bijzondere deposities.

Het interpreteren van de betekenis van het menselijk gedrag in het verleden vereist ofwel toegang tot rechtstreekse historische bronnen, ofwel het gebruik van analogieën. Omdat rechtstreekse historische bronnen voor de exacte plaatsen en perioden in dit onderzoek afwezig waren, moesten analogieën worden gebruikt tijdens het proces van de interpretatie. Historische analogieën werden gebruikt om verschillende deposities te interpreteren, evenals etnografische analogieën, hoewel relevante, gedetailleerde etnografische voorbeelden van ritueel gedrag, waaronder gedragingen met dierlijke beenderen, minder vaak voorkwamen dan was verwacht. Eigentijdse analogieën speelden hoogstwaarschijnlijk een rol bij de interpretatie van de deposities, aangezien deze zijn geïntegreerd in de algemene menselijke gedachtegangen. Ik heb echter geprobeerd de onbewuste en algemene invloed vanuit de moderne samenleving zo veel mogelijk te beperken.

Ook al ligt de focus van dit proefschrift op de methoden voor de interpretatie van deposities, toch is het interessant om te analyseren voor welk doel en in welke contexten de speciale deposities ontstonden. De meest gangbare rituele interpretatie bleek het deeloffer te zijn, iets wat niet zo verwonderlijk is als je bedenkt wat we weten uit historische en etnografische bronnen. Het lijkt erop dat het gebruikelijk was geen volledige dieren als offer aan te bieden aan de bovennatuurlijke machten, maar het meeste vlees zelf te consumeren. Alleen als het ritueel ook als een indicator voor welvaart en status functioneerde, werden complete dieren aan de bovennatuurlijke krachten gegeven. De interpretaties van de speciale deposities tonen in het algemeen dat we de samenstelling van deposities zorgvuldig moeten bestuderen. Het is altijd belangrijk om een stap terug te doen en na te denken over wat werkelijk werd gevonden en waarom dit materiaal op die plek belandde.

De resultaten van de analyses en de interpretaties leiden tot een methodologisch kader voor het begrijpen van individuele deposities, dat is gebaseerd op een holistisch perspectief, waarin alle informatie als potentieel waardevol moet worden beschouwd en waarbij elk mogelijk stukje informatie en elk idee op zijn waarde moet worden geschat. Dit betekent niet dat de onderzoeker ongelooflijk gedetailleerde studies moet produceren en gehinderd wordt door een eindeloos zoeken naar informatie. Wel suggereert dit kader openheid in de aanpak, waarbij verschillende methoden en ideeën in overweging worden genomen en waarbij formalisatie, simplificatie en stigmatisering worden voorkomen.

Voor een holistische benadering moeten de onderzoekers die zijn betrokken bij het archeologische proces, voldoen aan enkele verantwoordelijkheden. De archeoloog die de opgraving verricht, is de meest belangrijke persoon om de voorgestelde aanpak in dit hoofdstuk mogelijk te maken. Informatie die niet in het veld wordt genoteerd en materiaal dat niet wordt gelicht, zijn onherroepelijk verloren, hoe gedegen en methodisch de latere analyses ook zijn. De ervaring met de vier opgravingsmaterialen die werden geanalyseerd in dit proefschrift, is dat er een directe correlatie bestaat tussen de hoeveelheid informatie die tijdens de opgraving over een spoor wordt geregistreerd, en de omvang en betrouwbaarheid van de latere interpretaties. De volgende punten zijn de belangrijkste taken van de opgravers om gedetailleerde analyses van deposities binnen een vindplaats mogelijk te maken:

- Een grondige archeologische methodologie en kennis van de theoretische mogelijkheden en beperkingen van de discipline.

- Een erkenning van het belang van afzonderlijke grondsporen voor inzicht in het menselijke gedrag in het verleden.

- Grote aandacht voor de vondstomstandigheden, ook wanneer grote materialen worden opgegraven.

- Een goede velddocumentatie die zo vroeg mogelijk beschikbaar is voor de (archeozoölogische) expert.

- Een openheid naar en de integratie van de expert in een project vanaf een vroeg stadium, bij voorkeur al in de planning van de opgraving, en de bereidheid om gebruik te maken van zijn of haar vaardigheden, zowel in het veld als in de fase van de archeologische interpretatie.

In aanvulling op de plichten van het veldarcheoloog, heeft de (archeozoölogische) expert die het materiaal analyseert, overeenkomstige verantwoordelijkheden. De volgende punten zijn de belangrijkste taken van de expert, zowel als een potentiële deelnemer aan het archeologisch veldwerk, als tijdens de latere fasen van de analyse en interpretatie:

- Een bereidheid om de theorie, methodologie en beperkingen van archeologie als een discipline te leren, en niet alleen de specificaties van de eigen subdiscipline van de expert.

- Een bereidheid tot het bestuderen van het gebied, de tijdsperiode en de samenleving waarvan het materiaal afkomstig is, ook buiten de subdiscipline van de expert.

- Een bereidwilligheid om zowel deel te nemen aan de planningsfase als aan het opgravingsdeel van een project ter ondersteuning van de veldarcheologen.

- Het niet alleen leveren van een technisch rapport, maar vooral van een verslag van de materiaalanalyse toegespitst op de archeologische vraagstellingen van het onderzoek, waarin de interpretaties op basis van de kennis van de expert zijn opgenomen.

Acknowledgements

When an endeavour as long and extensive as the writing of a thesis is finally coming to an end and one looks back at it all trying to remember the people that inspired and helped in its creation, it suddenly dawns on you how many these actually were, and how easy it is to forget someone. Therefore I thank all my old teachers, colleagues and friends in Lund who taught me archaeology and osteology, encouraged me to apply to the position in Groningen and helped me with the analysis of the Uppåkra material.

Furthermore I thank all the colleagues and friends at the GIA who both helped me with the thesis, but also made the family feel at home in Groningen, guided us through the bureaucratic maze of moving to a new country, and helped us with all the practical problems we faced, large and small.

I would also like to specifically thank my supervisors Daan and Wietske, who have been so very helpful and patient with me.

Finally I give my gratitude and love to Elin who came with me to the Netherlands and who has read, commented, criticised and encouraged me without end, and to Ida and Erik, my two specialists in the art of distraction and in reminding me of what is really important in life.